Simulation of ODE/PDE Models with MATLAB®, OCTAVE and SCILAB

Alain Vande Wouwer · Philippe Saucez
Carlos Vilas

Simulation of ODE/PDE Models with MATLAB®, OCTAVE and SCILAB

Scientific and Engineering Applications

Springer

Alain Vande Wouwer
Service d'Automatique
Université de Mons
Mons
Belgium

Carlos Vilas
(Bio)Process Engineering Group
Instituto de Investigaciones Marinas (CSIC)
Vigo
Spain

Philippe Saucez
Service de Mathématique et Recherche Opérationnelle
Université de Mons
Mons
Belgium

Additional material to this book can be downloaded from http://extras.springer.com

ISBN 978-3-319-06789-6 ISBN 978-3-319-06790-2 (eBook)
DOI 10.1007/978-3-319-06790-2
Springer Cham Heidelberg New York Dordrecht London

Library of Congress Control Number: 2014939391

Printed on acid-free paper

Springer is part of Springer Science+Business Media (www.springer.com)

To V. L. E.

A.

To Daniel Druart

Philippe Saucez

Foreword

With the availability of computers of increasing power and lower cost, computer-based modeling is now a widespread approach to the analysis of complex scientific and engineering systems. First-principles models and numerical simulation can be used to investigate system dynamics, to perform sensitivity analysis, to estimate unknown parameters or state variables, and to design model-based control schemes. However, due to the usual gap between research and common practice, scientists and engineers still often resort to conventional tools (e.g. low-order approximate solutions), and do not make use of the full array of readily available numerical methods.

Many systems from science and engineering are *distributed parameter systems*, i.e., systems characterized by state variables (or dependent variables) in two or more coordinates (or independent variables). Time and space are the most frequent combination of independent variables, as is the case of the following (*time-varying, transient, or unsteady state*) examples:

- temperature profiles in a heat exchanger
- concentration profiles in a sorptive packed column
- temperature and concentration profiles in a tubular reactor
- car density along a highway
- deflection profile of a beam subject to external forces
- shape and velocity of a water wave
- distribution of a disease in a population (spread of epidemics)

but other combinations of independent variables are possible as well. For instance, time and individual size (or another characteristic such as age) occur in population models used in ecology, or to describe some important industrial processes such as polymerization, crystallization, or material grinding. In these models, space can also be required to represent the distribution of individuals (of various sizes) in a spatial region or in a nonhomogeneous reactor medium (due to nonideal mixing conditions in a batch reactor, or to continuous operation in a tubular reactor).

The preceding examples show that there exists a great variety of distributed parameter systems, arising from different areas of science and engineering, which are characterized by time-varying distributions of dependent variables. In view of the system complexity, a mathematical model, i.e., a mathematical description of the physical (chemical, biological, mechanical, electrical, etc.) phenomena taking place in the system, is often a prerequisite to system analysis and control. Such a model consists of partial differential equations (PDEs), boundary conditions (BCs), and initial conditions (ICs) describing the evolution of the state variables. In addition, distributed parameter systems can interact with lumped parameter systems, whose state variables are described by ordinary differential equations (ODEs), and supplementary algebraic equations (AEs) can be used to express phenomena such as thermodynamic equilibria, heat and mass transfer, and reaction kinetics (combinations of AEs and ODEs are also frequently termed differential-algebraic equations, or DAEs). Hence, a distributed parameter model is usually described by a mixed set of nonlinear AE/ODE/PDEs or PDAEs. Most PDAEs models are derived from first principles, i.e., conservation of mass, energy, and momentum, and are given in a state space representation which is the basis for system analysis.

This book is dedicated to numerical simulation of distributed parameter systems described by mixed systems of PDAEs. Special attention is paid to the numerical method of lines (MOL), a popular approach to the solution of time-dependent PDEs, which proceeds in two basic steps. First, spatial derivatives are approximated using finite difference, element, or volume approximations. Second, the resulting system of semi-discrete (discrete in space continuous in time) equations is integrated in time using an available solver. Besides conventional finite difference, element, and volume techniques, which are of high practical value, more advanced spatial approximation techniques are examined in some detail, including finite element and finite volume approaches.

Although the MOL has attracted considerable attention and several general-purpose libraries or specific software packages have been developed, there is still a need for basic, introductory, yet efficient, tools for the simulation of distributed parameter systems, i.e., software tools that can be easily used by practicing scientists and engineers, and that provide up-to-date numerical algorithms.

Consequently, a MOL toolbox has been developed within MATLAB/OCTAVE/SCILAB. These environments conveniently demonstrate the usefulness and effectiveness of the above-mentioned techniques and provide high-quality mathematical libraries, e.g., ODE solvers that can be used advantageously in combination with the proposed toolbox. In addition to a set of spatial approximations and time integrators, this toolbox includes a library of application examples, in specific areas, which can serve as templates for developing new programs. The idea here is that a simple code template is often more comprehensible and flexible than a software environment with specific user interfaces. This way, various problems including coupled systems of AEs, ODEs, and PDEs in one or more spatial dimensions can easily be developed, modified, and tested.

This text, which provides an introduction to some advanced computational techniques for dynamic system simulation, is suitable as a final year undergraduate course or at the graduate level. It can also be used for self-study by practicing scientists and engineers.

Preface

Our initial objective in developing this book was to report on our experience in numerical techniques for solving partial differential equation problems, using simple programming environments such as MATLAB, OCTAVE, or SCILAB. Computational tools and numerical simulation are particularly important for engineers, but the specialized literature on numerical analysis is sometimes too dense or too difficult to explore due to a gap in the mathematical background. This book is intended to provide an accessible introduction to the field of dynamic simulation, with emphasis on practical methods, yet including a few advanced topics that find an increasing number of engineering applications. At the origin of this book project, some years ago, we were teaming up with Bill Schiesser (Lehigh University) with whom we had completed a collective book on Adaptive Method of Lines. Unfortunately, this previous work had taken too much of our energy, and the project faded away, at least for the time being.

Time passed, and the book idea got a revival at the time of the post-doctoral stay of Carlos in the Control Group of the University of Mons. Carlos had just achieved a doctoral work at the University of Vigo, involving partial differential equation models, finite element techniques, and the proper orthogonal decomposition, ingredients, which all were excellent complements to our background material.

The three of us then decided to join our forces to develop a manuscript with an emphasis on practical implementation of numerical methods for ordinary and partial differential equation problems, mixing introductory material to numerical methods, a variety of illustrative examples from science and engineering, and a collection of codes that can be reused for the fast prototyping of new simulation codes.

All in one, the book material is based on past research activities, literature review, as well as courses taught at the University of Mons, especially introductory numerical analysis courses for engineering students. As a complement to the text, a website (www.matmol.org) has been set up to provide a convenient platform for downloading codes and method tutorials.

Writing a book is definitely a delicate exercise, and we would like to seize this opportunity to thank Bill for his support in the initial phase of this project. Many of his insightful suggestions are still present in the current manuscript, which has definitely benefited from our discussions and nice collaboration.

Of course, we also would like to express our gratitude to our colleagues at UMONS and at IMM-CSIC (Vigo), and particularly Marcel, Christine, Antonio, Julio, Eva, and Míriam, and all the former and current research teams, for the nice working environment and for the research work achieved together, which was a source of inspiration in developing this material. We are also grateful to a number of colleagues in other universities for the nice collaboration, fruitful exchanges at several conferences, or insightful comments on some of our developments: Michael Zeitz, Achim Kienle, Paul Zegeling, Gerd Steinebach, Keith Miller, Skip Thompson, Larry Shampine, Ken Anselmo, Filip Logist, to just name a few.

In addition, we acknowledge the support of the Belgian Science Policy Office (BELSPO), which through the Interuniversity Attraction Program Dynamical Systems, Control and Optimization (DYSCO) supported part of this research work and made possible several mutual visits and research stays at both institutions (UMONS and IIM-CSIC) over the past several years.

Finally, we would like to stress the excellent collaboration with Oliver Jackson, Editor in Engineering at Springer, with whom we had the initial contact for the publication of this manuscript and who guided us in the review process and selection of a suitable book series. In the same way, we would like to thank Charlotte Cross, Senior editorial assistant at Springer, for the timely publication process, and for her help and patience in the difficult manuscript completion phase.

Mons, March 2014 — Alain Vande Wouwer
Vigo — Philippe Saucez
Carlos Vilas

Contents

Acronyms

AE	Algebraic Equation
ALE	Algebraic Lagrangian Eulerian
BC	Boundary Conditions
BDF	Backward Differentiation Formula
DAE	Differential Algebraic Equation
FD	Finite Differences
FEM	Finite Element Method
IBVP	Initial Boundary Value Problem
IC	Initial Conditions
IVP	Initial Value Problem
LHS	Left Hand Side
MOL	Method of Lines
MC	Monotonized Central
ODE	Ordinary Differential Equation
PDAE	Partial Differential Algebraic Equation
PDE	Partial Differential Equation
POD	Proper Orthogonal Decomposition
RHS	Right-Hand Side
RK	Runge–Kutta
WRM	Weighted Residuals Methods

Chapter 1
An Introductory Tour

The computational study of mathematical models is an indispensable methodology in science and engineering which is now placed on an equal footing with theoretical and experimental studies. The most widely used mathematical framework for such studies consists of systems of algebraic equations (AEs), ordinary differential equations (ODEs), and partial differential equations (PDEs). In this introductory chapter, we successively consider

- Some ODE applications, including the development of physical models and ODEs; MATLAB programming; derivation of simple explicit ODE solvers (including the Euler, leapfrog, and modified Euler methods), stability, and accuracy of numerical methods with graphs of stability regions;
- An ODE/DAE application;
- A PDE application (briefly introducing the method of lines–MOL).

MATLAB has a number of high-quality library ODE integrators forming the so-called MATLAB ODE SUITE [1]. Most of the application examples considered in this book make use of these integrators. However, in this chapter and the following, we describe some ODE solvers, including Euler, leap frog, modified Euler, Runge-Kutta, and Rosenbrock methods, to facilitate the understanding of ODE integration algorithms. These time integrators will also easily be translated into other software environments such as OCTAVE and SCILAB.

We start the discussion of AE/ODE/PDE models with some modest examples that illustrate basic concepts and their implementation within MATLAB.

1.1 Some ODE Applications

ODEs are characterized by a single independent variable, typically an initial value variable such as time. For example, the ODE

$$\frac{\mathrm{d}X}{\mathrm{d}t} = \mu_{\max} X \tag{1.1}$$

A. Vande Wouwer et al., *Simulation of ODE/PDE Models with MATLAB®, OCTAVE and SCILAB*, DOI: 10.1007/978-3-319-06790-2_1,

where X is the dependent variable, t the independent variable and μ_{max} a given constant.

Equation (1.1) also requires an initial condition

$$X(t_0) = X_0 \tag{1.2}$$

where t_0 is the specified (initial) value of t and X_0 the specified (initial) value of X.

The solution to Eq. (1.1) is the dependent variable, X, as a function of the independent variable, t. This function is

$$X(t) = X_0 e^{(\mu_{max}(t-t_0))} \tag{1.3}$$

Equation (1.3) can be easily confirmed as the solution of Eq. (1.1) subject to initial condition (1.2).

We can note some interesting properties of the solution, Eq. (1.3):

- If $\mu_{max} < 0$, then $X(t)$ decreases exponentially with t
- If $\mu_{max} = 0$, then $X(t)$ is constant at X_0
- If $\mu_{max} > 0$, then $X(t)$ increases exponentially with t

Thus, the value of μ_{max} leads to markedly different solutions. For example, for the case of bacterial growth, $\mu_{max} > 0$, while for radioactive decay, $\mu_{max} < 0$.

Another important property of Eq. (1.1) is that it is *linear*. In other words, X and dX/dt are to the first power. Generally, linear ODEs are relatively easy to solve in the sense that analytical solutions such as Eq. (1.3) can be derived. Conversely, nonlinear ODEs (for which X and/or dX/dt are not to the first power) generally are not easily solved analytically (these conclusions, of course, depend on the number of ODEs and the particular form of their nonlinearities).

Because analytical solutions are generally difficult, if not impossible, to derive, we have to resort to numerical methods of solution which, in principle, can be used to compute solutions to any system of ODEs, no matter how large (how many equations) and how nonlinear the equations are. However, we will not obtain an analytical solution (as a mathematical function, e.g., Eq. (1.3)). Rather, we will obtain the dependent variables as a function of the independent one in numerical form. This is the central topic of this book, i.e., the computation of numerical solutions. The challenge then is to compute numerical solutions in reasonable time with acceptable numerical accuracy.

As an illustration of how Eqs. (1.1) and (1.2) might be extended, we consider the growth of bacteria (X) on a single substrate (S) in a culture (see for instance, [2])

$$\nu_S S \rightarrow X$$

ν_S is a pseudo-stoichiometric coefficient (or yield coefficient representing the mass of substrate consumed/number of cells produced).

For the culture, which we consider as the contents of a reactor operated in batch mode (no feed or product streams), the dynamic model is described by two coupled ODEs derived from two mass balances for X and S

Table 1.1 Bacterial growth laws

Case	Growth rate model	Law
1	$\mu = \mu_{\max}$	Zero-order (constant) law
2	$\mu(S) = \mu_{\max} \frac{S}{K_m+S}$	Monod law: substrate limitation
3	$\mu(S) = \mu_{\max} \frac{S}{K_m+S+S^2/K_i}$	Haldane law: substrate limitation and inhibition
4	$\mu(S, X) = \mu_{\max} \frac{S}{K_c X+S}$	Contois law: substrate limitation and biomass inhibition

$$\frac{dX}{dt} = \varphi(S, X) \tag{1.4}$$

$$\frac{dS}{dt} = -\nu_S \varphi(S, X) \tag{1.5}$$

where S denotes the concentration of substrate in the reactor (mass of substrate/volume reactor) and X the concentration of biomass (number of cells/volume reactor). The solution to Eqs. (1.4) and (1.5) will be the dependent variables, X and S, as a function of t in numerical form.

The kinetics (rate equation for the reaction) can be written as

$$\varphi(S, X) = \mu(S, X)X(t) \tag{1.6}$$

where $\mu(S, X)$ denotes the specific growth rate. Depending on the expression of $\mu(S, X)$, the two mass balance equations for X and S can be linear or nonlinear. For example, the specific growth rate models in Table 1.1 are used to model bacterial growth. Case 1 in Table 1.1 corresponds to a constant specific growth rate, and therefore to the linear Eq. (1.1). Cases 2 to 4 are nonlinear specific growth rates which if used in Eqs. (1.4) and (1.5) will preclude analytical solutions. Thus, we will develop numerical solutions to Eqs. (1.4) and (1.5) when using cases 2–4.

The MATLAB function `bacteria_odes` and the MATLAB script `Main_bacteria` can be used to compute numerical solutions to Eqs. (1.4) and (1.5) for the four cases of Table 1.1. We can note the following points about the function `bacteria_odes`:

1. The function named `bacteria_odes` is defined by `function xt = bacteria_odes(t,x)` The inputs to the function are `t`, the independent variable of Eqs. (1.4) and (1.5) (time), and `x`, the vector of dependent variables (X, S). The function computes the derivatives from the RHS of Eqs. (1.4) and (1.5), and returns these derivatives as a vector `xt`. Initially (at $t = 0$), X and S are available from the initial conditions for Eqs. (1.4) and (1.5) (not yet specified). After the integration of Eqs. (1.4) and (1.5) has progressed (by a numerical integration still to be explained), the values of X and S are inputs to `bacteria_odes` from the ODE integrator.
2. A set of global variables is defined, which can be shared between functions (such as between the function `bacteria_odes` and the main program that will be

described next). Note, also, that a line beginning with a % is a comment (and therefore has no effect on the execution of the MATLAB code; rather, it is included only for informational purposes).

3. Since ODE integrators, such as the one called by the main program, generally require that the dependent variables be stored as a vector, the two dependent variables of Eqs. (1.4) and (1.5), X and S, are extracted from the dependent variable vector `x` (again, recall that `x` is an input to the function `bacteria_odes`).
4. One of the growth functions in Table 1.1 is then selected by the *switch* utility of MATLAB, that is controlled by the value of the variable kinetics which is set in the main program and passed to the function `bacteria_odess` as a global variable. This coding illustrates a very important feature of numerical ODE solutions: The ODE complexity (in this case, the nonlinearity of the growth function) is not a limitation. In other words, ODEs of any complexity can be handled numerically; this is certainly not the case if we are attempting to derive an analytical solution.
5. The RHS of Eqs. (1.4) and (1.5) are then computed. Note that we adopted a convenient convention for naming derivatives: We add a character (such as `t`) to the dependent variable name, so that, for example, the derivative dX/dt becomes `Xt`.
6. Finally, the derivative vector consisting of the two elements `Xt` and `St` is returned to the ODE integrator as a column vector. Note that the derivative vector `[Xt St]` (two elements of the vector in a row delimited by square brackets) is interpreted by MATLAB as a row vector. To make it a column vector (as required by the ODE integrator), we apply the transpose operator, which in MATLAB can be denoted simply with an apostrophe (`'`). Other option would be to separate both elements by a semicolon, i.e., `[Xt; St]`.

```
function xt = bacteria_odes(t,x)

% Global variables
global nu mumax Km Ki Kc kinetics

% Transfer dependent variables
X = x(1);
S = x(2);

% Select growth function
switch kinetics
    % Constant
    case('constant')
        mu = mumax;
    % Monod
    case('monod')
        mu = mumax*S/(Km+S);
    % Haldane
    case('haldane')
        mu = mumax*S/(Km+S+(S^2)/Ki);
    % Contois
    case('contois')
        mu = mumax*S/(Kc*X+S);
end
```

```
% ODEs
phi = mu*X;
Xt  = phi;
St  = -nu*phi;

% Transfer temporal derivatives
xt = [Xt St]';
```

Function bacteria_odes Function defining the RHS of Eqs. (1.4) and (1.5)

```
% Culture of bacteria

close all;
clear all;

% Global variables
global nu mumax Km Ki Kc kinetics

% Model parameters
nu    = 0.5e-11;
mumax = 1.4;
Km    = 12;
Ki    = 3;
Kc    = 3e-11;

% Select growth function (comment/decomment one of the
% specific growth rate function)
%
% kinetics = 'constant'
% kinetics = 'monod'
% kinetics = 'haldane'
  kinetics = 'contois'

% Initial, final time, plot interval
t0     = 0;
tf     = 20;
Dtplot = 0.1;

% Initial conditions
ICs = 'case_1'
% ICs = 'case_2'
switch ICs
    % case 1
    case('case_1')
        X = 1.4e11;
        S = 9;
    % case 2
    case('case_2')
        X = 1.4e11;
        S = 12;
end
x0 = [X S]';
% Call to ODE solver
% method = 'ode45'
method = 'ode15s'
```

```
switch method
    % ode45
    case('ode45')
        options      = odeset('RelTol',1e-3,'AbsTol',1e-3);
        options      = odeset(options,'Stats','on',...
                              'Events',@events);
        t            = [t0:Dtplot:tf];
        [tout,xout] = ode45(@bacteria_odes,t,x0,options);
    % ode15s
    case('ode15s')
        options      = odeset('RelTol',1e-3,'AbsTol',1e-3);
        options      = odeset(options,'Stats','on',...
                              'Events',@events);
        t            = [t0:Dtplot:tf];
        [tout,xout] = ode15s(@bacteria_odes,t,x0,options);
end

% Plot results
figure(1)
plot(tout,xout(:,1));
xlabel('t');
ylabel('X(t)');
title('Biomass concentration');
figure(2)
plot(tout,xout(:,2));
xlabel('t');
ylabel('S(t)');
title('Substrate concentration')
```

Script Main_bacteria Main program that calls integrator `ode15s` (or `ode45`)

This completes the programming of Eqs. (1.4) and (1.5). Now we consider the main program `Main_bacteria` that calls the function `bacteria_odes` and two integrators from the ODE SUITE:

1. A defining statement is not required at the beginning.
2. The same global variables are again defined so that they can be shared between the main program `Main_bacteria` and the subordinate function `bacteria_ode`.
3. The model parameters used in Eqs. (1.4) and (1.5) and the growth functions are then defined numerically. The parameter values are taken from [3] and are given in Table 1.2. The units are g (grams), l (liters), and h (hours). When these values are substituted into the RHS of Eqs. (1.4) and (1.5) and the growth functions of Table 1.1, the net units must be those of X and S per hour (the units of the LHS derivatives of Eqs. (1.4) and (1.5)). Then, when these differential equations are integrated, the resulting solution will have the units of X and S as a function of time in hours.
4. The growth function is selected using the variable *kinetics* which is then passed as a global variable to the function `bacteria_odes` to compute a particular growth function. Note in this case the *Contois* growth function is selected since it is not in a comment.

Table 1.2 Bacterial growth model parameters

Parameter	Value	units
ν_S	0.5	$g(10^{11}\text{bacteria})^{-1}$
μ_{max}	1.4	h^{-1}
K_i	3	gl^{-1}
K_c	3	$g(10^{11}\text{bacteria})^{-1}$
K_m	12	gl^{-1}

5. The initial and final values of time and the ICs for Eqs. (1.4) and (1.5) are then defined (again, using the MATLAB *switch* utility). Note in this case that the first set of ICs is selected since the second set is within a comment (i.e., "commented out"). Also, the ICs are converted from a row vector to a column vector using the transpose, which is required by the library integrator.
6. An ODE integrator, either `ode45` (a nonstiff integrator) or `ode15s` (a stiff integrator for ODEs with widely separated eigenvalues or time scales), can be selected. Before actually proceeding with time integration, several parameters and options can be defined using the MATLAB utility `odeset`. In this case, `odeset` first defines a relative error, `RelTol`, and gives it a numerical value of 10^{-3}. Similarly, an absolute error is defined through `AbsTol` with a value of 10^{-3}. A second call to `odeset` defines some additional options:

    a. *Stats* with "on" activates a count of the computational statistics when integrating the ODEs, e.g., number of steps taken along the solution, number of times the RHS of Eqs. (1.4) and (1.5) are evaluated, and in the case of a stiff integrator, the number of times the Jacobian matrix of the ODEs is computed.
    b. *Events* with the user supplied function `@events` (note the `@` which designates another function named `events`) in this case is used to avoid negative concentrations; function `events` is discussed subsequently.

7. A vector, t, is then defined to contain the times at which the numerical solution is to be stored for subsequent output. In this case, t runs from 0 to `tf` in steps of `Dt plot`= 0.1. Since `tf` = 20 (as set previously), a total of 201 output values will be returned by the ODE integrator. This may seem like an excessive number, which might be true if numerical values of the solution are the primary result; however, the solution will be plotted, and 201 values are not excessive, i.e., they will produce a smooth plot with significant detail.
8. Integrator `ode15s` is then called. The function `bacteria_ode` is an input to `ode15s` (`@bacteria_odes`, recall again that a MATLAB function is denoted with an `@`). Also, `x0` is the initial condition vector set previously which is an input to `ode15s` (the solution starts at this initial condition); `ode15s` knows that two ODEs are to be integrated because the initial condition vector x_0 has two elements, the values $X(0)$ and $S(0)$. The options set previously are also an input to `ode15s`. The output is the numerical solution with the 201 values of t stored in tout and the two dependent variables of Eqs. (1.4) and (1.5), X and S, stored in `xout`. Thus, `xout` is a two-dimensional array with the first subscript running from 1 to 201

for the 201 values of t and the second subscript running from 1 to 2 for the two dependent variables X and S. As an example, `x(21,2)` would contain $S(t = 2)$.

9. The solution to Eqs. (1.4) and (1.5) is now computed, so the next section of code plots this solution. Two figures (plots) are drawn as specified by the MATLAB statement `figure`. The plot command uses the vector of 201 output times in array tout as the abscissa variable, and the array `xout` as the ordinate value. For the latter, the particular dependent variable is specified as the second subscript (1 for X, 2 for S). The colon used as the first subscript specifies all values of the first subscript are to be used, in this case 1 to 201. The x and y axis labels are written as well as a title at the top of the plot.

The function `events` (called by the integrators `ode45` and `ode15s` through their argument list, i.e., `options`) is listed below

```
function [value,isterminal,direction] = events(t,x)

% Transfer dependent variables
X = x(1);
S = x(2);

% Check if substrate concentration becomes negative
value      = S;    % monitor S and see if it vanishes
isterminal = 1;    % stop integration when S vanishes
direction  = -1;   % S decreases from initial positive values
```

Function events Function events to monitor S

If S (from the integration of Eq. (1.5)) should become negative (which physically is impossible since a concentration cannot be negative), the execution of the main program `Main_bacteria` will stop. To this end, the integrators `ode45` and `ode15s` use the function `events` to monitor the variable value (which is assigned to S in the example), and to check whether value is decreasing through zero (`direction=-1`), increasing through zero (`direction=+1`), or vanishing in either direction (`direction=0`). In the example, the substrate concentration starts at an initial positive value, decreases, and could possibly cross zero as time evolves, so that the former option is selected, i.e., `direction=-1`. When value becomes zero, the integration is halted (`isterminal=1`), since negative concentrations are unacceptable. In other examples, an event detection could be less critical (e.g., we could simply be interested in detecting and recording the zero crossings of a variable taking positive and negative values), so that the option `isterminal=0` would be selected (the event is recorded while time integration continues).

In fact, the use of an *event* function is required only when the first growth law of Table 1.1 is considered. In this linear case, the biomass concentration can grow unbounded and the substrate concentration can become negative, which is physically impossible. However, the nonlinear growth laws, cases 2–4 in Table 1.1, intrinsically avoid such situations and do not require the use of an *event* function. This stresses the development of meaningful models, guaranteeing bounded input–bounded output

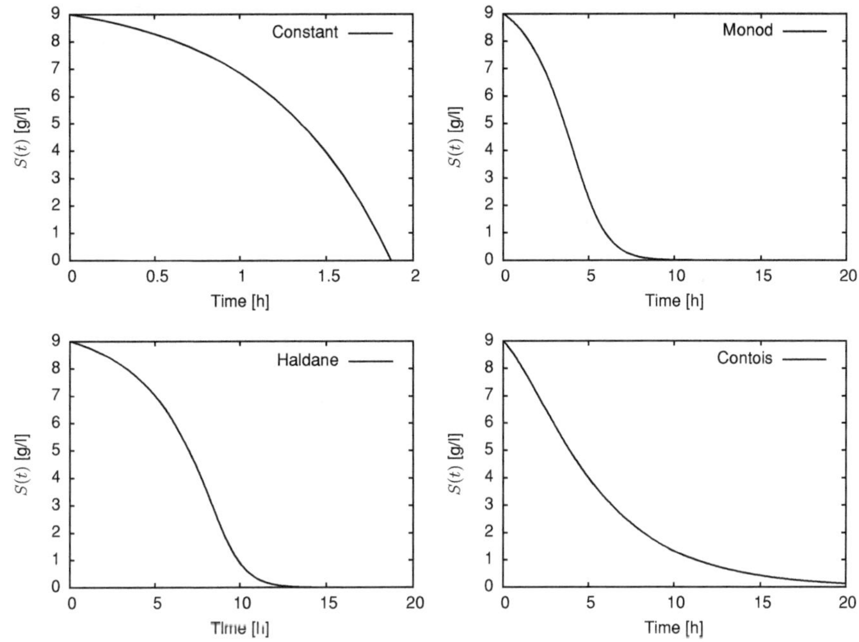

Fig. 1.1 Evolution of the substrate (S) obtained by means of Eqs. (1.4) and (1.5) for the growth laws in Table 1.1

behavior, or at least the use of simplified models (such as the linear growth model) in restricted operating ranges only (a constant specific growth law can reproduce data well when substrate is in abundance).

This completes the programming of Eqs. (1.4) and (1.5), and associated initial conditions. The plots of $S(t)$ vs. t are indicated in Fig. 1.1 for the four growth functions. Note that the plots all start at the correct initial condition $S(0) = 9$ (a good check for any numerical solution). Also, the time scale for the three nonlinear growth functions (Cases 2, 3, 4 in Table 1.1) is $0 \leq t \leq 20$ while for the linear growth function (Case 1 in Table 1.1) it is $0 \leq t \leq 2$. This is an important point, i.e., the time scale for an ODE integration must be selected by the analyst since the initial value independent variable is essentially open-ended. The time scale can change for the same ODE system with changes in the structure and parameters of the ODEs; for example, the time scale changes by a factor of 10 when changing the growth law from linear to nonlinear (which is logical since the nonlinear laws express that growth slows down as substrate resources become scarce). If the time scale is not selected carefully, the numerical solution may not be defined (only a small portion of the solution is computed if the time scale it too short, or the essence of the solution may be missed completely if the time scale is too long). Clearly, the choice of the growth function has a significant effect on the solution. The challenge then would be to select a growth function, and associated numerical parameters, that would give a plot of $S(t)$ vs. t which is in good agreement with experimental data.

Table 1.3 Numerical parameters for Eqs. (1.7) to (1.11)

Parameter	Value	Units
C_1	5×10^3	s^{-1}
C_{-1}	10^6	s^{-1}
E_1	10	$kcal \cdot mol^{-1}$
E_{-1}	10	$kcal \cdot mol^{-1}$
ρC_p	1	$kcal \cdot l^{-1} \cdot K^{-1}$
$-\Delta H_R$	5	$kcal \cdot mol^{-1}$
R	1.987	$cal \cdot mol^{-1} \cdot K^{-1}$

We described the MATLAB programming of Eqs. (1.4) and (1.5) in detail for the reader who may not have programmed ODE solutions previously. In the subsequent example applications, we will add a discussion of only those details which are specific to a particular application.

We next consider another ODE example that will then be extended to include other types of equations. The energy and material balances for a batch stirred tank reactor (BSTR), taken from [4], include a reversible exothermic reaction $A \leftrightarrow B$. The dynamic model is described by two mass balance ODEs and one energy balance ODE.

$$\frac{dA}{dt} = -k_1 A + k_{-1} B; \qquad A(0) = A_0 \tag{1.7}$$

$$\frac{dB}{dt} = k_1 A - k_{-1} B; \qquad B(0) = B_0 \tag{1.8}$$

Note that Eq. (1.8) can also be expressed in algebraic form as

$$B(t) = A_0 + B_0 - A(t) \tag{1.9}$$

since $dA/dt + dB/dt = 0$, and therefore $A(t) + B(t) = A_0 + B_0 = constant$.

$$\frac{dT}{dt} = -\frac{\Delta H_R}{\rho C_p}(k_1 A - k_{-1} B); \qquad T(0) = T_0 \tag{1.10}$$

with the Arrhenius temperature dependency of the reaction rate constants

$$k_1 = C_1 \exp\left(\frac{-E_1}{RT}\right); \qquad k_{-1} = C_{-1} \exp\left(\frac{-E_{-1}}{RT}\right) \tag{1.11}$$

where t is time, $A(t)$, $B(t)$ are the molar concentrations of components A and B, respectively, and $T(t)$ is the temperature of the liquid in the reactor. The constant parameter values taken from [4] are given in Table 1.3.

A function to define the RHS of Eqs. (1.7), (1.9), (1.10), and (1.11) is listed in Function `bstr_odes_ae` which has essentially the same format as the function `bacteria_odes`.

```
function xt = bstr_odes_ae(t,x)

% Global variables
global A0 B0 rhocp Cd Cr Ed Er DH R B

% Transfer dependent variables
A = x(1);
T = x(2);

% Algebraic equations
B  = A0 + B0 - A;
kd = Cd*exp(-Ed/(R*T));
kr = Cr*exp(-Er/(R*T));

% ODE temporal derivatives
At = -kd*A+kr*B;
Tt = (kd*A-kr*B)*DH/(rhocp);

% Transfer temporal derivatives
xt = [At Tt]';
```

Function bstr_odes_ae Implementation of the ODEs (1.7) and (1.10) using the algebraic relations (1.9) and (1.11)

We can note the following additional details:

1. Again, `t` and `x` are the two input arguments of `bstr_odes_ae`. Thus, since the concentration A is set to $x(1)$ and the temperature T is set to $x(2)$, they are available for use in the algebraic equations for `B`, `kd` and `kr`. In other words, *all of the algebraic variables that are used in the calculation of the derivatives must be computed first before the derivative vector* `[At Tt]` *is computed.*
2. As discussed before (for function `bacteria_ode`), the final output from `bstr_odes_ae` must be a vector with the derivatives for all of the ODEs, in this case, the vector containing `At`, `Tt`. In other words, *the input to a derivative function is the vector of dependent variables* (x) *and the independent variable* (t). *The output from the function is the vector of derivatives in column format* (`xt`).

The main program that calls Function `bstr_odes_ae` is listed in the MATLAB script `Main_bstr` where, for the sake of brevity, the problem presentation, i.e., the comments at the beginning of the main program, are not reproduced:

```
% Global variables
global A0 B0 rhocp Cd Cr Ed Er DH R B

% Model parameters
rhocp = 1000;
Cd    = 5000;
Cr    = 1000000;
Ed    = 10000;
```

```
Er     = 15000;
DH     = 5000;
R      = 1.987;

% Initial conditions
A0   = 1;
B0   = 0;
T0   = 430;
x0   = [A0 T0]';

% In-line Euler integrator
h          = 0.5;
t          = [0 : h : 100];
nout       = 200;
tk         = 0;
xout       = [A0 B0 T0];
fprintf('h = %5.2f\n\n',h);
fprintf(' %8.1f%10.4f%10.4f%8.1f\n',tk, A0,B0,T0);
for iout = 1:nout
    [xt]      = bstr_odes_ae(tk,x0);
    x0        = x0+xt*h;
    tk        = tk+h;
    xout      = [xout;x0(1) 1-x0(1) x0(2)];
    fprintf(' %8.1f%10.4f%10.4f%8.1f\n',tk,x0(1),...
               A0+B0-x0(1),x0(2));
end

% Plot the solution
subplot(3,1,1)
plot(t,xout(:,1));
ylabel('A(t)');

subplot(3,1,2)
plot(t,xout(:,2));
ylabel('B(t)');

subplot(3,1,3)
plot(t,xout(:,3));
xlabel('t');
ylabel('T(t)');
```

Script Main_bstr Main program that calls function `bstr_odes_ae`

Note that:

1. The same features are at the beginning as in the main program `Main_bacteria`, i.e., definition of the model parameters (which are passed as global variables) and the model initial conditions. An initial condition for B of Eq. (1.8) is set, but B will actually be computed algebraically using (1.9) (not by integrating Eq. (1.8)).
2. Numerical integration of Eqs. (1.7) and (1.10) is performed by the *Euler method*. To start the integration, the value of t (the independent variable), `tk`, is initialized and the *integration step*, h, along the solution is set. The number of Euler steps `nout` is defined numerically and an output vector `xout` is initialized with the initial conditions set previously.

3. `nout` steps are then taken along the solution using a `for` loop: the derivative function, `bstr_odes_ae`, is first called to evaluate the derivative vector `xt`. A step is next taken along the solution by the Euler method (note that the base value of the dependent variable, x in the RHS, is replaced by the new or advanced value of x at the next step along the solution, i.e., `x0 = x0 + xt*h;`) the independent variable is then incremented to reflect the next point along the solution. Finally, the solution vector at the advanced point along the solution is added to the solution vector to this point (in three-column format), i.e., `xout = [xout; x0(1) A0+B0-x0(1) x0(2)];`
4. Execution of the `for` loop for `nout` steps along the solution generates the complete solution which is now ready for plotting (since the entire solution has been collected in array `xout`). The plotting is essentially self explanatory. `subplot` subdivides a figure into separate graphs, i.e., `subplot(3,1,p)` defines a 3 by 1 array of graphs, and p is the handle to the current graph (e.g., $p = 1, 2, 3$). Note that a particular dependent variable is selected for all t using the notation of MATLAB, e.g., `plot(t,xout(:,1))`. This call to plot has as inputs the vector of the independent variable set previously with `t=[0:0.5:100];` followed by the three-column vector `xout` (for which the first column $A(t)$ is selected); the colon symbol (`:`) indicates all rows (for all of the values of t) of the first column are to be plotted. The resulting graphs are presented in Fig. 1.2 and the tabular output from the program is summarized in Table 1.4.

We now use this output to demonstrate some of the essential features of the Euler method. First, we note that the initial conditions are correct as set in the script `Main_bstr`. Displaying the initial conditions is important to ensure that the solution has the right starting point. Then, some physical reasoning can be applied to confirm that the solutions have reasonable form. In this case, as expected, $A(t)$ decreases with t, while $B(t)$ and $T(t)$ increase (as shown in Fig. 1.2). Finally, the dependent variables approach an equilibrium or a steady state ($A(t) = 0.3753$, $B(t) = 0.6247$, $T = 433.1$). We might then ask the question, why not $A(t) = 0$, $B(t) = A0 + B0$, the answer is because the chemical reaction is equilibrium limited.

While these checks may seem obvious, they are nonetheless important, especially when the size and complexity (numbers of ODEs) increase. If the solution does not make sense at this point, there is no sense in continuing to work with it until the apparent inconsistencies (with physical reasoning) are resolved.

At this point we ask an important question: *How do we know that the numerical solution is correct?*, or in other words, since the Euler method provides a numerical approximation to the exact solution, what accuracy does the numerical solution have? If we have an exact solution, e.g., Eq. (1.3), then we can determine the exact error in the numerical solution. However, this usually will not be the case, i.e., if we have an exact solution, there is no need to calculate a numerical solution. In fact, numerical methods are most useful for difficult problems for which an analytical solution is not available.

The question then is how do we ascertain the accuracy of a numerical solution when we do not have an exact solution. To answer this question, we start with the

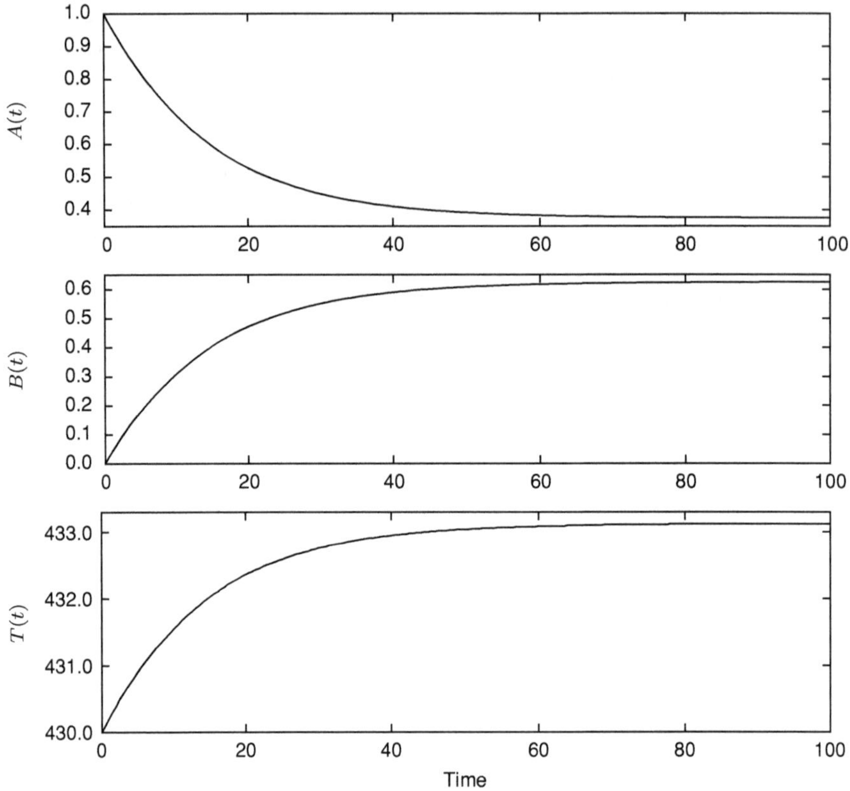

Fig. 1.2 Graphical output from the main program `Main_bstr`

Table 1.4 Partial output from the script `Main_bstr` and function `bstr_odes_ae`

h=0.50			
0.0	1.0000	0.0000	430.0
0.5	0.9793	0.0207	430.1
1.0	0.9593	0.0407	430.2
1.5	0.9399	0.0601	430.3
2.0	0.9210	0.0790	430.4
2.5	0.9028	0.0972	430.5
3.0	0.8850	0.1150	430.6
3.5	0.8679	0.1321	430.7
4.0	0.8512	0.1488	430.7
4.5	0.8351	0.1649	430.8
5.0	0.8195	0.1805	430.9
⋮	⋮	⋮	⋮
98.0	0.3753	0.6247	433.1
98.5	0.3753	0.6247	433.1
99.0	0.3753	0.6247	433.1
99.5	0.3753	0.6247	433.1

Taylor series. Thus, if we have an ODE

$$\frac{\mathrm{d}x}{\mathrm{d}t} = f(x, t) \tag{1.12}$$

with initial condition

$$x(t_0) = x_0 \tag{1.13}$$

the solution to Eq. (1.12) can be expanded around the initial condition

$$x(t_1) = x(t_0) + \frac{h}{1!}\left.\frac{\mathrm{d}x(t)}{\mathrm{d}t}\right|_{t=t_0} + \frac{h}{2!}\left.\frac{\mathrm{d}^2x(t)}{\mathrm{d}t^2}\right|_{t=t_0} + \ldots \tag{1.14}$$

where $h = t_1 - t_0$. If h is small, the higher order terms in Eq. (1.14) can be neglected, and therefore, it reduces to

$$x_1 = x(t_0) + \frac{h}{1!}\left.\frac{\mathrm{d}x(t)}{\mathrm{d}t}\right|_{t=t_0} \tag{1.15}$$

which is *Euler's method* (note that we use x_1 to denote the numerical solution at $t_1 = t_0 + h$ resulting from truncation of the Taylor series, while $x(t_1)$ denotes the exact solution at $t_1 = t_0 + h$).

In other words, we can step from the initial condition $x(t_0)$ to a nearby point along the solution, x_1, using Eq. (1.15) since the derivative $\mathrm{d}x(t)/\mathrm{d}t|_{t=t_0}$ is available from the ODE, Eq. (1.12). The only other required arithmetic operations (in addition to the evaluation of the derivative) in using Eq. (1.15) are multiplication (by h) and addition (of $x(t_0)$ which is the initial condition).

Then we can use x_1 in the RHS of Eq. (1.15) to compute x_2, x_2 to compute x_3, etc.,

$$x_{k+1} = x_k + \frac{h}{1!}\left.\frac{\mathrm{d}x}{\mathrm{d}t}\right|_{t=t_k} = x_k + hf(x_k, t_k) \tag{1.16}$$

until we have computed the entire numerical solution to some final value of t that we specify. This stepping process is illustrated by the `for` loop in script `Main_bstr` for 200 steps. An important point is to note that *this stepping procedure is entirely numerical* (which a computer can do very well); knowledge of mathematical methods for the analytical integration of ODEs is not required.

Note also that Eq. (1.16) can be applied to a system of ODEs if x is interpreted as a vector of dependent variables. In the case of Eqs. (1.7) and (1.10), this vector has the two components A and T. The vector of derivatives, $\mathrm{d}x/\mathrm{d}t$, is computed as required by Eq. (1.16) by calling a derivative function, in the case of script `Main_bstr`, function `bstr_odes_ae`.

The preceding discussion covers all of the essential details of integrating a system of ODEs by the fixed step, explicit Euler method of Eq. (1.16). One additional point we can note is that, during the integration, we evaluated three algebraic variables,

B, kd, and kr, which were then used in the calculation of the derivative vector in function `bstr_odes_ae`. These algebraic variables could be computed explicitly using only the dependent variables A and T. However, we might also have a situation where the calculation of the ODE derivatives and the algebraic variables cannot be separated, and therefore they must be calculated simultaneously. Generally, this is the requirement of a differential-algebraic (DAE) system. This simultaneous solution requires a more sophisticated algorithm than the Euler method of Eq. (1.16). We will subsequently consider some DAE algorithms.

We now return to the basic question of how do we determine the accuracy of the numerical solution of Table 1.4 (and we have to assess the accuracy without having an analytical solution). To address this question, one line of reasoning would be to ask if the integration step h is small enough so that the truncation of Eq. (1.14) to (1.15) provides sufficient accuracy when the numerical solution is computed by Eq. (1.15). In other words, we might reduce h and observe any change in the numerical solution. To this end, we modify the script `Main_bstr` by changing two statements

```
h = 0.25
```

and

```
fprintf(' %8.2f%10.4f%10.4f%8.1f\n',tk,x(1),A0+B0-x(1),x(2));
```

The only change in the second statement is to replace the format for t, `8.1f` by `8.2f` so that the additional figures in t due to a smaller change in t (from 0.5 to 0.25) will be observed in the output. The output from the modified program is summarized in Table 1.5 `bstr_odes_ae`.

Now, to assess the effect of the change in h, we can compare the outputs in Tables 1.4 and 1.5. For example, at $t = 1.0$ (see Table 1.6).

From these results, we can infer that the accuracy of the numerical solution is approximately *three significant figures*. If we were unable to come to this conclusion (because the results did not agree to three figures), we could execute the script `Main_bstr` again with a smaller h and compare solutions. Hopefully we would arrive at a small enough h that we could conclude that some required accuracy has been achieved, e.g., three figures. This process of reducing h to establish an apparent accuracy is termed *h refinement*. The process of arriving at a solution of a given accuracy is termed *convergence*.

The procedure of stepping along the solution with a prescribed h is illustrated in Fig. 1.3.

Table 1.5 Partial output from the script `Main_bstr` and function `bstr_odes_ae` with $h = 0.25$

h=0.25			
0.00	1.0000	0.0000	430.0
0.25	0.9897	0.0103	430.1
0.50	0.9795	0.0205	430.1
0.75	0.9695	0.0305	430.2
1.00	0.9596	0.0404	430.2
1.25	0.9499	0.0501	430.3
1.50	0.9403	0.0597	430.3
1.75	0.9309	0.0691	430.3
2.00	0.9216	0.0784	430.4
2.25	0.9125	0.0875	430.4
2.50	0.9034	0.0966	430.5
⋮	⋮	⋮	⋮
47.75	0.3953	0.6047	433.0
48.00	0.3950	0.6050	433.0
48.25	0.3946	0.6054	433.0
48.50	0.3942	0.6058	433.0
48.75	0.3939	0.6061	433.0
49.00	0.3935	0.6065	433.0
49.25	0.3932	0.6068	433.0
49.50	0.3929	0.6071	433.0
49.75	0.3925	0.6075	433.0

Table 1.6 Partial output from the script `Main_bstr` and function bstr_odes_ae for $h = 0.25, 0.5$

h=0.25				
h=0.5	1.0	0.9593	0.0407	430.2
h=0.25	1.00	0.9596	0.0404	430.2

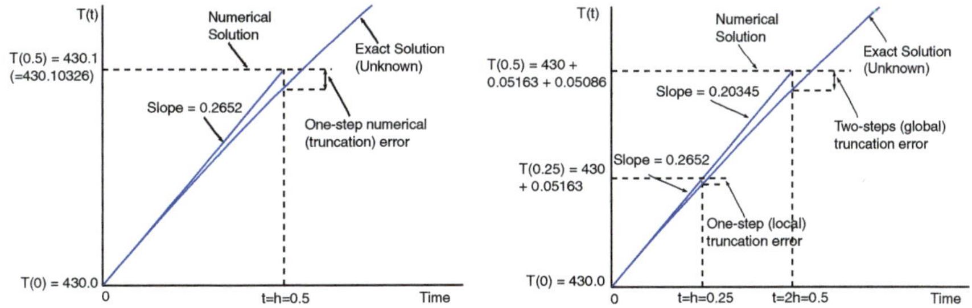

Fig. 1.3 Time stepping along the solution using Euler method

Table 1.7 Partial output from the script `Main_bstr` (modified) with $h = 0.5$

h=0.5			
0.00	1.0000	0.0000	430.0
-0.04130	-0.02065	0.20652	0.10326
0.50	0.9793	0.0207	430.1
-0.04007	-0.02004	0.20036	0.10018
1.0	0.9593	0.0407	430.2
-0.03887	-0.01943	0.19435	0.09717
1.5	0.9399	0.0601	430.3
2.0	0.9210	0.0790	430.4
2.5	0.9028	0.0972	430.5
3.0	0.8850	0.1150	430.6
⋮	⋮	⋮	⋮
98.0	0.3753	0.6247	433.1
98.5	0.3753	0.6247	433.1
99.0	0.3753	0.6247	433.1
99.5	0.3753	0.6247	433.1

The numbers required for Fig. 1.3 were produced by modifying the script `Main_bstr` slightly:

```
for iout = 1:nout
    fprintf(' %10.1f%10.4f%10.4f%10.1f\n',tk,x(1),A0+B0-...
            x(1), x(2));
    [xt] = bstr_odes_ae(tk,x);
    if tk<=1.0
        fprintf(' %10.5f%10.5f%10.5f%10.5f\n\n', xt(1),...
                xt(1)*h,xt(2),xt(2)*h);
    end
end
```

Now, in addition to displaying the solution vector, `tk`, `x(1)`, `A0+B0-x(1)`, `x(2)`, we also display the derivative vector, `xt(1)`, `xt(2)`, and the steps taken along the solution according to Eq. (1.15), `xt(1)*h`, `xt(2)*h`, which are usually called Euler steps. A sample of the output from this for loop is listed in Table 1.7, for $h = 0.5$.

Note that the output for $t = 0$ is plotted in Fig. 1.3a. For $h = 0.25$, a sample of the output from the program is given in Table 1.8.

The output for $t = 0$ and 0.25 is plotted in Fig. 1.3b. We can note the following important points in comparing the two plots:

1. The error that results from the application of Eq. (1.15) is a *one-step or local error* (one step is taken along the numerical solution) as illustrated in Fig. 1.3a and the first step (from $t = 0$) in Fig. 1.3b.
2. This local error can accumulate (as inferred in the second step of Fig. 1.3b) to produce a global error. Thus, at $t = 0.5$ in Fig. 1.3b, the global error is the result

Table 1.8 Partial output from the script `Main_bstr` (modified) with $h = 0.25$

h=0.25			
0.00	1.0000	0.0000	430.0
-0.04130	-0.02065	0.20652	0.10326
0.50	0.9793	0.0207	430.1
-0.04007	-0.02004	0.20036	0.10018
1.0	0.9593	0.0407	430.2
-0.03887	-0.01943	0.19435	0.09717
1.5	0.9399	0.0601	430.3
2.0	0.9210	0.0790	430.4
2.5	0.9028	0.0972	430.5
3.0	0.8850	0.1150	430.6
⋮	⋮	⋮	⋮
98.0	0.3753	0.6247	433.1
98.5	0.3753	0.6247	433.1
99.0	0.3753	0.6247	433.1
99.5	0.3753	0.6247	433.1

of taking two steps of length $h = 0.25$. Clearly as the number of steps increases, this accumulation of the one-step errors to produce the global error could lead to inaccurate solutions. In other words, we are really interested in *controlling the global error* since this is the actual error in the numerical solution at any point along the solution. Fortunately, this is generally possible by taking a small enough integration step h.

3. We can infer from Eq. (1.14) that the local error is proportional to h^2 if the error from the truncation of the Taylor series in Eq. (1.14) to produce the Euler method of Eq. (1.15) is *essentially limited to the leading term after the point of truncation*, i.e., the term $\frac{h^2}{2!}\frac{d^2x(t_0)}{dt^2}$. This property of the local error proportional to h^2 is generally termed *of order h^2 or second order correct*, and is denoted as $O(h^2)$ where "O" (big-oh) denotes "of order".
4. However, we are particularly interested in the global error since this is the actual error after a series of n integration steps. As the neglected term for each step from t_k to t_{k+1} has the form $\frac{h^2}{2!}\frac{d^2x}{dt^2}$, the global error will grow like the sum of the h^2 terms, i.e.:

$$\frac{h^2}{2!}\sum_{k=1}^{n}\left.\frac{\mathrm{d}^2x(t_0)}{\mathrm{d}t^2}\right|_t = \frac{nh^2}{2!}\left.\frac{\mathrm{d}^2x}{\mathrm{d}t^2}\right|_t = \frac{t_f - t_0}{2}\left.\frac{\mathrm{d}^2x}{\mathrm{d}t^2}\right|_t h = O(h)$$

an expression in which we assume that the derivative is smooth and the intermediate value theorem can be used. Hence, *the Euler method is first order correct with respect to the global error*. In general, for the ODE integration algorithms we will consider subsequently, the *order of the global error is one less than the order of the local error*, e.g., first and second order, respectively, for the Euler method. A practical way to observe the first-order behavior of the Euler method would be to compute numerical solutions to a problem with a known analytical solution (so that the exact error is available) using a series of integration steps, then plot the exact error versus the integration step size h at a particular value of the independent variable, t. Generally, we would observe a straight line with a slope of 1 (we cannot guarantee this first-order behavior for all problems since it will depend on the properties of the problem, but generally the first-order behavior will be observed).

This latter procedure suggests a method for computing a high accuracy numerical solution to an ODE problem. Solutions for a series of integration steps can be computed. Then the solutions can be plotted as a function of h (for a particular t) and an extrapolation to $h = 0$ can be performed to arrive at the high accuracy solution. This is known as a *Richardson extrapolation*. Its effectiveness is largely dependent on the reliability of the extrapolation to $h = 0$ (which would be an extrapolation using a straight line for a first-order method such as the Euler method). However, this procedure does not require the use of an analytical solution, so it is quite general. Also, the procedure can be efficiently organized in a computer program, and, in fact, ODE integrators that use the Richardson extrapolation are available [5].

We can now appreciate the limited accuracy of the Euler method due to the truncation of the Taylor series of Eq. (1.14) to arrive at Eq. (1.15). Thus, the resulting integration error is generally termed the *truncation error*. More generally, the preceding discussion demonstrates the importance of an *error analysis* whenever we compute a numerical solution to a problem; this is essential so that we have some confidence that the solution is accurate and therefore reliable and useful. The challenge then is to do an error analysis without having an analytical solution so that we can judge the error of the numerical solution. This might seem like an impossible requirement, but the secret is to *estimate the error* with sufficient accuracy that we have some confidence the estimated error does in fact indicate the accuracy of the numerical solution. We have actually gone through this process for an ODE solution, i.e., compute numerical solutions for a series of integration steps and observe if the solutions appear to be converging to a certain number of significant figures.

The next logical step would be to look for a more accurate algorithm. From Eq. (1.14), we see that Euler's method uses a forward finite difference approximation of the time derivative:

$$\left.\frac{\mathrm{d}x}{\mathrm{d}t}\right|_k = \frac{x_{k+1} - x_k}{h} + O(h) \tag{1.17}$$

If we now expand both x_{k+1} and x_{k-1} around x_k

$$x_{k+1} = x_k + \frac{h}{1!}\left.\frac{\mathrm{d}x}{\mathrm{d}t}\right|_k + \frac{h^2}{2!}\left.\frac{\mathrm{d}^2x}{\mathrm{d}t^2}\right|_k + \cdots$$

$$x_{k-1} = x_k - \frac{h}{1!}\left.\frac{\mathrm{d}x}{\mathrm{d}t}\right|_k + \frac{h^2}{2!}\left.\frac{\mathrm{d}^2x}{\mathrm{d}t^2}\right|_k + \cdots$$

and subtract the two expressions, we obtain a centered approximation

$$\left.\frac{\mathrm{d}x}{\mathrm{d}t}\right|_k = \frac{x_{k+1} - x_{k-1}}{2h} + O(h^2) \tag{1.18}$$

which is more accurate ($O(h^2)$ rather than $O(h)$).

The leapfrog method corresponds to this more accurate approximation, i.e.

$$x_{k+1} = x_{k-1} + 2h\left.\frac{\mathrm{d}x}{\mathrm{d}t}\right|_k + 0(h^3) = x_{k-1} + 2hf(x_k, t_k) + 0(h^3) \tag{1.19}$$

This method is third order correct with respect to the local error, and second order correct with respect to the global error. This is a substantial improvement in accuracy over Euler's method. For example, if the integration step h is reduced by a factor of two, the local error will decrease by a factor of four (leapfrog method) rather than two (Euler's method).

An apparent drawback of formula (1.19) however is that it requires two initial values, e.g., x_0 and x_1, to start the stepping process (so that in practice it is necessary to start integration with another method, e.g., Euler method, in order to get x_1 from x_0). If we apply this algorithm without initialization procedure (i.e., we simply assume $x_0 = x_1$) and with $h = 0.05$ (i.e., a small time step compared to the ones we have used with the Euler method) to our batch reactor example (1.7)–(1.10), we obtain the graphs of Fig. 1.4. Obviously, the numerical solution becomes unstable (and reducing h cannot cure the problem).

Let us now initialize the stepping process using Euler's method to compute the second initial condition required by the leapfrog method. The integration results, also with $h = 0.05$, are shown in Fig. 1.5. The solution has clearly improved, however, we can observe high-frequency and low amplitude oscillations at the end of the integration time. In both cases (with and without a proper initialization procedure), the algorithm is unstable. In some sense, an error in the initial condition acts as a perturbation which triggers the instability at an earlier stage. We will analyze stability in more detail later in this chapter.

Rather than considering the centered approximation (1.18), which involves the three points t_{k-1}, t_k and t_{k+1}, let us try an alternative centered approximation evaluated in the midpoint $t_{k+1/2}$ of the interval $[t_k, t_{k+1}]$

$$\left.\frac{\mathrm{d}x}{\mathrm{d}t}\right|_{k+\frac{1}{2}} = \frac{x_{k+1} - x_k}{h} + O(h^2) \tag{1.20}$$

This approximation, which is more accurate than (1.17) and has the same level of accuracy than (1.18) can be obtained by subtracting two Taylor series expansions around $x_{k+1/2}$

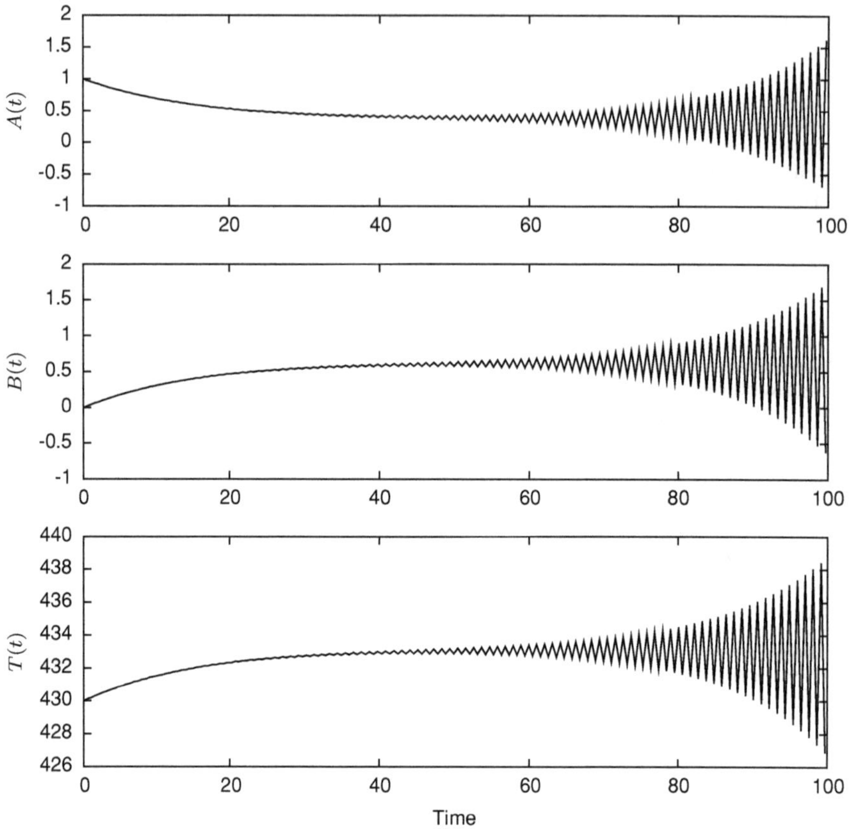

Fig. 1.4 Solution of the batch reactor problem (1.7)–(1.10) using the leapfrog method with $h = 0.05$ and $x_0 = x_1$

$$x_{k+1} = x_{k+\frac{1}{2}} + \frac{h}{2}\left.\frac{\mathrm{d}x}{\mathrm{d}t}\right|_{k+\frac{1}{2}} + \frac{h^2}{4}\left.\frac{\mathrm{d}^2x}{\mathrm{d}t^2}\right|_{k+\frac{1}{2}} + O(h^3)$$

$$x_k = x_{k+\frac{1}{2}} - \frac{h}{2}\left.\frac{\mathrm{d}x}{\mathrm{d}t}\right|_{k+\frac{1}{2}} + \frac{h^2}{4}\left.\frac{\mathrm{d}^2x}{\mathrm{d}t^2}\right|_{k+\frac{1}{2}} + O(h^3)$$

This yields the following numerical scheme

$$x_{k+1} = x_k + h\left.\frac{\mathrm{d}x}{\mathrm{d}t}\right|_{k+\frac{1}{2}} + O(h^3) = x_k + hf(x_{k+\frac{1}{2}}, t_{k+\frac{1}{2}}) + O(h^3) \qquad (1.21)$$

where the term $f(x_{k+1/2}, t_{k+1/2})$ is however unknown.

To estimate the solution at the midpoint $t_{k+1/2}$, we can simply use Euler's method as a *predictor*

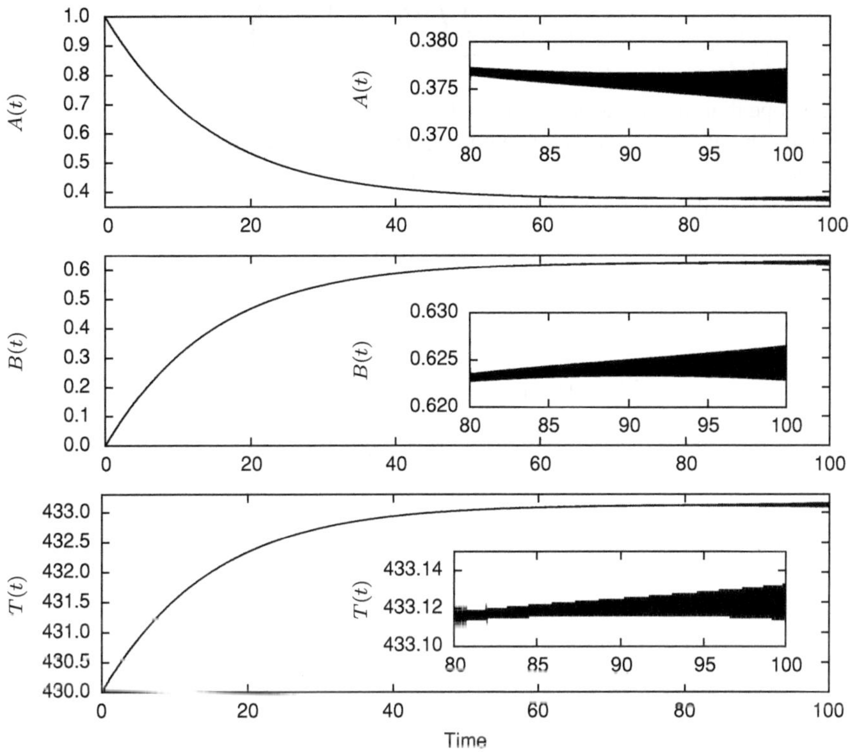

Fig. 1.5 Solution of the batch reactor problem (1.7)–(1.10) using the leap frog method with $h = 0.05$ and a first step of the Euler method to start the stepping process

$$\tilde{x}_{k+\frac{1}{2}} = x_k + \frac{h}{2} f(x_k, t_k) \tag{1.22}$$

which, when substituted in (1.21), gives

$$x_{k+1} = x_k + \frac{h}{2} f(\tilde{x}_{k+\frac{1}{2}}, t_{k+\frac{1}{2}}) \tag{1.23}$$

This latter expression is known as the *midpoint method* since it uses the slope of the solution evaluated in the midpoint $t_{k+1/2}$ instead of the base point t_k, as in the original Euler's method. This method, which is by one order more accurate than the Euler method, is an example of *predictor-corrector formula*, i.e., Euler's method (1.22) is used to predict the solution in the midpoint $t_{k+1/2}$, and this prediction is then used in the corrector (1.23).

Instead of estimating the slope of the solution in the midpoint $t_{k+1/2}$, we could also try to use an average slope over the interval $[t_k, t_{k+1}]$

$$x_{k+1} = x_k + \frac{1}{2}\left(\left.\frac{dx}{dt}\right|_k + \left.\frac{dx}{dt}\right|_{k+1}\right) \tag{1.24}$$

in the hope that this representation would lead to a more accurate result than Euler's method. This is indeed an interesting idea, which can be interpreted as an attempt to include the second-order derivative term in the truncated Taylor series of the solution, or at least a finite difference of this latter term

$$\begin{aligned} x_{k+1} &= x_k + \frac{h}{1!}\left.\frac{dx}{dt}\right|_k + \frac{h^2}{2!}\left.\frac{d^2x}{dt^2}\right|_k + O(h^3) \\ &= x_k + \frac{h}{1!}\left.\frac{dx}{dt}\right|_k + \frac{h^2}{2!}\frac{\left(\left.\frac{dx}{dt}\right|_{k+1} - \left.\frac{dx}{dt}\right|_k\right)}{h} + O(h^3) \\ &= x_k + \frac{h^2}{2!}\frac{\left(\left.\frac{dx}{dt}\right|_{k+1} + \left.\frac{dx}{dt}\right|_k\right)}{h} + O(h^3) \end{aligned} \tag{1.25}$$

As for the midpoint method, a predictor step is needed in the form

$$\tilde{x}_{k+1} = x_k + hf(x_k, t_k) \tag{1.26}$$

which can then be used in a corrector step

$$\tilde{x}_{k+1} = x_k + \frac{h}{2}\left(f(x_k, t_k) + f(\tilde{x}_{k+1}, t_{k+1})\right) \tag{1.27}$$

Equations (1.26)–(1.27) constitute what is generally referred to as *Heun's method* (in some texts, this predictor-corrector formula is also referred to as the *modified Euler method* or *the extended Euler method*, whereas other authors use these terms to designate the midpoint method). As with the leapfrog method, Heun's method is third order correct with respect to the local error, and second order correct with respect to the global error. Note, however, that there is a computational price for the improved accuracy. Whereas the Euler method (1.16) requires only one derivative evaluation for each integration step, Heun's method requires two derivative evaluations. In general, this additional computational effort is well worth doing to improve the accuracy of the numerical solution. If we apply this algorithm to our batch reactor example (1.7)–(1.10), we obtain essentially the same results as with Euler's method, but we can now use much larger time steps, e.g., $h = 2$.

We could pursue the idea of including higher order derivative terms in the truncated Taylor series expansion in order to improve the accuracy of the numerical solution. This is actually what is done in a family of method called *Taylor series methods*, in which the ODE function $f(x, t)$ is successively differentiated with respect to t (at this stage, it is necessary to use the chain rule to take account of the fact that x is also a function of t). While this might be feasible for one or a few ODEs, it would not be practical for large systems of ODEs, e.g., several hundreds or thousands of ODEs. Thus, we need an approach by which we can get the higher order derivatives of the Taylor series without actually differentiating the ODEs. In fact, we have just seen such an approach in Heun's method, where the computation of the second-order derivative has been replaced by the computation of a predicted point. This approach is the basis of the ingenious *Runge-Kutta* (RK) methods [6], which can be of any order of accuracy. In practice, fourth- and fifth-order RK methods are widely used. In subsequent applications, we will demonstrate the use of general purpose MATLAB functions that implement higher order ODE integration methods.

Finally, since we now have several methods for computing an ODE solution, e.g., Euler's method, the midpoint method, Heun's method (as well as the leapfrog method, but we do not know yet in which cases this solver can be applied), we could compare the results from two different methods to assess the accuracy of the solution. If the numerical solutions produced by these different methods agree to within a certain number of figures, we have some confidence that the solutions are accurate to this number of figures. More generally, we could estimate the numerical error by comparing the solutions from two methods which are $O(h^p)$ and $O(h^{p+q})$, respectively, where q is the increase in order between the low- and high-order methods ($p = 1$ and $q = 1$ for the comparison of the solutions from Euler's and Heun's method). Since the basic order is denoted with p, the increased accuracy that can be obtained by using higher order integration methods is generally termed *p-refinement*. The important feature of this approach is again that we can infer the accuracy of numerical solutions without knowing the analytical solution.

The preceding discussion is devoted primarily to the *accuracy of a numerical ODE solution*. There is, however, another important consideration, the *stability of the solution*. We have observed that stability of the numerical algorithm can be a severe limitation when applying the leapfrog method to the batch reactor example. We will not go into this subject in detail, but rather, just cover a few basic concepts. Much more complete discussions of numerical ODE stability can be found in numerous texts, e.g., [6].

The usual approach to the stability analysis of numerical ODE solutions is to use the *model equation*

$$\frac{\mathrm{d}x}{\mathrm{d}t} = \lambda x; \quad x(t_0) = x_0 \tag{1.28}$$

Note that Eq. (1.28) is linear with a parameter λ (which is an *eigenvalue*). The analytical solution is

$$x(t) = x_0 e^{\lambda(t-t_0)} \tag{1.29}$$

An important feature of this solution is that it is *stable (bounded) for $Re(\lambda) \leq 0$ and asymptotically stable if $Re(\lambda) < 0$ (λ can be a complex number)*. Thus, we would expect that a numerical solution to Eq. (1.28) would also be bounded for this condition. However, it is not the case as the following reasoning shows:

(a) If $\lambda < 0 \Rightarrow \lim_{t\to\infty} x(t) = 0$ from (1.29).
(b) The numerical solution is asymptotically stable if $\lim_{k\to\infty} x_k = 0$

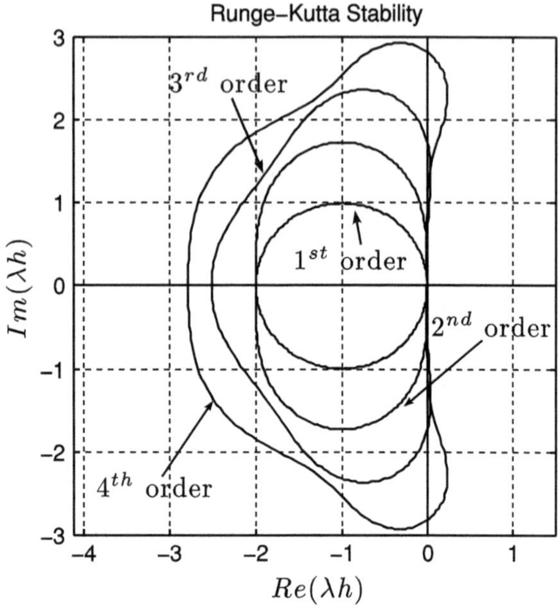

Fig. 1.6 Stability diagrams for Euler, Heun, and higher order RK methods (orders 3 and 4)

(c) From Eqs. (1.16) (Euler's method) and (1.28), $x_{k+1} = x_k + hf(x_k, t_k) = x_k + h\lambda x_k$, so that $x_{k+1} = (1 + h\lambda)x_k = (1 + h\lambda)^{k+1}x_0$. In turn, $\lim_{k\to\infty} x_k = 0$ if

$$|1 + h\lambda| < 1 \tag{1.30}$$

This latter condition defines a disk of unit radius centered on the point $(-1, 0j)$. The boundary circle $|1 + h\lambda| < 1$ corresponds to stable (but not asymptotically stable) solutions. This region is represented in Fig. 1.6. Note that if λ is a real negative number (and h a positive time step), then condition (1.30) simplifies to $-2 < h\lambda < 0$, i.e., a segment on the real axis. Inequality (1.30) places an upper limit on the integration step h, above which the numerical integration by Euler's method will become unstable; in other words, the region outside the circle corresponds to an unstable Euler integration. This stability limit on h was not a significant constraint for the ODE problems considered previously since h was constrained by accuracy. However, there is a class of problems for which the integration step is constrained not by accuracy, but rather by stability. This class of problems is called *stiff*. For example, a system of n linear ODEs will have n eigenvalues. If one (or more) of these eigenvalues is large as compared to the other eigenvalues, it will require h to be small according to inequality (1.30). This small h then requires many integration steps for a complete numerical solution, and thus leads to a lengthy calculation. This is a defining characteristic of a stiff system, i.e., a *large separation in eigenvalues and a prohibitively long calculation to maintain stability in the numerical integration.*

In Fig. 1.6, which has been drawn using an elegant MATLAB code described in [7], the stability contour just outside the circle is for second-order Runge-Kutta

(RK) methods such as Heun's method. While this region is somewhat larger than for Euler's method, using a second-order RK does not produce much of an improvement in stability, only an improvement in accuracy. The remaining stability contours are for the third- and fourth-order RK methods. Again, improvement in stability is marginal (but improvement in accuracy can be very substantial).

We now turn to the stability analysis of the leapfrog method (1.19). If the numerical solution of the model Eq. (1.28) is stable, we expect it to be of the form $x_{k+1} = \alpha x_k$ with $|\alpha| < 1$, i.e., to decay exponentially. If this expression is substituted into Eq. (1.19), then

$$x_{k+1} = \alpha x_k = \alpha^2 x_{k-1} = x_{k-1} + f(x_k, t_k)2h = x_{k-1} + \lambda x_k 2h = x_{k-1} + \lambda\alpha x_{k-1} 2h$$

or

$$\alpha^2 x_{k-1} - 2h\lambda\alpha x_{k-1} - x_{k-1} = 0$$

This second-order difference equation has the second-order characteristic equation in α

$$\alpha^2 - 2h\lambda\alpha - 1 = 0 \tag{1.31}$$

The solutions $\alpha_{1,2}$ of this equation determine the exponential decay of the numerical solution x_k, and for this latter solution to be stable, we expect $|\alpha_{1,2}| \leq 1$. However, the product $\alpha_1\alpha_2 = -1$ (i.e., the value of the constant term in the second-order equation), so that if α_1 is one solution, $\alpha_2 = -1/\alpha_1$ is the other solution. In turn, if $|\alpha_1| < 1$, then $|\alpha_2| > 1$ and the numerical solution will be unstable (since the numerical solution has a component $\alpha_2 x_k$ which grows unbounded). Thus, stability requires that the solutions lie on the unit circle and are distinct, i.e., $|\alpha_{1,2}| = 1$ and $\alpha_1 \neq \alpha_2$, so that $\alpha_{1,2} \neq \pm j$. As the solutions to Eq. (1.31) are given by $\alpha_{1,2} = h\lambda \pm \sqrt{(h\lambda)^2 + 1}$, equation parameter λ (eigenvalue) must be purely imaginary, i.e., $\lambda = \pm j\omega$, and $-j < h\lambda < j$, so that $\alpha_{1,2} = \sqrt{1 - (h\omega)^2}$ lie on the unit circle and $|\alpha_{1,2}| \pm j$; see Fig. 1.7a. The stability region, i.e., a segment on the imaginary axis $-j < h\lambda < j$, is drawn in Fig. 1.7b.

Hence, the leapfrog method is stable when applied to problems with pure imaginary eigenvalues. The simplest example of this class of systems is a spring-mass system without damping and friction (i.e., an ideal mechanical oscillator), e.g.,

$$m\frac{d^2 z}{dt^2} + kz = 0 \tag{1.32}$$

where z is the horizontal mass position (it is assumed here that the mass is moving in a horizontal plane, and that there are no external forces) and k is the spring constant. This equation can equivalently be reformulated as two first-order ODEs

$$\frac{dz}{dt} = v, \quad m\frac{dv}{dt} = -kz \tag{1.33}$$

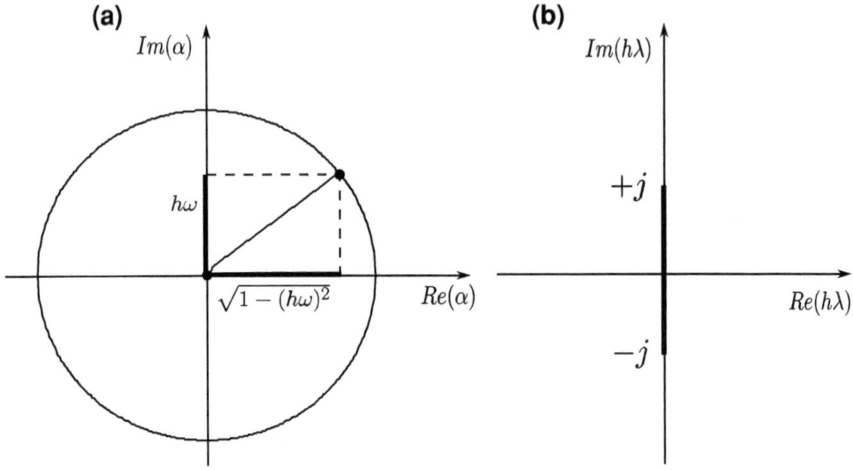

Fig. 1.7 **a** Sketch of the solution to Eq. (1.31) and **b** stability region of the leapfrog method

We do not detail the code here, which is similar to the previous examples (the interested reader can find the code in the companion library), but concentrate on the graphical outputs presented in Figs. 1.8 and 1.9. Figure 1.8 shows the periodic evolution of the mass position and velocity in the time interval $t \in [0, 90]$, whereas Fig. 1.9 represents one system state against the other (the velocity against the position), i.e., a phase portrait. These numerical results correspond to $m = 1$, $k = 0.1$, i.e., $\omega^2 = k/m = 0.1$. Stability therefore requires $h < 1/\sqrt{0.1} = 3.16$. Accuracy however requires a much smaller time step, e.g., $h = 0.01$. Note that if there is any damping, i.e., a term $+cdx/dz$ in the RHS of Eq. (1.32), then the numerical results would become unstable (since the eigenvalues are complex conjugates with a nonzero real part).

As a general conclusion, using higher order explicit ODE integration methods does very little to enhance the stability of a numerical integration. By an *explicit method* we mean that we can step forward along the solution using only past values of the dependent variable. For example, in the case of Eqs. (1.16), (1.19), (1.22), (1.23), (1.26), (1.27), we can compute the solution x_{k+1} using only past values x_k and x_{k-1}. While computationally this is an important convenience, it also means that explicit methods are stability limited, i.e., *we must limit h so that we are in the stable region of the stability diagram*. While this may not seem like a significant constraint based on the algorithms and applications discussed previously, it is very important for stiff ODEs to the extent that it can preclude the calculation of a stable numerical solution with reasonable effort.

To circumvent the stability limit of explicit algorithms, we can consider implicit algorithms. For example, a backward finite difference approximation could be used instead of a forward or a centered finite difference approximation as in Eq. (1.16) or Eq. (1.19), yielding

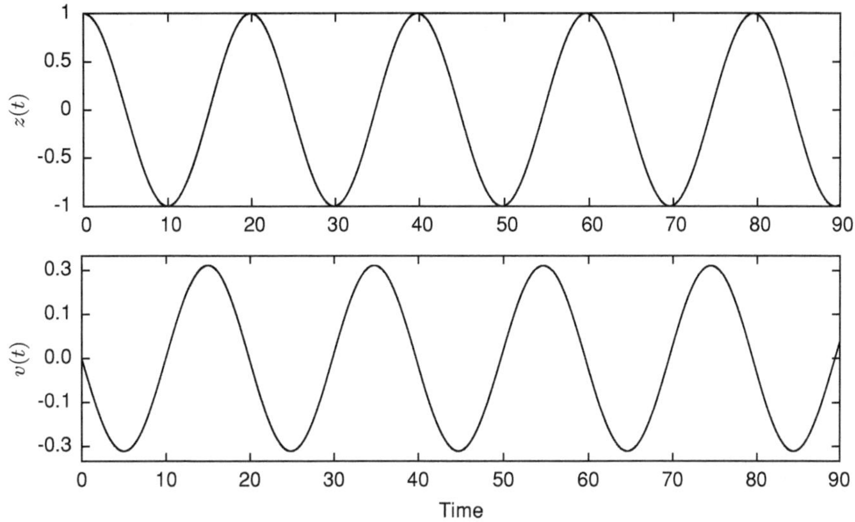

Fig. 1.8 Position and velocity of the spring mass system with $m = 1, k = 0.1$, as predicted by the leapfrog method with $h = 0.01$

Fig. 1.9 Velocity versus position, i.e., phase portrait, of the spring-mass system

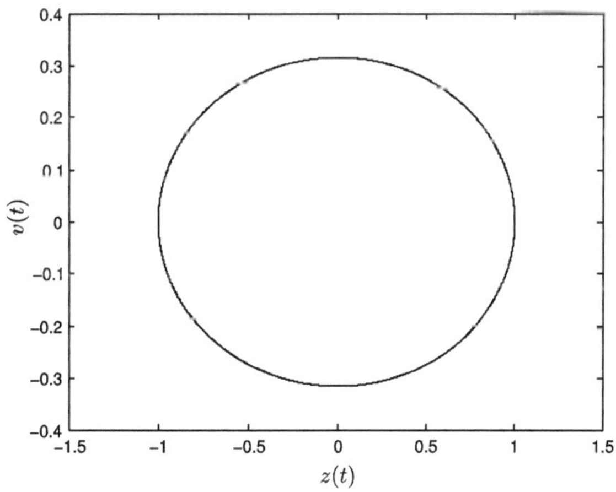

$$x_{k+1} = x_k + \left.\frac{dx}{dt}\right|_{k+1} h = x_k + f(x_{k+1}, t_{k+1})h \tag{1.34}$$

Note that the solution at the advanced point x_{k+1} appears on both sides of the equation, thus the name *implicit Euler method*.

If the derivative function $f(x, t)$ is nonlinear, the solution of Eq. (1.34) for x_{k+1} requires the *solution of a nonlinear equation (or systems of nonlinear equations for systems of simultaneous ODEs)*. The question then arises whether this additional computational requirement is worthwhile. The answer is *YES if the ODE system is stiff*. The reason for this conclusion is that the implicit Euler method possesses a much larger stability domain than Euler's method. From Eqs. (1.34) and (1.28) we obtain:

$$x_{k+1} = x_k + f(x_{k+1}, t_{k+1})h = x_k + h\lambda x_{k+1}$$

so that

$$x_{k+1} = \frac{x_k}{1 - h\lambda} = \frac{x_0}{(1 - h\lambda)^{k+1}}$$

In turn, $\lim_{k \to \infty} (x_k) = 0$ if

$$|1 - h\lambda| > 1 \tag{1.35}$$

The region of stability of the implicit Euler method is much larger than the region of stability of the explicit Euler method since it corresponds to almost all the complex plane, with the exception of the disk of unit radius centered on the point $(+1, 0j)$. The circle $|1 - h\lambda| = 1$ corresponds to stable (but not asymptotically stable) solutions. Note that if the real part of λ is a negative number (i.e., the physical system under consideration is asymptotically stable) and h is a positive time step, then condition (1.35) simplifies to $Re(h\lambda) < 0$, i.e., the method is stable over the entire left half of the complex plane and is said to be *unconditionally stable*. In this case $(Re(h\lambda) < 0)$, there is *no limit on h due to stability* (only accuracy).

However, the implicit Euler method has the same accuracy limitation as the explicit Euler method (1.16), i.e., it is *first order accurate*. Thus, for stiff ODE problems, we would like to have a *higher order method (for good accuracy) that is also implicit (for good stability)*. Several classes of such algorithms are available, and probably the best known are the *Backward Differentiation Formulas (BDF)* [8]. Their principle is easy to understand. If we consider Eq. (1.12)

$$\frac{dx}{dt} = f(x, t)$$

and assume that the numerical solution (t_i, x_i) is known for $i = 0, \ldots, k$, then the solution x_{k+1} at t_{k+1} is obtained by defining an $(m + 1)^{\text{th}}$ order interpolation polynomial $p_{m+1,\,k+1}(t)$ based on the points $(t_{k-m}, x_{k-m}), \ldots, (t_{k+1}, x_{k+1})$ and by requiring that

$$\frac{dp_{m+1},\ k + 1(t_{k+1})(t_{k+1})}{dt} = f(x_{k+1},\ t_{k+1}) \tag{1.36}$$

This expression leads to the general form (BDF formula)

$$\alpha_{m+1,k+1}x_{k+1} + \alpha_{m,\,k+1}x_k + \cdots + \alpha_{0,k+1}x_{k-m} = f(x_{k+1}, t_{k+1}) \tag{1.37}$$

which has to be solved for x_{k+1} using a nonlinear solver such as Newton's method.

The stability diagrams of BDF methods of orders 1–6 are represented in Fig. 1.10 (again this figure has been graphed using a MATLAB code from [7]).

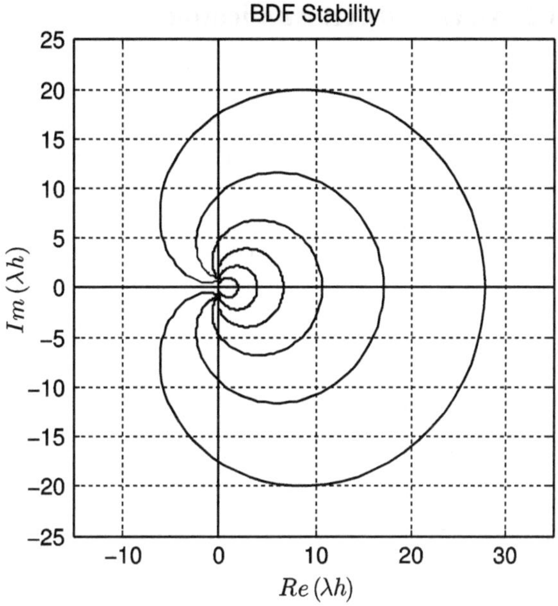

Fig. 1.10 Stability diagram for the BDF methods of orders 1 to 6

The small circle of unit radius centered at $(1, 0j)$ is for the implicit Euler method (1.34), which is also the first-order BDF. The region of stability is *outside* this circle. Thus, as can be observed, most of the complex plane is the stable region for the implicit Euler method, including the entire left half of the complex plane (for which the ODE is stable). The next contour is for the second-order BDF; again the region of stability is outside the contour. The next four contours are for BDFs of orders three to six; the stability regions are again outside the contours. One detail, in particular, should be observed. The BDFs of orders three to six have a section of the left half of the complex plane along the imaginary axis which is *unstable*. Thus, these BDFs are not unconditionally stable in the entire left half plane. While this may seem like a minor point, in fact it can become quite important if the ODEs have any eigenvalues near the imaginary axis, e.g., highly oscillatory systems. In this case, the integration step h is limited by stability and if the ODE system is stiff, this can impose a significant constraint on the step size, resulting in a lengthy calculation. One approach to circumventing this problem is to limit the order of the BDF to two or less. Another approach is to modify or extend the BDF methods to include in their stability region more of the region along the imaginary axis [9, 10].

This completes the introductory discussion of the numerical integration of ODEs. We now consider a second class of problems, *differential-algebraic equations* (DAEs).

1.2 An ODE/DAE Application

The basic ODE we considered previously, Eq. (1.12) was discussed essentially as a single (scalar) equation. However, all of the discussion in Sect. 1.1 also applies to systems of ODEs, and in fact, this reflects the power of the numerical methods that were discussed in Sect. 1.1, i.e., *they can be applied to ODE systems of essentially any size (number or dimension) and complexity (e.g., nonlinearity).*

To emphasize this point, we can write Eq. (1.12) as

$$\frac{\mathrm{d}\mathbf{x}}{\mathrm{d}t} = \mathbf{f}(\mathbf{x}, t) \tag{1.38}$$

where a bold character now denotes a vector (or a matrix). Also, Eq. (1.38) can be written as

$$\mathbf{I}\frac{\mathrm{d}\mathbf{x}}{\mathrm{d}t} = \mathbf{f}(\mathbf{x}, t) \tag{1.39}$$

where $\mathbf{I}$ is the identity matrix (ones on the main diagonal and zeros everywhere else). Note that (1.38), (1.39) denote a system of ODEs with *only one derivative in each equation*, and is therefore called an explicit *ODE system* (not to be confused with an explicit integration algorithm as discussed previously).

Equation (1.39) can then be generalized to

$$\mathbf{M}\frac{d\mathbf{x}}{dt} = \mathbf{f}(\mathbf{x}, t) \tag{1.40}$$

where, if $\mathbf{x}$ is an n-vector (column vector of length n), $\mathbf{M}$ is a $n \times n$ matrix that can bestow interesting properties on the differential equation system. For example, if one or more off-diagonal elements of $\mathbf{M}$ is nonzero, the corresponding differential equation will have *more than one derivative*. For example, if the fourth row in $\mathbf{M}$ has two nonzero off-diagonal elements (and the diagonal element is not zero), the fourth ODE in (1.40) will have three derivatives, one corresponding to the diagonal element and two corresponding to the two off-diagonal elements. Such a differential equation is called *linearly implicit* since the derivatives appear as linear combinations (sums of derivatives with each derivative multiplied by a weighting coefficient from $\mathbf{M}$ rather than having only one derivative in each differential equation). Another description we could then use for such a differential equation is *linearly coupled* (through the derivatives), and $\mathbf{M}$ is therefore termed a *coupling matrix* (or a *mass matrix*, in analogy with model equations describing mechanical systems). We shall shortly consider such a linearly implicit system of differential equations.

Another interesting case is when $\mathbf{M}$ has a row of all zero elements. The corresponding equation actually has *no derivatives* and is therefore *algebraic*. Thus, depending on the structure (coefficients) of $\mathbf{M}$, Eq. (1.40) can be a system of *differential-algebraic (DAE) equations*, with the algebraic equations resulting from rows of zeros in $\mathbf{M}$, and the differential equations resulting from rows of $\mathbf{M}$ with some nonzeros

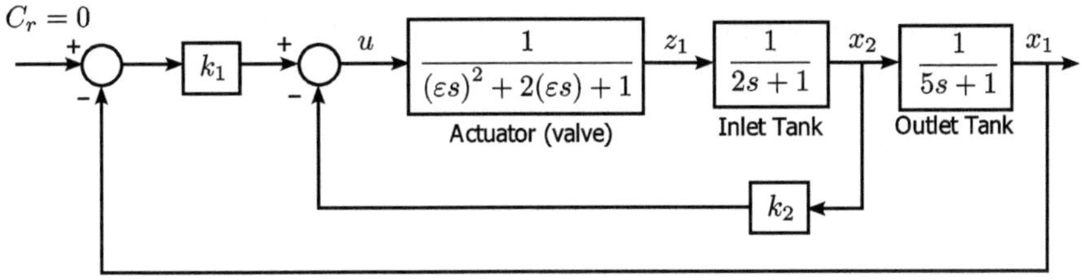

Fig. 1.11 Linearized model of the control scheme (x_1: deviation in the concentration of the outlet tank, x_2: deviation in the concentration of the inlet tank, z_1: actuator first state variable, z_2: actuator second state variable, u: control signal, k_1, k_2: controller gains)

(and the latter can be linearly coupled). The numerical solution of such DAE systems is substantially more difficult than the explicit ODE system (1.39), but library integrators are available for such systems, as illustrated in the example that follows.

Consider a two-tank chemical reaction system, where it is required to maintain the concentration in the outlet tank at a desired level by the addition of reactant in the inlet tank through a control valve [11]. A linearized model of the feedback control system is illustrated in Fig. 1.11, where $C_r = 0$ is the deviation in the desired concentration. The actuator block, i.e., the control valve, is represented by a transfer function, using the Laplace transform variable s

$$\frac{1}{(\varepsilon s)^2 + 2(\varepsilon s) + 1} \tag{1.41}$$

which has two real poles ($s_{1,2} = -\frac{1}{\varepsilon}$) that can be scaled using the parameter ε.

Using a change of variables, $z_2 = \varepsilon \dot{z}_1$, the complete model takes the *singularly perturbed form*

$$\begin{bmatrix} \dot{x}_1 \\ \dot{x}_2 \\ \varepsilon \dot{z}_1 \\ \varepsilon \dot{z}_2 \end{bmatrix} = \begin{bmatrix} -0.2 & 0.2 & 0 & 0 \\ 0 & -0.5 & 0.5 & 0 \\ 0 & 0 & 0 & 1 \\ 0 & 0 & -1 & -2 \end{bmatrix} \begin{bmatrix} x_1 \\ x_2 \\ z_1 \\ z_2 \end{bmatrix} + \begin{bmatrix} 0 \\ 0 \\ 0 \\ 1 \end{bmatrix} u \tag{1.42}$$

The design objective is to select the state feedback controller gains k_1 and k_2 so that the departure of x_1 from the set point, C_r, is in some sense minimized.

The state equations (1.42) have a coupling matrix

$$\mathbf{M} = \begin{bmatrix} 1 & 0 & 0 & 0 \\ 0 & 1 & 0 & 0 \\ 0 & 0 & \varepsilon & 0 \\ 0 & 0 & 0 & \varepsilon \end{bmatrix} \tag{1.43}$$

If the parameter ε is set to zero, the third and fourth rows of $\mathbf{M}$ have entirely zeros. In other words, the third and four equations of (1.42) are algebraic. Physically, this means that the response of the actuator is instantaneous, as compared to the slow dynamics of the tank system. For this case, a solver for explicit ODEs cannot be used; rather, a DAE solver is required. The MATLAB integrator `ode15s` will be used for this purpose as explained in the MATLAB code to follow (Script `Main_two_tanks`; see also the script `Main_bacteria` for a previous use of `ode15s`).

```
% Global variables
global A B eps K

% Model parameters (matrices and vectors)
A = [-0.2    0.2    0.0    0.0;
      0.0   -0.5    0.5    0.0;
      0.0    0.0    0.0    1.0;
      0.0    0.0   -1.0   -2.0];
B = [0 0 0 1]';
K = [7 1.5];

% Initial conditions
t = [0:0.1:10];
x = [2.5 2.0 0 0]';

% Call to ODE solver with eps = 0.1
eps     = 0.1;
options = odeset('Mass',@mass,'MassSingular','no',...
                 'RelTol',1e-3,'AbsTol',1e-3);
[tout,xout] = ode15s(@two_tanks_odes,t,x,options);

% Plot for eps = 0.1
figure(1)
subplot(2,1,1)
plot(t,xout(:,1:2));
xlabel('t');
ylabel('x1(t) and x2(t)');
subplot(2,1,2)
plot(t,xout(:,3:4));
xlabel('t');
ylabel('x3(t) and x4(t)');

% Next case: eps = 0.05
eps = 0.05;

% Call to ODE solver with eps = 0.05
options = odeset('Mass',@mass,'MassSingular','no',...
                 'RelTol',1e-3,'AbsTol',1e-3);
[tout,xout] = ode15s(@two_Tanks_odes,t,x,options);

% Plot for eps = 0.05 (superimposed on first plot)
figure(1)
subplot(2,1,1)
hold on
plot(t,xout(:,1:2),'k');
xlabel('t');
ylabel('x_1(t) and x_2(t)');
title('Singular perturbation problem');
subplot(2,1,2)
hold on
```

```
plot(t,xout(:,3:4),'k');
xlabel('t');
ylabel('z_1(t) and z_2(t)');
```

Script Main_two_tanks Main program that calls the function `two_tanks_odes` for the solution of Eq. (1.42)

If ε is small (the actuator dynamics is fast compared to the tank dynamics), we would expect the corresponding solution would be close to that for $\varepsilon = 0$. This conclusion will be confirmed by the numerical output from `ode15s` to be subsequently discussed.

Reversely, this suggests another way to compute a solution to a DAE system, i.e., add derivatives to the algebraic equations with a small parameter, ε, to produce a *singularly perturbed system*, then integrate the resulting system of ODEs (rather than solve the original DAE system directly). This is an established approach for solving DAE systems. However, since ε is small, the ODEs will be stiff, and therefore a stiff solver (an implicit ODE integrator) will generally be required.

The function `two_tanks_odes` implements the RHS terms of Eq. (1.42) using the vector-matrix facilities of `MATLAB`. K is a 1×2 vector (1 row, 2 columns) of the controller gains k_1, k_1 is set in the main program that calls `two_tanks_odes` and is passed as a global variable. A is a 4×4 matrix and B is a 4×1 matrix (column vector) with the RHS coefficients of Eq. (1.42) set in the main program (discussed subsequently) and passed as global variables. Note that `xt` is then a 4×1 column vector defining the four derivatives of the state variables.

```
function xt = two_tanks_odes(t,x)

% Global variables
global A B eps K

% Temporal derivatives
u  = -K*x(1:2);
xt = A*x+B*u;
```

Function two_tanks_odes Function `xt` to define the RHS of Eq. (1.42)

The function `mass` implements the mass matrix `M` with `eps` set in the main program for the singular ($\varepsilon = 0$) and nonsingular ($\varepsilon \neq 0$) cases.

```
function M = mass(t,x)

% Global variables
global A B eps K

% Mass matrix
```

```
M = [1 0  0    0;
     0 1  0    0;
     0 0 eps   0;
     0 0  0  eps];
```

Function mass MATLAB programming of Eq. (1.43)

Functions `two_tanks_odes` and `mass` are called by the main program `Main_two_tanks` (in which the problem presentation, i.e., the comments at the beginning of the main program, are not reproduced).

The main program `Main_two_tanks` shares many of the features of the script `Main_bacteria`. The main additional feature is the use of the function `mass` to compute the mass matrix `M`, and the specification of its characteristics (singular or nonsingular) via `odeset`.

Note that two plots are produced (Fig. 1.12). The first is for the two concentrations, x_1, x_2, as a function of t. The second is for the two control states (or singular perturbation) variables, z_1, z_2. The coding can be repeated for $\varepsilon = 0.05$ and then for the DAE case, $\varepsilon = 0$. If the cases $\varepsilon = 0.1, 0.05$ do correspond to "small" values of the perturbation variable, all three solutions should be nearly the same. On inspection of Fig. 1.12, it is clear that ε has little influence on the dynamics of the slow variables x_1, x_2. On the other hand, the trajectories of the actuator variables z_1, z_2 exhibit a two-time-scale behavior. They start with a fast transient and then settle down slowly.

To conclude this example, we considered the extension of the explicit ODE problem (1.38) to the linearly implicit and DAE problem (1.40). The later includes the coupling matrix **M** that can be singular (the DAE problem). Solvers for the more general problem (1.40) are available, such as `ode15s`. Beyond this, we can consider the fully *implicit ODE/DAE system*

$$\mathbf{f}\left(\mathbf{x}, \frac{\mathrm{d}\mathbf{x}}{\mathrm{d}t}, t\right) = \mathbf{0} \tag{1.44}$$

which can be significantly more difficult to solve than Eq. (1.40). The difficulty arises from both the mathematical generality of (1.44) and the computational requirements. We will not go into the details of solving Eq. (1.44) at this point other than to mention that solvers are available that can handle certain classes of problems [12]. In recent MATLAB versions, a new solver called `ode15i` has been included to handle fully implicit systems of differential equations. Besides, the library SUNDIALS [13] developed at LLNL provides several DAE solvers for explicit and implicit DAE problems and has an interface to MATLAB (SUNDIALSTB). The solver DASSL [12] is also available in OCTAVE and SCILAB.

The remaining application example in this chapter is a partial differential equation (PDE) reformulated as a system of ODEs. By using this approach, we are able to solve the PDE using the ODE methods discussed previously.

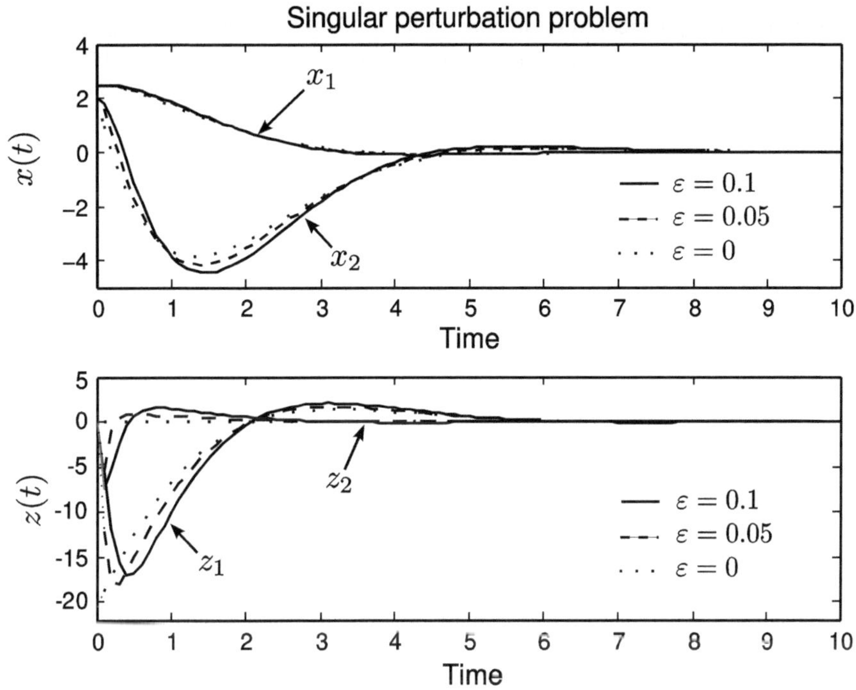

Fig. 1.12 Solution of Eq. (1.42) for $\varepsilon = 0$ (*dotted line*), $\varepsilon = 0.05$ (*dashed line*), $\varepsilon = 0.1$ (*solid line*)

1.3 A PDE Application

The PDE to be numerically integrated in this section is known as Burgers' equation and it presents the following form:

$$\frac{\partial u}{\partial t} = -u\frac{\partial u}{\partial z} + \mu\frac{\partial^2 u}{\partial z^2} \tag{1.45}$$

where t is the independent (initial value) variable, z represents the spatial (boundary value) independent variable, and μ is a given constant parameter. The usual notation in the literature for the dependent variable of Burgers equation is u, therefore, the notation employed so far in this book will be slightly modified in this particular example so that $x = u$.

We again mean by the solution of a PDE (as with an ODE), the *dependent variable as a function of the independent variables*. Note that *variables* is plural since the feature that distinguishes a PDE from an ODE is having *more than one independent variable*, in this case, z and t. In other words, we seek $u(z, t)$ in numerical form.

Since t is an initial value variable and appears in a first-order derivative $\partial u/\partial t$, we require one initial condition (to be specified). z is a boundary value variable. This name comes from the typical application of specification of conditions at the boundaries of a physical system. Since Eq. (1.45) is second order in z (from the derivative $\partial^2 u/\partial z^2$), we require two boundary conditions (to be specified).

Equation (1.45) is a special case of the Navier Stokes equations of fluid mechanics. It is also a widely used problem for testing numerical methods for the following reasons:

1. It is a *nonlinear* PDE (due to the term $u\frac{\partial u}{\partial z}$) which has exact solutions, for instance:

$$u_a(z,t) = \frac{0.1e^a + 0.5e^b + e^c}{e^a + e^b + e^c} \tag{1.46}$$

   where $a = -(0.05/\mu)(z - 0.5 + 4.95t)$, $b = -(0.25/\mu)(z - 0.5 + 0.75t)$, $c = -(0.5/\mu)(z - 0.375)$. This exact solution can be used to assess the accuracy of numerical solution (as we will do subsequently).
2. The difficulty of computing a numerical solution can be increased by reducing the value of the parameter μ. In the limit as $\mu \rightarrow 0$, Burgers' equation produces moving discontinuities that provide a very stringent test of any numerical procedure. We will discuss more problems with moving fronts in Chap. 5.
3. Equation (1.45) can easily be extended to more than one spatial dimension, and the analytical solutions for several dimensions are also available. Thus, Burgers' equation can be used to test numerical methods in one, two, and three dimensions.

For the initial and boundary conditions for Eq. (1.45), we use the analytical solution, i.e.,

$$u(z,0) = u_a(z,0) \tag{1.47}$$

$$u(0,t) = u_a(0,t) \tag{1.48}$$

$$u(z_L,t) = u_a(z_L,t) \tag{1.49}$$

where z_L is the right boundary of the spatial domain.

A function to implement Eqs. (1.45) and (1.47)–(1.49) is listed in `burgers_pde`.

```
function ut = burgers_pde(t,u)

% Global variables
global mu z0 zL n D1 D2;

% Boundary condition at z = 0
u(1) = burgers_exact(z0,t);

% Boundary condition at z = zL
u(n) = burgers_exact(zL,t);

% Second-order spatial derivative
uzz = D2*u;

% First-order spatial derivative
```

```
fz = D1*(0.5*u.^2);

% Temporal derivatives
ut    = -fz + mu*uzz;
ut(1) = 0;
ut(n) = 0;
```

Function burgers_pde Function for the MOL solution of Burgers' equation (1.45)

This programming includes calls to the function `burgers_exact` that implements the exact solution (1.46).

```
function [u] = burgers_exact(z,t)
% This function computes an exact solution to Burgers'
% equation

% Global variables
global mu

% Analytical (exact) solution
a  = (0.05/mu)*(z-0.5+4.95*t);
b  = (0.25/mu)*(z-0.5+0.75*t);
c  = (0.5/mu)*(z-0.375);
ea = exp(-a);
eb = exp(-b);
ec = exp(-c);
u  = (0.1*ea+0.5*eb+ec)/(ea+eb+ec);
```

Function burgers_exact Function for computing the exact solution to Burgers' equation (1.46)

The second derivative in Eq. (1.45), $\partial^2 u/\partial z^2$, is computed as `uzz`. We follow a subscript notation in expressing PDEs that facilitates their programming. With this notation, Eq. (1.45) can be written as

$$u_t = -uu_z + \mu u_{zz} \tag{1.50}$$

The subscript denotes the particular independent variable. The number of times the subscript is repeated indicates the order of the derivative. For example

$$u_t \equiv \frac{\partial u}{\partial t}, u_{zz} \equiv \frac{\partial^2 u}{\partial z^2}$$

Then it follows in the programming that `ut` $= u_t$, `uz` $= u_z$ and `uzz` $= u_{zz}$.

The second partial derivative u_{zz} is computed by `uzz=D2*u`. `D2` is a *second-order differentiation matrix* (`u` is a vector and `*` denotes a matrix-vector multiply using the special MATLAB utility for this operation). The theory and programming of `D2` will be explained in detail in subsequent chapters. The net result of applying `D2` to a vector is to compute a numerical second-order derivative of the vector (element by element); in this case, the derivative is with respect to z.

The nonlinear (convective) term $uu_z = (1/2)(u^2)_z$ is also computed by the application of a differentiation matrix `D1`. The square of a given variable x, i.e., x^2, is computed in MATLAB as `x.^2`, where x is a vector (one-dimensional array) and its elements are squared individually. The additional (`.`) in `x.^2` is the MATLAB notation for an *element by element operation*; in this case the operation is squaring.

Now that all the partial derivatives in Eq. (1.45) -or Eq. (1.50)- have been computed, it can be programmed as a system of ODEs, for which we can note the following details:

- The close resemblance of the PDE, $u_t = -uu_z + \mu u_{zz}$, and the programming of the PDE `ut=-fz+mu*uzz` is apparent. In fact, this is one of the major advantages of this approach of approximating a PDE as a system of ODEs (called the *method of lines* or abbreviated as *MOL*).
- The RHS of `ut=-fz+mu*uzz` is a column vector (as a result of applying the differentiation matrix `D2`). Of course, this means that the LHS, `ut`, is also a column vector and, therefore, a transpose from a row vector to a column vector is not required.
- Since the left and right hand boundary values of u are specified by constant Dirichlet boundary conditions (1.48) and (1.49), their t derivatives are zeroed so that they will not change, i.e., `ut(1)=0; ut(n)=0;`
- The variation of the dependent variable u with respect to z is accomplished by using a grid with 201 points, i.e., `u` is an array of 201 values. At each of the grid points (points in `z`), the derivative is calculated according to the RHS of Eq. (1.45); the result is 201 values of u_t stored in the array `ut`. These 201 values of `ut` can then be integrated by an ODE solver (in this case `ode15s`) to produce the solution vector `u`. This is the essence of the MOL.

In summary, the MOL programming of a PDE is a straightforward extension of the programming of ODEs. The only additional requirement is the calculation of the spatial (boundary value) derivatives in the PDE. This can usually be accomplished by library utilities such as `D1` and `D2`. The theory and programming of these *spatial differentiation* utilities will be covered in detail in subsequent chapters. We include the Burgers' equation example here just to demonstrate the general MOL approach.

```
% Global variables
global mu z0 zL n D1 D2;

% Spatial grid
z0 = 0.0;
zL = 1.0;
n  = 201;
dz = (zL-z0)/(n-1);
z  = [z0:dz:zL]';
```

```
% Model parameter
mu = 0.001;

% Initial condition
for i=1:n
    u(i) = burgers_exact(z(i),0);
end

% Define differentiation matrix (convective term)
v  = 1;
D1 = five_point_biased_upwind_uni_D1(z0,zL,n,v);

% Define differentiation matrix (diffusive term)
D2 = five_point_centered_uni_D2(z0,zL,n);

% Call to ODE solver
t            = [0:0.1:1];
options      = odeset('RelTol',1e-3,'AbsTol',1e-3,...
                      'Stats','on');
[tout,uout] = ode15s('burgers_pde',t,u,options);

% Plot numerical solution
plot(z,uout,'k');
xlabel('z');
ylabel('u(z,t)');
title('Burgers equation')
hold on

% Superimpose plot of exact solution
for k = 1:length(tout)
    for i = 1:n
        uexact(i) = burgers_exact(z(i),tout(k));
    end
    plot(z,uexact,':k')
end
```

Script Main_burgers Script for the MOL solution of Burgers' equation (1.45)

The main program that calls function `burgers_pde` is listed in the script `burgers` (again, to save space, the introductory comments are not listed). This main program has most of the features of the previous main programs, with a few additional details:

1. The spatial grid in z is defined. First, the boundary values of the spatial domain $z_0 \leq z \leq z_L$ are set. For a grid of 201 points, the grid spacing `dz` is computed. Finally, the 201 values of z are stored in a one-dimensional array `z` (a column vector).
2. The parameter μ in Eq. (1.45) is then defined. μ is a critical parameter in determining the characteristics (features) of the solution. For large μ, the diffusion term μu_{zz} dominates and Eq. (1.45) behaves much like a *diffusion* or *parabolic equation*, including smoothed (and therefore easily computed) solutions. For small μ, the convective term $-uu_z$ dominates and Eq. (1.45) behaves much like a *convective* or *hyperbolic equation*, including the propagation of steep moving

fronts (which makes the calculation of numerical solutions relatively difficult). The value used in the current case, $\mu = 0.001$, is sufficiently small to make Eq. (1.45) *strongly convective* and therefore the numerical solution will exhibit steep moving fronts as we shall observe.

3. The analytical solution (1.46) is used to set the initial condition over the grid of `n` (= 201) points in z.
4. A *five-point biased upwind FD*, which is especially well suited for strongly convective PDEs, is selected for the approximation of the convective term $-uu_z$ in Eq. (1.45). The differentiation matrix, `D1`, can be defined numerically using a function `D1 = five_point_biased_upwind_uni_D1(z0,zL,n,v)` (included in the companion software). The boundaries of the spatial domain $(z0, zL)$ are defined over a `n` point grid. The last argument basically defines the sign of the term $-uu_z$ (this convective term is for flow in the positive z direction, so the fourth argument is given a positive value corresponding to a positive velocity `v = 1`).
5. The differentiation matrix for the diffusion derivative, u_{zz}, is also defined by a call to a function `D2 = five_point_centered_uni_D2(z0,zL,n)` (also included in the companion software). Note that in the use of centered approximations, a fourth argument is not required since the centered approximation is applied to diffusion (not convection) and a specification of a direction of flow is not required.
6. The spatial differentiation matrices are now defined `(D1, D2)`, so the ODEs on the 201 point grid in z can be computed through function `burgers_pde`. These 201 ODEs can then be integrated by `ode15s` and the solution can be plotted.

Note that this code produces a plot of the solution $u(z, t)$ vs z with t as a parameter (this will be clear when considering the plotted output that follows). The analytical solution is then superimposed on the plot for comparison with the numerical solution. The plot of the numerical and analytical solutions appears in Fig. 1.13. The solution has 11 curves corresponding to the output times defined by the statement `t=[0:0.1:1];` just before the call to `ode15s`. The leftmost curve corresponds to the initial condition at $t = 0$. The rightmost curve is for $t = 1$. The numerical and analytical solutions are essentially indistinguishable, with the exception of small spurious oscillations near the front (as shown in the enlarged picture). We will come back later on, in Chap. 5, to this observation. Thus, the numerical procedures implemented in the programs `burgers_pde` and `burgers` have performed satisfactorily for Burgers' equation, and we have some confidence that they could be applied effectively to other PDEs.

An important reason why the numerical solution has good accuracy is the choice of the number of grid points, $n = 201$. Clearly if we reduce this number substantially, the grid in z will become so coarse that the solution cannot be refined, particularly where there are sharp spatial variations in z. The choice of 201 grid points was selected essentially by trial and error, i.e., enough grid points to accurately resolve the solution, but not an excessive number which would lead to a lengthy calculation (because more ODEs would be integrated). Thus, the choice of the number of grid

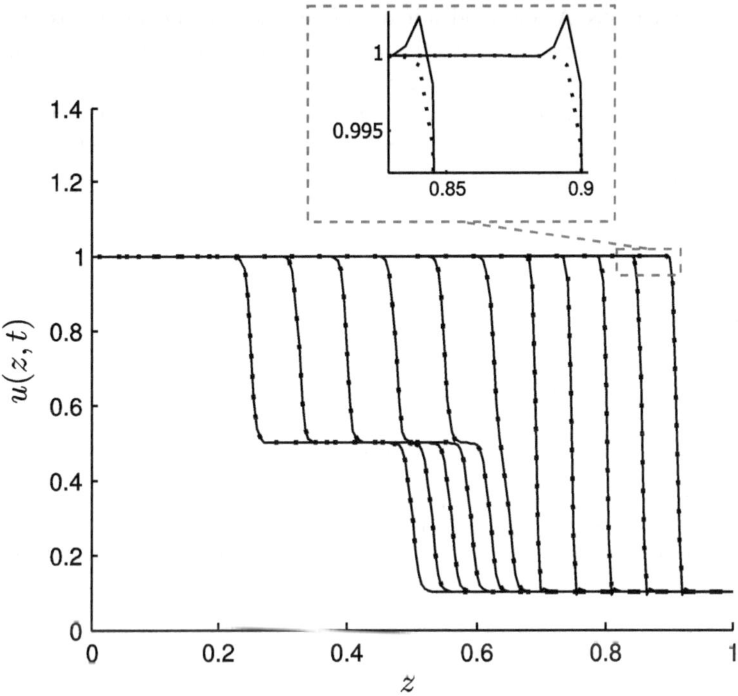

Fig. 1.13 Numerical (*lines*) and analytical (*dots*) solutions of Burgers' equation (1.45)

points in a PDE solution is typically guided by some knowledge of the characteristics (features) of the solution and by experience. However, in subsequent chapters, we shall develop methods by which the number of grid points, and their placement, is done by numerical algorithms, i.e., by the code itself. These are termed *adaptive grid methods*.

Note that as the curves move from left to right (with increasing t), they *sharpen spatially* (become more vertical). In fact, if $\mu = 0$ and t progressed far enough, a shock (discontinuity) would develop. The important point is that, due to the nonlinear hyperbolic term $-uu_z$ in Eq. (1.50), the solution becomes progressively more difficult to compute. This is handled to a degree by the use of the five-point biased upwind approximation, but ultimately, it would fail for large t.

We can then come to the general conclusion that *hyperbolic or strongly convective (parabolic) PDEs are generally the most difficult PDEs to solve numerically (of the three geometric classes, hyperbolic, parabolic, and elliptic).*

1.4 Summary

In this chapter, we have given examples of the numerical solution of the three major classes of differential equations, ODEs, DAEs, and PDEs. Within each of these classes, we have observed special computational requirements as illustrated through example applications. A central idea in discussing these applications was the use of library utilities that implement effective numerical methods. A few basic time integrators, including Euler, modified Euler, and leapfrog methods, have been discussed, whereas more sophisticated solvers, i.e., higher order Runge-Kutta (RK) and backward differentiation formulas (BDF) methods, have been briefly sketched. The very important concepts of accuracy, stability, and ODE stiffness have been introduced. In the next chapters, some of these ideas are pursued further and illustrated with additional example applications.

References

1. Shampine LF, Gladwell I, Thompson S (2003) Solving ODEs with MATLAB. Cambridge University Press, Cambridge
2. Bastin G, Dochain D (1990) On-line estimation and adaptive control bioreactors. Elsevier, Amsterdam
3. Bogaerts PH (1999) Contribution à la modélisation mathématique pour la simulation et l'observation d'états des bioprocédés. PhD thesis, Université Libre de Bruxelles, Belgium
4. Economou CG, Morari M, Palsson B (1986) Internal model control. 5. extension to nonlinear systems. Ind Eng Chem Process Des Dev 25(2):403–411
5. Press WH, Teukolsky SA, Vetterling WT, Flannery BP (1986) Numerical recipes in fortran 77: the art of scientific computing. Cambridge University Press, Cambridge (1986–1992)
6. Hairer E, Wanner G (1996) Solving ordinary differential equations II: stiff and differential algebraic problems. Springer, Heidelberg
7. Trefethen LN (2000) Spectral methods in Matlab. SIAM, Philadelphia
8. Gear CW (1971) Numerical initial value problems in ordinary differential equations. Prentice-Hall, Engelwood Cliffs
9. Cash JR (1979) Stable recursions with applications to the numerical solution of stiff systems. Academic Press, London
10. Cash JR, Considine S (1992) An MEBDF code for stiff initial value problems. ACM Trans Math Softw 18(2):142–158
11. Kokotovic P, Khalil HK, O'Reilly J (1986) Singular perturbation methods in control: analysis and design. Academic Press, London
12. Ascher UM, Petzold LR (1998) Computer methods for ordinary differential equations and differential-algebraic equations. SIAM, Philadelphia
13. Hindmarsh AC, Brown PN, Grant KE, Lee SL, Serban R, Shumaker DE, Woodward CS (2005) SUNDIALS: suite of nonlinear and differential/algebraic equation solvers. ACM Trans Math Softw 31(3):363–396

Further Reading

14. Hairer E, Norsett SP, Wanner G (1993) Solving ordinary differential equations I: nonstiff problems. Springer, Berlin
15. Higham DJ, Higham NN (2000) MATLAB guide. SIAM, Philadelphia
16. Lee HJ, Schiesser WE (2004) Ordinary and partial differential equation routines in C, C++, fortran, Java, maple and MATLAB. Chapman Hall/CRC Press, Boca Raton
17. Petzold L, Ascher U (1998) Computer methods for ordinary differential equations and differential-algebraic equations. SIAM, Philadelphia

Chapter 2
More on ODE Integration

In Chap. 1, we considered through examples several numerical ODE integrators, e.g., Euler, leap-frog, Heun, Runge Kutta, and BDF methods. As we might expect, the development and use of ODE integrators is an extensive subject, and in this introductory text, we will only try to gain some overall perspective. To this end, we will first focus our attention to the `MATLAB` coding of a few basic integrators, i.e., a fixed-step Euler's method, a variable-step (nonstiff) Runge–Kutta (RK) method and a variable-step (stiff) Rosenbrock's method, and then turn our attention to more sophisticated library integrators. Typically, library integrators are written by experts who have included features that make the integrators robust and reliable. However, the coding of library integrators is generally long and relatively complicated, so that they are often used without a detailed understanding of how they work. A popular library is the `MATLAB ODE SUITE` developed by Shampine et al. [1]. Other options are provided by the use of `MATLAB` MEX-files and of readily available integrators originally written in `FORTRAN` or `C/C++` language. On the other hand, SCILAB provides a library of solvers based on ODEPACK [2] whereas OCTAVE includes LSODE [3] and DASSL [4]. We will explore these several options and discuss a few more ODE applications.

2.1 A Basic Fixed Step ODE Integrator

We considered previously the Euler's method—see Eqs. (1.15) and (1.16). A function that implements Euler's method with a fixed or constant integration step is listed in function `Euler_solver`.

A. Vande Wouwer et al., *Simulation of ODE/PDE Models with MATLAB®, OCTAVE and SCILAB*, DOI: 10.1007/978-3-319-06790-2_2,

```
function [tout,xout] = euler_solver(odefunction, t0, tf, x0,...
                                    Dt, Dtplot)

% This function solves first-order differential equations using
% Euler's method.
% [tout,xout] = euler_solver(@f,t0,tf,x0,Dt,Dtplot)
% integrates the system of differential equations xt=f(t,x) from
% t0 to tf with initial conditions x0.  f is a string containing
% the name of an ODE file.  Function f(t,x) must return a column
% vector.  Each row in solution array x corresponds to a value
% returned in column vector t.
%
% Argument list
%
% f - string containing the name of the user-supplied problem
%     call: xt = problem_name(t,x) where f = 'problem_name'
%     t - independent variable (scalar)
%     x - solution vector
%     xt - returned derivative vector; xt(i) = dx(i)/dt
% t0 - initial value of t
% tf - final value of t
% x0 - initial value vector
% Dt - time step size
% Dtplot - plot interval
% tout - returned integration points (column-vector).
% xout - returned solution, one solution row-vector per tout-value

% Initialization
plotgap = round(Dtplot/Dt);      % number of computation

% steps within a plot interval
Dt     = Dtplot/plotgap;
nplots = round((tf-t0)/Dtplot);  % number of plots
t      = t0;                     % initialize t
x      = x0;                     % initialize x
tout   = t0;                     % initialize output value
xout   = x0';                    % initialize output value

% Implement Euler's method
for i = 1: nplots
    for j = 1: plotgap
        % Use MATLAB's feval function to access the function
        % file, then take Euler step
        xnew = x + feval(odefunction,t,x)*Dt;
        t    = t + Dt;
        x    = xnew;
    end
    % Add latest result to the output arrays
    tout = [tout;t];
    xout = [xout;x'];
end
```

Function Euler_solver Basic Euler integrator

We can note the following points about this code:

1. The operation of function `Euler_solver`, and its input and output arguments are generally explained and defined in a block of comments.
2. The number of computation steps within a plot (output) interval, `plotgap`, the fixed-length Euler step, `Dt`, and the number of plot points (outputs), `nplots`, are first computed to take the integration from the initial value `t0` to the final value `tf`.
3. The initial conditions corresponding to Eq. (1.13) are then stored. Note that `t`, `t0` and `tout` are scalars while `x`, `x0` and `xout` are one-dimensional vectors.
4. `nplots` plot (output) steps are then taken using an outer `for` loop, and `plotgap` Euler steps are taken within each plot step using an inner `for` loop.
5. The solution is then advanced by Euler's method (1.16). Note that a `MATLAB` function can be evaluated using the utility `feval`. In the present case, the RHS of (1.12) is coded in function `odefunction`. The solution is then updated for the next Euler step. Finally, this code completes the inner for `loop` to give the solution at a plot point after `plotgap` Euler steps.
6. The solution is then stored in the arrays `tout` and `xout`. Thus, `tout` becomes a column vector and `xout` is a two-dimensional array with a column dimension the same as `tout` and each row containing the dependent variable vector for the corresponding `tout` (note that the latest solution is transposed into a row of `xout`).

The basic Euler integrator (see function `Euler_solver`) can now be called in any new application, with the particular ODEs defined in a function. To illustrate this approach, we consider the *logistic equation* which models population growth (N represents the number of individuals):

$$\frac{dN}{dt} = (a - bN)N = aN - bN^2 \tag{2.1}$$

first proposed by Verhulst in 1838 [5]. Equation (2.1) is a generalization of the ODE

$$\frac{dN}{dt} = aN \tag{2.2}$$

which for the Malthusian rate $a > 0$ gives unlimited exponential growth, i.e., the solution to Eq. (2.2) is

$$N(t) = N_0 e^{at} \tag{2.3}$$

where N_0 is an initial value of N for $t = 0$.

Since, realistically, unlimited growth is not possible in any physical process, we now consider Eq. (2.1) as an extension of Eq. (2.2) to reflect limited growth. Thus, instead of $\frac{dN}{dt} \to \infty$ as $N \to \infty$ according to Eq. (2.2), the solution now reaches the *equilibrium condition*

$$\frac{dN}{dt} = 0 = (a - bN)N$$

corresponding to $N = \frac{a}{b}$.

The approach to this equilibrium can be understood by considering the RHS of Eq. (2.1). At the beginning of the solution, for small t and N, the term aN dominates, and the solution grows according to Eqs. (2.2) and (2.3). For large t, N increases so that $-bN^2$ begins to offset aN and eventually the two terms become equal (but opposite in sign), at which point $\frac{dN}{dt} = 0$. Thus, the solution to Eq. (2.1) shows rapid growth initially, then slower growth, to produce a S-shaped solution.

These features are demonstrated by the analytical solution to Eq. (2.1), which can be derived as follows. First, *separation of variables* can be applied:

$$\frac{dN}{(a - bN)N} = dt \tag{2.4}$$

The LHS of Eq. (2.4) can then be expanded into two terms by *partial fractions*

$$\frac{dN}{(a - bN)N} = \frac{\frac{dN}{a}}{N} + \frac{\frac{bdN}{a}}{a - bN} = dt \tag{2.5}$$

Integration of Eq. (2.5) gives

$$\frac{1}{a}\ln(N) - \frac{b}{a}\ln(|a - bN|) = t + c \tag{2.6}$$

where c is a constant of integration that can be evaluated from the initial condition

$$N(0) = N_0 \tag{2.7}$$

so that Eq. (2.6) can be rewritten as

$$\frac{1}{a}\ln(N) - \frac{1}{a}\ln(|a - bN|) = t + \frac{1}{a}\ln(N_0) - \frac{b}{a}\ln(|a - bN_0|)$$

or, since $a - bN_0$ and $a - bN$ have the same sign,

$$\ln\left(\frac{N}{N_0}\right) + \ln\left(\frac{a - bN_0}{a - bN}\right) = at$$

and thus the analytical solution of Eq. (2.1) is

$$\left(\frac{N}{N_0}\right)\left(\frac{a - bN_0}{a - bN}\right) = e^{at}$$

or solving for N

$$N = \frac{\frac{a}{b}}{1 + \left(\frac{a - bN_0}{bN_0}\right) e^{-at}} = \frac{K}{1 + \left(\frac{K}{N_0} - 1\right) e^{-at}}, \quad K = \frac{a}{b} \tag{2.8}$$

Note again from Eq. (2.8) that for $a > 0, b > 0$, we have that as $t \to \infty, N \to \frac{a}{b} = K$, where K is called the *carrying capacity*.

Equation (2.8) can be used to evaluate the numerical solution from function `Euler_solver`. To do this, a function must be provided to define the ODE.

```
function Nt = logistic_ode(t,N)

% Global variables
global a b K N0

% ODE
Nt = (a-b*N)*N;
```

Function logistic_ode Function to define the ODE of the logistic equation (2.1)

The coding of function `logistic_ode` follows directly from Eq. (2.1). Note that this ODE function must return the time derivative of the state variable (`Nt` in this case). The ODE function is called by `Euler_solver`, which is called by the main program `Main_logistic`.

```
clear all
close all

% Set global variables
global a b K N0

% Model parameters
a = 1;
b = 0.5e-4;
K = a/b;

% Initial conditions
t0     = 0;
N0     = 1000;
tf     = 15;
Dt     = 0.1;
Dtplot = 0.5;

% Call to ODE solver
[tout,xout] = euler_solver(@logistic_ode,t0,tf,N0,Dt,Dtplot);

% Plot results
figure(1)
```

```
plot(tout,xout,'k');
hold on
Nexact = logistic_exact(tout);
plot(tout,Nexact,':r')
xlabel('t');
xlabel('N(t)');
title('Logistic equation');
```

Script Main_logistic Main program that calls functions `Euler_solver` and `logistic_ode`

This program proceeds in several steps:

1. After clearing `MATLAB` for a new application (via `clear` and `close`), the initial and final value of the independent variable, `t`, and the initial condition for N (`N0`) are defined.
2. The fixed integration step and plot (output) intervals are then defined. Thus, function `Euler_solver` will take $(15-0)/0.1 = 150$ Euler steps and the solution will be plotted $(15-0)/0.5+1 = 31$ times (counting the initial condition). Within each `Dtplot` step (outer `for` loop), `Euler_solver` will take $0.5/0.1 = 5$ Euler steps (inner `for` loop).
3. Function `Euler_solver` is then called with the preceding parameters and the solution returned as the one-dimensional column vector `tout` and the matrix `xout` (one-dimensional with the number of rows equal to the number of elements of tout, and in each row, the corresponding value of `x`). Note that the name of the function that defines the ODE, i.e., function `logistic_ode`, in the call to `Euler_solver`, does not have to be the same as in function `Euler_solver`, i.e., `odefunction`; rather, the name of the ODE function is specified in the call to `Euler_solver` as `@logistic_ode` where `@` specifies an argument that is a function. This is an important detail since it means that the user of the script `Main_logistic` can select any convenient name for the function that defines the ODEs.

In order to evaluate the numerical solution from `Euler_solver`, the analytical solution (2.8) is evaluated in the function `logistic_exact`.

```
function N = logistic_exact(t)

% Global variables
global a b K N0

% Analytical solution
N = K./(1+(K/N0-1)*exp(-a*t));
```

Function logistic_exact Function to compute the analytical solution of the logistic equation.

The plot produced from the preceding program `Main_logistic` is shown in Fig. 2.1.

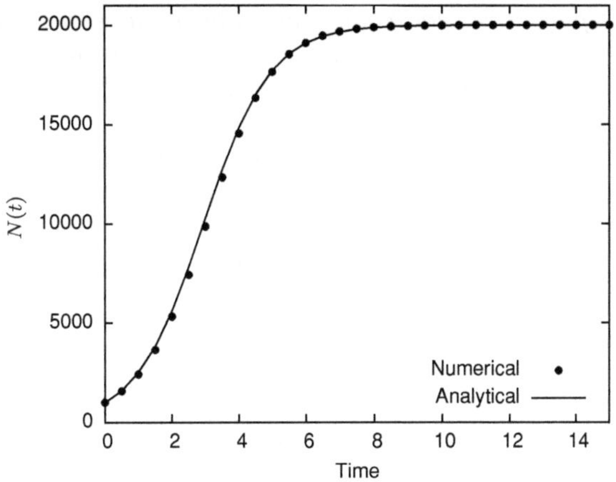

Fig. 2.1 Plot of the numerical and analytical solutions from `Main_logistic`

Note that there is a small difference between the numerical solution (dotted line) and the analytical solution (2.8) (solid line). This difference is due principally to the limited accuracy of the Euler's method and to the small number of Euler steps taken during each output interval (5 steps). Thus, to improve the accuracy of the numerical solution, we could use a more accurate (higher-order) integration algorithm (p refinement discussed in Chap. 1) and/or more integration steps between each output value. We will next consider a higher-order algorithm that generally gives good accuracy with a modest number of integration steps.

2.2 A Basic Variable-Step Nonstiff ODE Integrator

The preceding example illustrates the use of an ODE integrator that proceeds along the solution with a fixed step (constant h). Although this worked satisfactorily for the modest example of Eq. (2.1) (particularly if we used more than five Euler steps in each output interval), we can envision situations where varying the integration step would be advantageous, e.g., the integration step might be relatively small at the beginning of the solution when it changes most rapidly, then increased as the solution changes more slowly. In other words, the integration step will not be maintained at an excessively small value chosen to maintain accuracy where the solution is changing most rapidly, when we would like to use a larger step where the solution is changing less rapidly. The central requirement will then be to vary the integration step so as to maintain a given, specified accuracy.

It might, at first, seem impossible to vary the integration step to maintain a prescribed accuracy since this implies we know the exact solution in order to determine the integration error and thereby decide if the error is under control as the integration step is changed. In other words, if we require the exact solution to determine the integration error, there is no reason to do a numerical integration, e.g., we can just use the exact solution. However, *we can use an estimated error to adjust the integration step that does not require knowledge of the exact solution, provided the estimated error is accurate enough to serve as a basis for varying the integration step.* Note here the distinction between the exact integration error (generally unknown) and the estimated integration error (generally known).

To investigate the use of an estimated error to adjust the integration step, we will now consider RK methods. Representatives of this class of methods have already been presented in Chap. 1, i.e., the Euler's method, which is a first-order RK method, and the modified Euler or Heun's method, which is a second-order RK method. We are now looking at higher-order schemes, which are based on Taylors series expansions of the solution $x(t)$ of Eq. (1.12), $\frac{\mathrm{d}x}{\mathrm{d}t} = f(x, t)$, i.e.,

$$\begin{aligned} x_{k+1} &= x_k + h \left.\frac{\mathrm{d}x}{\mathrm{d}t}\right|_k + \frac{h^2}{2!} \left.\frac{\mathrm{d}^2x}{\mathrm{d}t^2}\right|_k + \cdots + \frac{h^q}{q!} \left.\frac{\mathrm{d}^qx}{\mathrm{d}t^q}\right|_k + O\left(h^{q+1}\right) \\ &= x_k + hf(x_k, t_k) + \frac{h^2}{2!} \left.\frac{\mathrm{d}f}{\mathrm{d}t}\right|_k + \cdots + \frac{h^q}{q!} \left.\frac{\mathrm{d}^{q-1}f}{\mathrm{d}t^{q-1}}\right|_k + O\left(h^{q+1}\right) \end{aligned} \tag{2.9}$$

The idea behind the RK methods is to evaluate the terms involving higher-order derivatives in (2.9), without actually differentiating the ODEs, but using q intermediate function evaluations $f(x_k, t_k)$ (also called *stages*) between $f(x_k, t_k)$ and $f(x_{k+1}, t_{k+1})$, and selecting coefficients w_i so as to match the Taylor series expansion (2.9) with

$$\begin{aligned} x_{k+1} &= x_k + h \sum_{i=1}^{q} w_i f(x_{k,i}, t_{k,i}) + O\left(h^{q+1}\right) \\ &= x_k + h \sum_{i=1}^{q} w_i k_i + O\left(h^{q+1}\right) \end{aligned} \tag{2.10}$$

The intermediate stages can generally be expressed as:

$$t_{k,i} = t_k + h\alpha_i, \quad x_{k,i} = x_k + h \sum_{j=1}^{q} \beta_{i,j} k_j \tag{2.11}$$

so that

$$k_i = f\left(x_k + h \sum_{j=1}^{q} \beta_{i,j} k_j, t_k + h\alpha_i\right) \tag{2.12}$$

with $\alpha_1 = 0$ and $\beta_{i,j} = 0\ \forall j \geq i$ an *explicit RK method* is derived (otherwise, the solution of a nonlinear system of Eq. (2.12) is required to get the values of k_i, yielding an *implicit RK method*).

Note that RK methods are *single-step* methods, i.e., x_{k+1} is given in terms of x_k only (in contrast to *multi-step* methods, such as the BDF formulas (1.37), which give

Table 2.1 Maximum order p_{max} that can be obtained with a q-stage explicit RK method

q	1	2	3	4	5	6	7	8	9
p	1	2	3	4	4	5	6	6	7

x_{k+1} in terms of $x_k, \ldots, x_{k-m}$). They are also called *self starting*, in the sense that only the initial condition x_0 is required to compute x_1.

As a specific example of derivation, consider the explicit, second-order RK methods ($q = 2$), for which Eqs. (2.10)–(2.12) reduce to

$$\begin{aligned} x_{k+1} &= x_k + h\left(w_1 f(x_k, t_k) + w_2 f(x_k + h\beta_{2,1} f(x_k, t_k), t_k + h\alpha_2)\right) \\ &= x_k + hw_1 f(x_k, t_k) + hw_2 \left(f(x_k, t_k) + h\alpha_2 \left.\frac{\partial f}{\partial t}\right|_k + h\beta_{2,1} \left.\frac{\partial f}{\partial x}\right|_k\right) + O\left(h^3\right) \\ &= x_k + h(w_1 + w_2) f(x_k, t_k) + h^2 w_2 \alpha_2 \left.\frac{\partial f}{\partial t}\right|_k \\ &\quad + h^2 w_2 \beta_{2,1} \left.\frac{\partial f}{\partial x}\right|_k f(x_k, t_k) + O\left(h^3\right) \end{aligned} \tag{2.13}$$

Upon comparison with the truncated Taylor series expansion

$$\begin{aligned} x_{k+1} &= x_k + hf(x_k, t_k) + \frac{h^2}{2!} \left.\frac{df}{dt}\right|_k + O\left(h^3\right) \\ &= x_k + hf(x_k, t_k) + \frac{h^2}{2!} \left.\frac{\partial f}{\partial t}\right|_k + \frac{h^2}{2!} \left.\frac{\partial f}{\partial x}\right|_k f(x_k, t_k) + O\left(h^3\right) \end{aligned} \tag{2.14}$$

we obtain

$$w_1 + w_2 = 1, \quad w_2 \alpha_2 = \frac{1}{2}, \quad w_2 \beta_{2,1} = \frac{1}{2} \tag{2.15}$$

This is a system of three equations and four unknowns which, solved for $w_2 = \lambda$, gives a one-parameter family of explicit methods

$$x_{k+1} = x_k + h\left((1-\lambda) f(x_k, t_k) + \lambda f\left(x_k + \frac{h}{2\lambda} f(x_k, t_k), t_k + \frac{h}{2\lambda}\right)\right) \tag{2.16}$$

For $\lambda = 0$, we find back the first-order Euler's method (1.16), and for $\lambda = 1/2$, the second-order modified Euler or Heun's method (1.26)–(1.27).

Deriving higher-order RK methods following the same line of thoughts would however require laborious algebraic manipulations. Another approach, based on *graph theory* (*rooted tree theory*), has been proposed by Butcher [6] among others [7, 8]. This approach enables a systematic and an efficient derivation of higher-order explicit and implicit methods. The maximum order p_{max} that can be obtained with a

q-stage explicit RK method is given in Table 2.1. For a q-stage implicit RK method, $p_{\max} = 2q$. Among the family of fourth- and fifth-order methods, a classical embedded pair of explicit RK formulas is given by

$$x_{k+1,4} = x_k + \frac{25}{216}k_1 + \frac{1408}{2565}k_3 + \frac{2197}{4104}k_4 - \frac{1}{5}k_5 \tag{2.17}$$

$$x_{k+1,5} = x_k + \frac{16}{135}k_1 + \frac{6656}{12825}k_3 + \frac{28561}{56430}k_4 - \frac{9}{50}k_5 + \frac{2}{55}k_6 \tag{2.18}$$

These formulas are said *embedded* as they share the same stages k_i, $i = 1, \ldots, 5$. Such stages can be computed as follows:

$$k_1 = f(x_k, t_k)h \tag{2.19a}$$

$$k_2 = f\left(x_k + 0.25k_1, t_k + 0.25h\right)h \tag{2.19b}$$

$$k_3 = f\left(x_k + \frac{3}{32}k_1 + \frac{9}{32}k_2, t_k + \frac{3}{8}h\right)h \tag{2.19c}$$

$$k_4 = f\left(x_k + \frac{1932}{2197}k_1 - \frac{7200}{2197}k_2 + \frac{7296}{2197}k_3, t_k + \frac{12}{13}h\right)h \tag{2.19d}$$

$$k_5 = f\left(x_k + \frac{439}{216}k_1 - 8k_2 + \frac{3680}{513}k_3 - \frac{845}{4104}k_4, t_k + h\right)h \tag{2.19e}$$

$$k_6 = f\left(x_k - \frac{8}{27}k_1 + 2k_2 - \frac{3544}{2565}k_3 + \frac{1859}{4104}k_4 - \frac{11}{40}k_5, t_k + 0.5h\right)h \tag{2.19f}$$

Returning to the idea of using an estimated error to adjust the integration step, we can obtain an error estimate for the fourth-order result from

$$\varepsilon_{k+1,4} = x_{k+1,5} - x_{k+1,4} \tag{2.20}$$

that can be then used to adjust the integration step h in accordance with a user-prescribed error tolerance.

Equation (2.17) fits the underlying Taylor series up to including the fourth-order term $\frac{d^4 x_k}{dt^4}\left(\frac{h^4}{4!}\right)$ while Eq. (2.18) fits the Taylor series up to an including the fifth-order term $\frac{d^5 x_k}{dt^5}\left(\frac{h^5}{5!}\right)$. Thus, the subtraction (2.20) provides an estimate of the fifth-order term.

The RK pair (2.17)–(2.18) is termed the *Runge–Kutta–Fehlberg method* [9], usually designated *RKF45*. Note that an essential feature of an embedded pair as illustrated by Eq. (2.19) is that the stages are the same for the lower and higher-order methods (k_1–k_6 in this case). Therefore, the subtraction (2.20) can be used to combine terms of the same stages (since they are the same for the lower- and higher-order methods and therefore have to be *calculated only once for both of them*).

Two functions for implementing the *RKF45* method, with the use of Eq. (2.20) to estimate the truncation error and adjust the integration step, are `ssrkf45` and `rkf45_solver` listed below. The first function, `ssrkf45`, takes a single step along the solution based on Eqs. (2.17)–(2.18):

1. After a block of comments explaining the operation of `ssrkf45`, the derivative vector is computed at the base point and the first stage, k_1, is evaluated according to (2.19a). Note that k_1 is a column vector with the row dimension equal to the number of first order ODEs to be integrated. Each stage is just the ODE derivative vector multiplied by the integration step h, so each `xt0` is calculated by a call to the derivative routine via `feval`.
2. The dependent variable vector and the independent variable are then evaluated to initiate the calculation of the next stage, k_2, according to (2.19b).
3. This basic procedure is repeated for the calculation of k_3, k_4, k_5 and k_6 according to (2.19c)–(2.19f). Then the fourth- and fifth-order solutions are computed at the next point by the application of Eqs. (2.17)–(2.18), and the error is estimated according to (2.20).
4. The independent variable, the fifth-order solution and the vector of estimated errors at the next point along the solution are then returned from `ssrkf45` as `t`, `x`, and `e` (the output or return arguments of `ssrkf45`).

```
function [t,x,e] = ssrkf45(odefunction,t0,x0,h)
%
% This function computes an ODE solution by the RK Fehlberg 45
% method for one step along the solution (by calls to
% 'odefunction' to define the ODE derivative vector).  It also
% estimates the truncation error of the solution, and applies
% this estimate as a correction to the solution vector.
%
% Argument list
%
% odefunction - string containing name of user-supplied problem
%   t0 - initial value of independent variable
%   x0 - initial condition vector
%   h  - integration step
%   t  - independent variable (scalar)
%   x  - solution vector after one rkf45 step
%   e  - estimate of truncation error of the solution vector

% Derivative vector at initial (base) point
[xt0] = feval(odefunction,t0,x0);

% k1, advance of dependent variable vector and independent
% variable for calculation of k2
k1 = h*xt0;
x = x0 + 0.25*k1;
t = t0 + 0.25*h;

% Derivative vector at new x, t
```

```
[xt] = feval(odefunction,t,x);

% k2, advance of dependent variable vector and independent
% variable for calculation of k3
k2 = h*xt;
x = x0 + (3.0/32.0)*k1...
       + (9.0/32.0)*k2;
t = t0 + (3.0/8.0)*h;

% Derivative vector at new x, t
[xt] = feval(odefunction,t,x);

% k3, advance of dependent variable vector and independent
% variable for calculation of k4
k3 = h*xt;
x = x0 + (1932.0/2197.0)*k1...
    - (7200.0/2197.0)*k2...
    + (7296.0/2197.0)*k3;
t = t0 + (12.00/13.0)*h;

% Derivative vector at new x, t
[xt] = feval(odefunction,t,x);

% k4, advance of dependent variable vector and independent
% variable for calculation of k5
k4 = h*xt;
x = x0 + ( 439.0/ 216.0)*k1...
    - (    8.0         )*k2...
    + (3680.0/ 513.0)*k3...
    - ( 845.0/4104.0)*k4;
t = t0 + h;

% Derivative vector at new x, t
[xt] = feval(odefunction,t,x);

% k5, advance of dependent variable vector and independent
% variable for calculation of k6
k5 = h*xt;
x = x0 - (    8.0/   27.0)*k1...
    + (    2.0         )*k2...
    - (3544.0/2565.0)*k3...
    + (1859.0/4104.0)*k4...
    - (   11.0/   40.0)*k5;
t = t0 + 0.5*h;

% Derivative vector at new u, t
[xt] = feval(odefunction,t,x);

% k6
k6 = h*xt;

% Fourth order step
sum4 = x0 + (      25.0/  216.0)*k1...
          + (    1408.0/2565.0)*k3...
          + (    2197.0/4104.0)*k4...
          - (       1.0/    5.0)*k5;

% Fifth order step
sum5 = x0 + (      16.0/   135.0)*k1...
          + (  6656.0/12825.0)*k3...
          + (28561.0/56430.0)*k4...
          - (       9.0/     50.0)*k5...
          + (       2.0/     55.0)*k6;
t = t0 + h;

% Truncation error estimate
```

```
e = sum5 - sum4;

% Fifth order solution vector (from 4,5 RK pair);
% two ways to the same result are listed
% x = sum4 + e;
x = sum5;
```

Function ssrkf45 Function for a single step along the solution of (1.12) based on (2.15)–(2.17)

We also require a function that calls `ssrkf45` to take a series of step along the solution. This function, `rkf45_solver`, is described below:

1. After an initial block of comments explaining the operation of `rkf45_solver` and listing its input and output arguments, the integration is initialized. In particular, the initial integration step is set to `h = Dtplot/2` and the independent variable `tout`, the dependent variable vector `xout` and the estimated error vector `eout` are initiated for subsequent output. The integration steps, `nsteps`, and the number of outputs, `nplots`, are also initialized.
2. Two loops are then executed. The outer loop steps through `nplots` outputs. Within each pass through this outer loop, the integration is continued while the independent variable `t` is less than the final value `tplot` for the current output interval. Before entering the inner `while` loop, the length of the output interval, `tplot`, is set.
3. Within each of the output intervals, a logic variable is initialized to specify a successful integration step and a check is made to determine if the integration step, `h`, should be reset to cover the remaining distance in the output interval.
4. `ssrkf45` is then called for one integration step of length `h`.
5. A series of tests then determines if the integration interval should be changed. First, if the estimated error (for any dependent variable) exceeds the error tolerances (note the use of a combination of the absolute and relative error tolerances), the integration step is halved. If the integration step is reduced, the logic variable `fin1` is set to 0 so that the integration step is repeated from the current base point.
6. Next (for `fin1 = 1`), if the estimated error for all of the dependent variables is less than $1/32$ of the composite error tolerance, the step is doubled. The factor $1/32$ is used in accordance with the fifth-order RK algorithm. Thus, if the integration step is doubled, the integration error will increase by a factor of $2^5 = 32$. If the estimated error for *any* dependent variable exceeds $1/32$ of the composite error tolerance, the integration step is unchanged (`fin1 = 0`).
7. Next, two checks are made to determine if user-specified limits have been reached. If the integration step has reached the specified minimum value, the integration interval is set to this minimum value and the integration continues from this point. If the maximum number of integration steps has been exceeded, an error message is displayed, execution of the `while` and `for` loops is terminated through the two `break`s, i.e., the ODE integration is terminated since the total number of integration steps has reached the maximum value.

8. Finally, at the end of the output interval, the solution is stored in arrays, which are the return arguments for `rkf45_solver`. Note that the final `end` statement terminates the `for` loop that steps the integration through the `nplots` outputs.

```
function [tout,xout,eout]=rkf45_solver(odefunction,t0,tf,x0,...
                                       hmin,nstepsmax,abstol,...
                                       reltol,Dtplot)
% This function solves first-order differential equations using
% the Runga-Kutta-Felhberg (4,5) method
% [tout, xout] = rkf45_solver(@f,t0,tf,x0,hmin,nstepsmax,...
%                             abstol,reltol,Dtplot)
% integrates the system of differential equations xt=f(t,x)
% from t0 to tf with initial conditions x0.  f is a string
% containing the name of an ODE file.  Function f(t,x) must
% return a column vector. Each row in solution array xout
% corresponds to a value returned in column vector t.
%
% rkf45_solver.m solves first-order differential equations
% using the variable step RK Fehlberg 45 method for a series of
% points along the solution by repeatedly calling function
% ssrkf45 for a single RK Fehlberg 45 step.  The truncation error
% is estimated along the solution to adjust the integration step
% according to a specified error tolerance.
%
% Argument list
%
% f - String containing name of user-supplied problem description
%       Call: xt = problem_name(t,x) where f = 'problem_name'
%             t  - independent variable (scalar)
%             x  - solution vector
%             xt - returned derivative vector; xt(i) = dx(i)/dt
%
% t0        - initial value of t
% tf        - final value of t
% x0        - initial value vector
% hmin      - minimum allowable time step
% nstepsmax - maximum number of steps
% abstol    - absolute error tolerance
% reltol    - relative error tolerance
% Dtplot    - plot interval
% tout      - returned integration points (column-vector)
% xout      - returned solution, one solution row-vector per
%             tout-value

% Start integration
t = t0;
tini = t0;
tout = t0;                          % initialize output value
xout = x0';                         % initialize output value
eout = zeros(size(xout));           % initialize output value
nsteps = 0;                         % initialize step counter
nplots = round((tf-t0)/Dtplot);     % number of outputs

% Initial integration step
h = 10*hmin;

% Step through nplots output points
for i = 1:nplots
    % Final (output) value of the independent variable
    tplot = tini+i*Dtplot;
    % While independent variable is less than the final value,
    % continue the integration
```

```
    while t <= tplot*0.9999
        % If the next step along the solution will go past the
        % final value of the independent variable, set the step
        % to the remaining distance to the final value
        if t+h > tplot, h = tplot-t; end

        % Single rkf45 step
        [t,x,e] = ssrkf45(odefunction,t0,x0,h);

        % Check if any of the ODEs have violated the error
        % criteria
        if max( abs(e) > (abs(x)*reltol + abstol) )
            % Error violation, so integration is not complete.
            % Reduce integration step because of error violation
            % and repeat integration from the base point.
            % Set logic variable for rejected integration step.
            h = h/2;

            % If the current step is less than the minimum
            % allowable step, set the step to the minimum
            % allowable value
            if h < hmin,  h = hmin; end

            % If there is no error violation, check if there is
            % enough "error margin" to increase the integration
            % step
        elseif max( abs(e) > (abs(x)*reltol + abstol)/32 )
            % The integration step cannot be increased, so leave
            % it unchanged and continue the integration from the
            % new base point.
            x0 = x; t0 = t;
            % There is no error violation and enough "security
            % margin"
        else
            % double integration step and continue integration
            % from new base point
            h = 2*h; x0 = x; t0 = t;
        end %if

        % Continue while and check total number of integration
        % steps taken
        nsteps=nsteps+1;
        if(nsteps > nstepsmax)
      fprintf(' \n nstepsmax exceeded; integration terminated\n');
      break;
        end
    end %while

    % add latest result to the output arrays and continue for
    % loop
    tout = [tout ; t];
    xout = [xout ; x'];
    eout = [eout ; e'];
end % for
% End of rkf45_solver
```

Function rkf45_solver Function for a variable step solution of an ODE system

In summary, we now have provision for increasing, decreasing or not changing the integration interval in accordance with a comparison of the estimated integration error (for each dependent variable) with the composite error tolerance, and for taking some special action if user-specified limits (minimum integration interval and maximum number of integration steps) are exceeded.

We can now apply `ssrkf45` an `rkf45_solver` to the logistic equation (2.1). We already have function `logistic_ode` and the exact solution in `logistic_exact`, so all we now require is a main program, which closely parallels the main program calling the fixed step Euler integrator (`Main_logistic`). The complete code is available in the companion library. Here, we just note a few points:

1. Since `rkf45_solver` is a variable step integrator, it requires error tolerances, a minimum integration step and the maximum number of steps

```
abstol     = 1e-3;
reltol     = 1e-3;
hmin       = 1e-3;
nstepsmax = 1000;
```

   These parameters are inputs to `rkf45_solver`.

```
% Call to ODE solver
[tout,xout] = rkf45_solver(@logistic_ode,t0,tf,N0,...
                             hmin,nstepsmax,abstol,...
                             reltol,Dtplot);
```

2. In order to assess the accuracy of the numerical solution, the absolute and relative errors are computed from the exact solution. These (exact) errors can then be compared with the absolute and relative error tolerances

```
% Print results
fprintf('    t     x(t)    xex(t)    abserr     relerr\n');
for i = 1:length(tout)
    fprintf('%7.1f%10.2f%10.2f%10.5f%10.5f\n',...
     tout(i),xout(i,1),Nexact(i),xout(i,1)-Nexact(i),...
      (xout(i,1)-Nexact(i))/Nexact(i));
end
```

The plot produced by the main program is shown in Fig. 2.2. The exact and numerical solutions are indistinguishable (as compared with Fig. 2.1). This issue is also confirmed by the numerical output reproduced in Table 2.2.

We can note the following features of the numerical solution:

1. The initial conditions of the numerical and exact solutions agree (a good check).
2. The maximum relative error is −0.00001 which is a factor of 0.01 better than the relative error tolerance set in the main program, i.e., `reltol = 1e-3;`
3. The maximum absolute error is −0.02847 which exceeds the absolute error tolerance, i.e., `abstol = 1e-3;` this absolute error tolerance can be considered excessively stringent since it specifies 0.001 for dependent variable values between 1,000 ($t = 0$) and 20,000 ($t = 15$). To explain why this absolute error tolerance is not observed in the numerical solution, consider the use of the absolute and relative error tolerances in function `rkf45_solver`, i.e., `if`

Table 2.2 Numerical output obtained with `rkf45_solver` (`abstol = 1e-3`)

t	x(t)	xex(t)	abserr	relerr
0.0	1000.00	1000.00	0.00000	0.00000
0.5	1596.92	1596.92	-0.00033	-0.00000
1.0	2503.21	2503.22	-0.00958	-0.00000
1.5	3817.16	3817.18	-0.01991	-0.00001
2.0	5600.06	5600.09	-0.02673	-0.00000
2.5	7813.65	7813.68	-0.02799	-0.00000
3.0	10277.71	10277.73	-0.02742	-0.00000
3.5	12708.47	12708.50	-0.02847	-0.00000
4.0	14836.80	14836.83	-0.02623	-0.00000
4.5	16514.30	16514.31	-0.01358	-0.00000
5.0	17730.17	17730.17	0.00552	0.00000
5.5	18558.95	18558.92	0.02088	0.00000
6.0	19100.47	19100.44	0.02777	0.00000
6.5	19444.59	19444.56	0.02766	0.00000
7.0	19659.41	19659.39	0.02385	0.00000
7.5	19792.03	19792.01	0.01888	0.00000
8.0	19873.35	19873.33	0.01416	0.00000
8.5	19922.99	19922.98	0.01023	0.00000
9.0	19953.22	19953.21	0.00720	0.00000
9.5	19971.60	19971.60	0.00497	0.00000
10.0	19982.77	19982.76	0.00338	0.00000
10.5	19989.54	19989.54	0.00227	0.00000
11.0	19993.66	19993.66	0.00151	0.00000
11.5	19996.15	19996.15	0.00100	0.00000
12.0	19997.67	19997.67	0.00065	0.00000
12.5	19998.58	19998.58	0.00043	0.00000
13.0	19999.14	19999.14	0.00028	0.00000
13.5	19999.48	19999.48	0.00018	0.00000
14.0	19999.68	19999.68	0.00012	0.00000
14.5	19999.81	19999.81	0.00007	0.00000
15.0	19999.88	19999.88	0.00005	0.00000

Table 2.3 Numerical output obtained with `rkf45_solver` (`abstol=1`)

t	x(t)	xex(t)	abserr	relerr
0.0	1000.00	1000.00	0.00000	0.00000
0.5	1596.92	1596.92	-0.00033	-0.00000
1.0	2503.21	2503.22	-0.00958	-0.00000
1.5	3817.16	3817.18	-0.01991	-0.00001
2.0	5600.06	5600.09	-0.02673	-0.00000
2.5	7813.65	7813.68	-0.02799	-0.00000
3.0	10277.71	10277.73	-0.02742	-0.00000
3.5	12708.47	12708.50	-0.02847	-0.00000
4.0	14836.80	14836.83	-0.02623	-0.00000
4.5	16514.30	16514.31	-0.01358	-0.00000
5.0	17730.17	17730.17	0.00552	0.00000

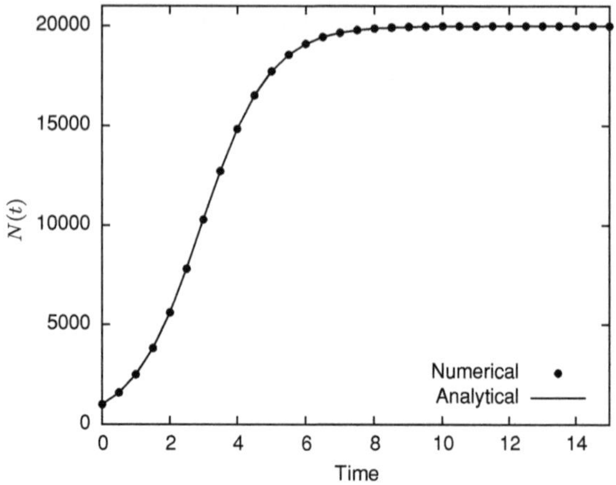

Fig. 2.2 Plot of the numerical (using the rkf45 IVP solver) and analytical solutions of the logistic equation

abs(e(i)) > (abs(x(i))*reltol + abstol). This composite error is typically (1,000)*1.0e-03 + 1.0e-03 and therefore the contribution of the absolute error tolerance (the second 1.0e-03) is small in comparison to the contribution of the relative error tolerance (the first 1.0e-03). In other words, the relative error tolerance controls the integration step in this case, which explains why the absolute error tolerance in the output of Table 2.3 exceeds the absolute error tolerance. This conclusion does emphasize the need to carefully understand and specify the error tolerances.

To explore this idea a little further, if the absolute error tolerance had been specified as abserr = 1, it would then be consistent with the relative error, i.e., both the absolute error (=1) and the relative error of 1.0e-03 are 1 part in 1,000 for the dependent variable N equal to its initial value of 1,000.

A sample of the output is given in Table 2.3. Note that this output is essentially identical to that of Table 2.2 (thus confirming the idea that the absolute error tolerance 1.0e-03 has essentially no effect on the numerical solution), but now both error criteria are satisfied. Specifically, the maximum absolute error −0.02847 is well below the specified absolute error (=1).

This example illustrates that the specification of the error tolerances requires some thought, including the use of representative values of the dependent variable, in this case N = 1,000 for deciding on appropriate error tolerances. Also, this example brings to mind the possibility that the dependent variable may have a value of zero in which case the relative error tolerance has no effect in the statement if abs(e(i)) > (abs(x(i))*reltol + abstol) (i.e., x(i) = 0) and therefore the absolute error tolerance completely controls the step changing algorithm. In other words, specifying an absolute error tolerance of zero may not be a good idea (if the dependent variable passes through a zero value).

This situation of some dependent variables passing through zero also suggests that having absolute and relative error tolerances that might be different for each dependent variable might be worthwhile, particularly if the dependent variables have typical values that are widely different. In fact, this idea is easy to implement since the integration algorithm returns an estimated integration error for each dependent variable as a vector (`e(i)` in the preceding program statement). Then, if absolute and relative error tolerances are specified as vectors (e.g., `reltol(i)` and `abstol(i)`), the comparison of the estimated error with the absolute and relative error tolerances for each dependent variable is easily accomplished, e.g., by using a `for` loop with index `i`. Some library ODE integrators have this feature (of specifying absolute and relative error vectors), and it could easily be added to `rkf45_solver`, for example. Also, the idea of a relative error brings widely different values of the dependent variables together on a common scale, while the absolute error tolerance should reflect these widely differing values, particularly for the situation when some of the dependent variables pass through zero.

4. As a related point, we have observed that the error estimator for the *RKF45* algorithm (i.e., the difference between the fourth- and fifth-order solutions) is conservative in the sense that it provides estimates that are substantially above the actual (exact) error. This is desirable since the overestimate of the error means that the integration step will be smaller than necessary to achieve the required accuracy in the solution as specified by the error tolerances. Of course, if the error estimate is too conservative, the integration step might be excessively small, but this is better than an error estimate that is not conservative (too small) and thereby allows the integration step to become so large the actual error is above the error tolerances. In other words, we want an error estimate that is reliable.
5. Returning to the output of Table 2.2, the errors in the numerical solution actually decrease after reaching maximum values. This is rather typical and very fortuitous, i.e., the errors do not accumulate as the solution proceeds.
6. The numerical solution approaches the correct final value of 20,000 (again, this is important in the sense that the errors do not cause the solution to approach an incorrect final value).

In summary, we now have an *RKF45* integrator that can be applied to a broad spectrum of initial value ODE problems. In the case of the preceding application to the logistic equation, *RKF45* was sufficiently accurate that only one integration step was required to cover the entire output interval of 0.5 (this was determined by putting some additional output statements in `rkf45_solver` to observe the integration step h, which illustrates an advantage of the basic integrators, i.e., experimentation with supplemental output is easily accomplished). Thus, the integration step adjustment in `rkf45_solver` was not really tested by the application to the logistic equation.

Therefore, we next consider another application which does require that the integration step is adjusted. At the same time with this application, we will investigate the notion of stiffness. Indeed, there is one important limitation of *RKF45*: it is an explicit integrator which does not perform well when applied to stiff problems.

The next application consists of a system of two linear, constant coefficient ODEs

$$\frac{dx_1}{dt} = -ax_1 + bx_2 \tag{2.21a}$$

$$\frac{dx_2}{dt} = bx_1 - ax_2 \tag{2.21b}$$

For the initial conditions

$$x_1(0) = 0; \quad x_2(0) = 2 \tag{2.22}$$

the analytical solution to Eq. (2.21) is

$$x_1(t) = e^{\lambda_1 t} - e^{\lambda_2 t} \tag{2.23a}$$

$$x_2(t) = e^{\lambda_1 t} + e^{\lambda_2 t} \tag{2.23b}$$

where

$$\lambda_1 = -(a-b); \quad \lambda_2 = -(a+b) \tag{2.24}$$

and a, b are constants to be specified.

We again use the *RKF45* algorithm implemented in functions `rkf45_solver` and `ssrkf45`. The main program, designated `stiff_main`, and associated functions `stiff_odes` and `stiff_odes_exact` (which compute the exact solution from Eq. (2.23) to (2.24)) can be found in the companion library and follows directly from the previous discussion. Note that the absolute and relative error tolerances are chosen as `abstol=1e-4` and `reltol=1e-4`, which is appropriate for the two dependent variables $x_1(t)$ and $x_2(t)$ since they have representative values of 1. Further, since $x_1(t)$ has an initial condition of zero, the specification of an absolute error tolerance is essential.

The numerical output from these functions is listed in abbreviated form in Table 2.4. As we can see in the table, two rows are printed at each time instant. The first row corresponds with the output for the first state variable (x_1) while the second row corresponds with the second state variable (x_2). The error tolerances are easily satisfied throughout this solution and the number of steps taken by the solver is `nsteps = 24`. The plotted solution is shown in Fig. 2.3. The eigenvalues for this solution with $a = 1.5$, $b = 0.5$ are $\lambda_1 = -(a-b) = -1$ and $\lambda_2 = -(a+b) = -2$. As t increases, e^{-2t} decays more rapidly than e^{-t} and eventually becomes negligibly small (in comparison to e^{-t}). Therefore, from Eq. (2.21), the two solutions merge (at about $t = 5$ from Fig. 2.3). In other words, the solution for both $x_1(t)$ and $x_2(t)$ becomes essentially e^{-t} for $t > 5$.

We can now vary a and b to investigate how these features of the solution change and affect the numerical solution. For example, if $a = 10.5$, $b = 9.5$, the eigenvalues are $\lambda_1 = -1$ and $\lambda_2 = -20$ so that the exponential $e^{\lambda_2 t} = e^{-20t}$ decays much more rapidly than $e^{\lambda_1 t} = e^{-t}$. The corresponding plotted solution is shown in Fig. 2.4. Note that the two solutions merge at about $t = 0.5$. Also, the specified error criteria, `abserr = 1.0e-04`, `relerr=1.0e-04`, are satisfied throughout the solution, but clearly, this is becoming more difficult for the variable-step algorithm to accomplish because of the rapid change in the solution just after the initial condition.

Table 2.4 Numerical output from `stiff_main` for $a = 1.5, b = 0.5$

t	x(t)	xex(t)	abserr	relerr
0.00	0.0000000	0.0000000	0.0000000	0.0000000
	2.0000000	2.0000000	0.0000000	0.0000000
0.10	0.0861067	0.0861067	0.0000000	0.0000000
	1.7235682	1.7235682	-0.0000000	-0.0000000
0.20	0.1484108	0.1484107	0.0000000	0.0000003
	1.4890508	1.4890508	-0.0000000	-0.0000000
0.30	0.1920067	0.1920066	0.0000001	0.0000004
	1.2896298	1.2896299	-0.0000001	-0.0000001
0.40	0.2209912	0.2209911	0.0000001	0.0000004
	1.1196489	1.1196490	-0.0000001	-0.0000001
0.50	0.2386513	0.2386512	0.0000001	0.0000004
	0.9744100	0.9744101	-0.0000001	-0.0000001
⋮	⋮	⋮	⋮	⋮
9.60	0.0000677	0.0000677	-0.0000000	-0.0000001
	0.0000677	0.0000677	-0.0000000	-0.0000001
9.70	0.0000613	0.0000613	-0.0000000	-0.0000001
	0.0000613	0.0000613	-0.0000000	-0.0000001
9.80	0.0000554	0.0000554	-0.0000000	-0.0000001
	0.0000555	0.0000555	-0.0000000	-0.0000001
9.90	0.0000502	0.0000502	-0.0000000	-0.0000001
	0.0000502	0.0000502	-0.0000000	-0.0000001
10.00	0.0000454	0.0000454	-0.0000000	-0.0000001
	0.0000454	0.0000454	-0.0000000	-0.0000001

Practically, this manifests in an increased number of integration steps required to meet the specified error tolerances, i.e., `nsteps = 78`.

If this process of separating the eigenvalues is continued, clearly the difficulty in computing a numerical solution will increase, due to two causes:

- The initial "transient" (or "boundary layer") in the solution just after the initial condition will become shorter and therefore more difficult for the variable step algorithm to resolve.
- As the eigenvalues separate, the problem becomes stiffer and the stability limit of the explicit *RKF45* integrator places an even smaller limit on the integration step to maintain stability. The maximum allowable integration step to maintain stability can be estimated from the approximate stability condition for *RKF45* (see Fig. 1.6, in particular the curve corresponding to the fourth-order method)

$$|\lambda h| < 2.7 \tag{2.25}$$

Thus, for $\lambda_2 = -20$, $h < \frac{2.7}{20} \approx 0.135$ which is still not very restrictive (relative to the time scale of $0 \leq t \leq 5$ in Fig. 2.4) so that for this case, accuracy still probably dictates the maximum integration step (to meet the error tolerances).

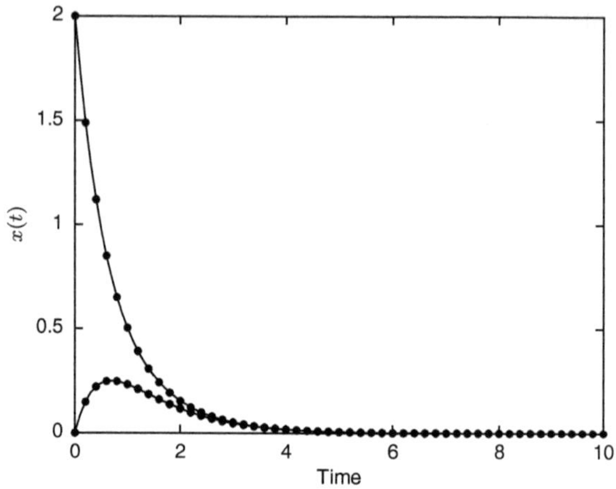

Fig. 2.3 Plot of the numerical and analytical solutions (2.21)–(2.24) with $a = 1.5, b = 0.5$

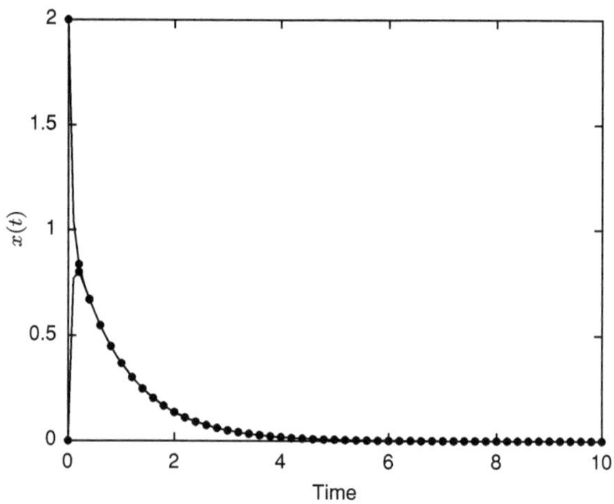

Fig. 2.4 Plot of the numerical and analytical solutions (2.21)–(2.24) with $a = 10.5, b = 9.5$

We consider one more case in separating the eigenvalues, $a = 5000.5, b = 4999.5$, for which $\lambda_1 = -1, \lambda_2 = -10{,}000$; thus, the *stiffness ratio* is $10{,}000/1$. The output for this case is incomplete since `rkf45_solver` fails at $t = 0.33$ with an error message indicating nsteps has exceeded the limit `nstepsmax=1,000`. This is to be expected since the maximum step size to still maintain stability is now determined by $\lambda_2 = -10{,}000$, $h < \frac{2.7}{10000} = 2.7 \times 10^{-4}$. In other words, to cover the total time interval $0 \leq t \leq 5$, which is set by $\lambda_1 = -1$ (so that the exponential for λ_1 decays to $e^{-(1)(5)}$), the integrator must take at least $\frac{5}{2.7\times 10^{-4}} \approx 2 \times 10^4$ steps which is greater than `nstepsmax=1,000` (i.e., the numerical solution only proceeded to $t = 0.33$ when `nstepsmax=1,000` was exceeded).

Note in general that a stiff system (with widely separated eigenvalues) has the characteristic that the total time scale is determined by the smallest eigenvalue (e.g., $\lambda_1 = -1$) while the maximum step allowed to still maintain stability in covering this total time interval is determined by the largest eigenvalue (e.g., $\lambda_2 = -10{,}000$); if these two extreme eigenvalues are widely separated, the combination makes for a large number of integration steps to cover the total interval (again, as demonstrated by *RKF45* which in this case could only get to $t = 0.33$ with 10,00 steps).

The preceding example illustrates the general limitation of the stability of *explicit ODE integrators* for the solution of stiff systems. We therefore now consider *implicit ODE integrators*, which generally circumvent this stability limit.

2.3 A Basic Variable Step Implicit ODE Integrator

Many implicit ODE integrators have been proposed and implemented in computer codes. Thus, to keep the discussion to a reasonable length, we consider here only one class of methods to illustrate some properties that clearly demonstrate the advantages of using an implicit integrator. This type of integrator is generally termed as *Rosenbrock* [10] or *linearly implicit Runge–Kutta (LIRK)* method.

Consider an autonomous function f (f does not depend explicitly on time) and the following equation:

$$\frac{dx}{dt} = f(x) \tag{2.26}$$

The explicit RK methods developed earlier in this chapter (see Eqs. 2.10–2.12) are given by

$$\begin{aligned} k_1 &= f(x_k) \\ k_2 &= f(x_k + h\beta_{2,1}k_1) \\ k_3 &= f(x_k + h(\beta_{3,1}k_1 + \beta_{3,2}k_2)) \\ &\vdots \\ k_q &= f(x_k + h(\beta_{q,1}k_1 + \beta_{q,2}k_2 + \cdots + \beta_{q,q-1}k_{q-1})) \\ x_{k+1} &= x_k + h\sum_{i=1}^{q} w_i k_i \end{aligned} \tag{2.27}$$

whereas the general implicit formulas can be expressed by

$$\begin{aligned} k_1 &= f(x_k + h(\beta_{1,1}k_1 + \beta_{1,2}k_2 + \cdots + \beta_{1,q}k_q)) \\ k_2 &= f(x_k + h(\beta_{2,1}k_1 + \beta_{2,2}k_2 + \cdots + \beta_{2,q}k_q)) \\ k_3 &= f(x_k + h(\beta_{3,1}k_1 + \beta_{3,2}k_2 + \cdots + \beta_{3,q}k_q)) \\ &\vdots \\ k_q &= f(x_k + h(\beta_{q,1}k_1 + \beta_{q,2}k_2 + \cdots + \beta_{q,q}k_q)) \end{aligned} \tag{2.28}$$

$$x_{k+1} = x_k + h \sum_{i=1}^{q} w_i k_i$$

Diagonally implicit Runge–Kutta (DIRK) is a particular case of this general formulation where $\beta_{i,j} = 0$ for $j > i$, i.e.,

$$\begin{aligned}
k_1 &= f(x_k + h\beta_{1,1}k_1) \\
k_2 &= f(x_k + h(\beta_{2,1}k_1 + \beta_{2,2}k_2)) \\
k_3 &= f(x_k + h(\beta_{3,1}k_1 + \beta_{3,2}k_2 + \beta_{3,3}k_3)) \\
&\vdots \\
k_q &= f(x_k + h(\beta_{q,1}k_1 + \beta_{q,2}k_2 + \cdots + \beta_{q,q}k_q)) \\
x_{k+1} &= x_k + h \sum_{i=1}^{q} w_i k_i
\end{aligned} \tag{2.29}$$

LIRK methods introduce a linearization of stage k_i

$$\begin{aligned}
k_i &= f(x_k + h(\beta_{i,1}k_1 + \cdots + \beta_{i,i-1}k_{i-1}) + h\beta_{i,i}k_i) \\
k_i &\approx f(x_k + h(\beta_{i,1}k_1 + \cdots + \beta_{i,i-1}k_{i-1})) \\
&\quad + h\beta_{i,i} \left.\frac{\partial f}{\partial x}\right|_{x_k + h(\beta_{i,1}k_1 + \cdots + \beta_{i,i-1}k_{i-1})} k_i
\end{aligned} \tag{2.30}$$

or, if we consider a system of ODEs $\frac{d\mathbf{x}}{dt} = \mathbf{f}(\mathbf{x})$ (instead of a single ODE)

$$\begin{aligned}
\mathbf{k}_i &= \mathbf{f}(\mathbf{x}_k + h(\beta_{i,1}\mathbf{k}_1 + \cdots + \beta_{i,i-1}\mathbf{k}_{i-1})) \\
&\quad + h\beta_{i,i} \left.\frac{\partial \mathbf{f}}{\partial \mathbf{x}}\right|_{\mathbf{x}_k + h(\beta_{i,1}\mathbf{k}_1 + \cdots + \beta_{i,i-1}\mathbf{k}_{i-1})} \mathbf{k}_i
\end{aligned} \tag{2.31}$$

where the Jacobian $\left.\frac{\partial \mathbf{f}}{\partial \mathbf{x}}\right|_{\mathbf{x}_k + h(\beta_{i,1}\mathbf{k}_1 + \cdots + \beta_{i,i-1}\mathbf{k}_{i-1})} \approx \left.\frac{\partial \mathbf{f}}{\partial \mathbf{x}}\right|_{\mathbf{x}_k}$ is usually not evaluated for each stage, but rather assumed constant across the stages, so as to keep the computational expense at a reasonable level.

In addition, Rosenbrock's methods replace stage $\mathbf{k}_i$ in the preceding expression by a linear combination of the previous stages (constructed so as to preserve the lower-triangular structure, i.e., $\gamma_{i,j} = 0$ for $j > i$)

$$\begin{aligned}
\mathbf{k}_1 &= \mathbf{f}(\mathbf{x}_k) + h\beta_{1,1}\mathbf{J}_k\mathbf{k}_1 \\
\mathbf{k}_2 &= \mathbf{f}(\mathbf{x}_k + h\beta_{2,1}\mathbf{k}_1) + h\mathbf{J}_k\left(\gamma_{2,1}\mathbf{k}_1 + \gamma_{2,2}\mathbf{k}_2\right) \\
&\vdots \\
\mathbf{k}_q &= \mathbf{f}(\mathbf{x}_k + h(\beta_{q,1}\mathbf{k}_1 + \beta_{q,2}\mathbf{k}_2 + \cdots + \beta_{q,q-1}\mathbf{k}_{q-1}))
\end{aligned} \tag{2.32}$$

$$
\begin{aligned}
&+ h\mathbf{J}_k \left(\gamma_{2,1}\mathbf{k}_1 + \cdots + \gamma_{q,q}\mathbf{k}_q\right) \\
\mathbf{x}_{k+1} &= \mathbf{x}_k + h\sum_{i=1}^{q} w_i \mathbf{k}_i
\end{aligned}
$$

Therefore, each stage $\mathbf{k}_i$ can be computed by solving a linear system of equations of the form

$$
\begin{aligned}
\left(\mathbf{I} - h\gamma_{i,i}\mathbf{J}_k\right)\mathbf{k}_i &= \mathbf{f}\left(\mathbf{x}_k + h(\beta_{i,1}\mathbf{k}_1 + \beta_{i,2}\mathbf{k}_2 + \cdots + \beta_{i,i-1}\mathbf{k}_{i-1})\right) \\
&\quad + h\mathbf{J}_k\left(\gamma_{i,1}\mathbf{k}_1 + \cdots + \gamma_{i,i-1}\mathbf{k}_{i-1}\right)
\end{aligned}
\tag{2.33}
$$

If the parameters $\gamma_{i,i}$ are all given the same numerical values

$$
\gamma_{1,1} = \cdots = \gamma_{q,q} = \gamma \tag{2.34}
$$

then, the same LU decomposition can be used for all the stages, thus saving computation time.

In short form, the preceding equations which define a q-stage Rosenbrock's method for an autonomous system are given by

$$
(\mathbf{I} - h\gamma_{i,i}\mathbf{J}_k)\mathbf{k}_i - f\left(\mathbf{x}_k + h\sum_{j=1}^{i-1}\beta_{i,j}\mathbf{k}_j\right) + h\mathbf{J}_k\sum_{j=1}^{i-1}\gamma_{i,j}\mathbf{k}_j; \quad \mathbf{x}_{k+1} = \mathbf{x}_k + h\sum_{i=1}^{q}\omega_i\mathbf{k}_i \tag{2.35}
$$

If we now consider a *nonautonomous* system of equations (explicit dependence on t)

$$
\frac{d\mathbf{x}}{dt} = \mathbf{f}(\mathbf{x}, t) \tag{2.36}
$$

then this system can be transformed into an autonomous form as

$$
\frac{d\mathbf{x}}{d\tau} = \mathbf{f}(\mathbf{x}, t); \qquad \frac{dt}{d\tau} = 1 \tag{2.37}
$$

and Eq. (2.35) lead to

$$
\begin{aligned}
(\mathbf{I} - h\gamma_{i,i}\mathbf{J}_k)\mathbf{k}_i &= f\left(\mathbf{x}_k + h\sum_{j=1}^{i-1}\beta_{i,j}\mathbf{k}_j, t_k + h\beta_i\right) \\
&\quad + h\mathbf{J}_k\sum_{j=1}^{i-1}\gamma_{i,j}\mathbf{k}_j + h\left.\frac{\partial \mathbf{f}}{\partial t}\right|_{t_k,\mathbf{x}_k}\gamma_i
\end{aligned}
\tag{2.38}
$$

with $\beta_i = \sum_{j=1}^{i-1}\beta_{i,j}$ and $\gamma_i = \sum_{j=1}^{i}\gamma_{i,j}$.

For instance, a first-order accurate Rosenbrock's scheme for nonautonomous equations is given by

$$(\mathbf{I} - h\mathbf{J}_k)\mathbf{k}_1 = \mathbf{f}(\mathbf{x}_k, t_k) + h \left.\frac{\partial \mathbf{f}}{\partial t}\right|_{t_k,\mathbf{x}_k} \tag{2.39}$$

$$\mathbf{x}_{k+1} = \mathbf{x}_k + h\mathbf{k}_1 \tag{2.40}$$

A second-order accurate Rosenbrock's method developed for autonomous equations in [11], and called ROS2, is as follows:

$$(\mathbf{I} - \gamma h\mathbf{J}_k)\mathbf{k}_1 = \mathbf{f}(\mathbf{x}_k) \tag{2.41}$$

$$(\mathbf{I} - \gamma h\mathbf{J}_k)\mathbf{k}_2 = \mathbf{f}(\mathbf{x}_k + h\mathbf{k}_1) - 2\mathbf{k}_1 \tag{2.42}$$

$$\mathbf{x}_{k+1} = \mathbf{x}_k + 1.5h\mathbf{k}_1 + 0.5h\mathbf{k}_2 \tag{2.43}$$

with desirable stability properties for $\gamma \geq 0.25$.

In the following, we will focus attention on a modified Rossenbrock's method originally proposed in [12] and specifically designed for the solution of nonlinear parabolic problems, which will be of interest to us in the following chapters dedicated to the method of lines solutions of partial differential equations. As we have just seen, the main advantage of Rosenbrock's methods is to avoid the solution of nonlinear equations, which naturally arise when formulating an implicit method. In [12], the authors establish an efficient third-order Rosenbrock's solver for nonlinear parabolic PDE problems, which requires only three stages. In mathematical terms, the method described in [12] is stated in a transformed form, which is used in practice to avoid matrix-vector operations

$$\left(\frac{\mathbf{I}}{h\gamma} - \frac{\partial \mathbf{f}}{\partial \mathbf{x}}(\mathbf{x}_k, t_k)\right)\mathbf{X}_{k,i} = \mathbf{f}\left(\mathbf{x}_k + \sum_{j=1}^{i-1} a_{i,j}\mathbf{X}_{k,j}, t_k + h\alpha_i\right) + \sum_{j=1}^{i-1} \frac{c_{i,j}}{h}\mathbf{X}_{k,j} + hd_i\frac{\partial \mathbf{f}}{\partial t}(\mathbf{x}_k, t_k) \quad i = 1, 2, 3 \tag{2.44}$$

Two stepping formulas for a third- and a second-order methods, respectively, are used to estimate the truncation error, i.e., can be computed by taking the difference between the two following solutions

$$\mathbf{x}_{k+1} = \mathbf{x}_k + \sum_{j=1}^{3} m_j\mathbf{X}_{k,j} \tag{2.45a}$$

Table 2.5 Parameters of the ROS23P algorithm

$\gamma = 0.5 + \sqrt{3}/6$	
$a_{21} = 1.267949192431123$	$c_{21} = -1.607695154586736$
$a_{31} = 1.267949192431123$	$c_{31} = -3.464101615137755$
$a_{32} = 0$	$c_{32} = -1.732050807567788$
$\alpha_1 = 0$	$d_1 = 0.7886751345948129$
$\alpha_2 = 1$	$d_2 = -0.2113248654051871$
$\alpha_3 = 1$	$d_3 = -1.077350269189626$
$m_1 = 2$	$\hat{m}_1 = 2.113248654051871$
$m_2 = 0.5773502691896258$	$\hat{m}_2 = 1$
$m_3 = 0.4226497308103742$	$\hat{m}_3 = 0.4226497308103742$

$$\hat{\mathbf{x}}_{k+1} = \mathbf{x}_k + \sum_{j=1}^{3} \hat{m}_j \mathbf{X}_{k,j} \tag{2.45b}$$

Note that the solutions at intermediate points, $\mathbf{X}_{k,j}$, are the same for both orders reflecting the embedding of the second-order method in the third-order method.

γ, $a_{i,j}$, $c_{i,j}$, α_i, d_i, m_i, $\hat{m}_i$ are parameters defined for a particular Rosenbrock's method. Particular values are listed in Table 2.5, which defines a method designated as *ROS23P*. *ROS* denotes *Rosenbrock*, 23 indicates a second-order method embedded in a third-order method in analogy with the fourth-order method embedded in a fifth-order method in *RKF45*, and *P* stands for *parabolic problems*. This solver will indeed be particularly useful for the time integration of ODE systems arising from the application of the method of lines to parabolic PDE problems. The selection of appropriate parameters, as in Table 2.5, confers desirable stability properties to the algorithm, which can therefore be applied to stiff ODEs. To reiterate, the system of Eq. (2.44) is *linear* in $\mathbf{X}_{k,j}$ which is a very favorable feature, i.e., a single application of a linear algebraic equation solver is all that is required to take the next step along the solution (from point k to $k + 1$). This is in contrast to implicit ODE methods which require the solution of *simultaneous nonlinear equations* that generally is a substantially more difficult calculation than solving linear equations (usually done by Newton's method, which must be iterated until convergence).

As in the case of *RKF45*, we program the algorithm in two steps: (1) a single step function analogous to `ssrkf45` and (2) a function that calls the single step routine to step along the solution and adjust the integration step that is analogous to `rkf45_solver`. We first list the single step routine, `ssros23p`

```
function [t,x,e] = ssros23p(odefunction,jacobian,...
                            time_derivative,t0,x0,h,gamma,a21,...
                            a31,a32,c21,c31,c32,d,alpha,m,mc)
%
% Function ssros3p computes an ODE solution by an implicit
% third-order Rosenbrock method for one step along the
```

```
% solution (by calls to 'odefunction' to define the ODE
% derivative vector, calls to 'jacobian' to define the
% Jacobian and calls to time_derivative if the problem is
% non autonomous).
%
% Argument list
%
% odefunction - string containing name of user-supplied problem
% jacobian - string containing name of user-supplied Jacobian
% time_derivative - sting containing name of user-supplied
%                   function time derivative
%
% t0 - initial value of independent variable
% x0 - initial condition vector
% h - integration step
% t - independent variable (scalar)
% x - solution vector after one rkf45 step
% e - estimate of truncation error of the solution vector
%
%   gamma,a21,a31,a32,c21,c31,c32,alpha,d,m,mc are the
%   method parameters

% Jacobian matrix at initial (base) point
[Jac] = feval(jacobian, t0, x0);

% Time derivative at initial (base) point
[Ft] = feval(time_derivative, t0, x0);

% Build coefficient matrix and perform L-U decomposition
CM = diag(1/(gamma*h)*ones(length(x0),1)) - Jac;
[L,U] = lu(CM);

% stage 1
xs = x0;
[xt] = feval(odefunction, t0+alpha(1)*h, xs);
rhs = xt + h*d(1)*Ft;
xk1 = U\(L\rhs);

% stage 2
xs = x0 + a21*xk1;
[xt] = feval(odefunction, t0+alpha(2)*h, xs);
rhs = xt + (c21/h)*xk1 + h*d(2)*Ft;
xk2 = U\(L\rhs);

% stage 3
xs = x0 + a31*xk1 + a32*xk2;
[xt] = feval(odefunction, t0+alpha(3)*h, xs);
rhs = xt + (c31/h)*xk1 + (c32/h)*xk2 + h*d(3)*Ft;
xk3 = U\(L\rhs);

% second-order step
x2 = x0 + mc(1)*xk1 + mc(2)*xk2 + mc(3)*xk3;

% third-order step
x3  = x0 + m(1)*xk1 + m(2)*xk2 + m(3)*xk3;

% error evaluation
t = t0 + h;
e = x3 - x2;
x = x3;
```

Function ssros23p Routine for a single step along the solution

We can note the following details about `ssros23p`:

1. The first statement defining function `ssros23p` has an argument list that includes the parameters (constants) used in the *ROS23P* algorithm as defined in Table 2.5.
2. After an initial set of comments explaining the operation of `ssros23p` and its arguments, the Jacobian matrix of the ODE system is first computed

```
% Jacobian matrix at initial (base) point
[Jac] = feval(jacobian,t0,x0);
```

   The user-supplied routine `jacobian` (to be discussed subsequently) is called to compute the Jacobian matrix in Eq. (2.44), $f_x(x_k, t_k)$, at the base point $t_k = t_0$, $x_k = x0$.
3. Then, the time derivative of the ODE function is computed

```
%...
%... Time derivative at initial (base) point
     [Ft] = feval(time_derivative, t0, x0);
```

   The user-supplied routine `time_derivative` is called to compute $f_t(x_k, t_k)$.
4. The LHS coefficient matrix of Eq. (2.44)

$$\left(\frac{I}{h\gamma} - f_x(\mathbf{x}_k, t_k)\right)$$

   is then constructed

```
% Build coefficient matrix and perform L-U decomposition
CM    = diag(1/(gamma*h)*ones(length(x0),1))-Jac;
[L,U] = lu(CM);
```

   Note that the main diagonal of the identity matrix is coded as `ones(length(x0),1)`. Then subtraction of the Jacobian matrix, `Jac`, results in a square coefficient matrix, `CM`, with dimensions equal to the length of the ODE dependent variable vector (`length(x0)`). After `CM` is constructed, it is factored (decomposed) into `L` and `U` lower and upper triangular factors using the MATLAB utility `lu`.

   Just to briefly review why this decomposition is advantageous, if the coefficient, A, of a linear algebraic system

$$Ax = b \tag{2.46a}$$

   is written in factored form, i.e., $A = LU$, Eq. (2.46a) can be written as

$$LUx = b \tag{2.46b}$$

Equation (2.46b) can be written as two algebraic equations

$$Ly = b \tag{2.47}$$

$$Ux = y \tag{2.48}$$

Note that Eq. (2.47) can be solved for y and Eq. (2.48) can then be solved for x (the solution of Eq. (2.46a)). If matrix A can be decomposed into a lower triangular matrix L and and upper triangular matrix U, then the solution of (2.47) can be easily done by elementary substitution starting from the first equation to the last one (forward substitution), whereas Eq. (2.48) can then be solved in the same way from the last to the first equation (backward substitution). This is a substantial simplification. Also, once the LU decomposition is performed, it can be used repeatedly for the solution of Eq. (2.46a) with different RHS vectors, b, as we shall observe in the coding of `ssros23p`. This reuse of the LU factorization is an important advantage since this factorization is the major part of the computational effort in the solution of linear algebraic equations.

5. We now step through the first of three stages, using $i = 1$ in Eq. (2.44)

```
% Stage 1
xs    = x0;
[xt]  = feval(odefunction,t0+alpha(1)*h,xs);
rhs   = xt;
xk1   = U\(L\rhs);
```

Thus, the RHS of Eq. (2.44) is simply $xt = f(x_k, t_k)$ (from Eq. 1.12). Finally, the solution of Eq. (2.44) for X_{k1} is obtained by using the MATLAB operator twice, corresponding to the solution of Eqs. (2.47) and (2.48).

6. The second stage is then executed essentially in the same way as the first stage, but with $i = 2$ in Eq. (2.44) and using X_{k1} from the first stage

```
% Stage 2
xs    = x0 + a21*xk1;
[xt]  = feval(odefunction,t0+alpha(2)*h,xs);
rhs   = xt + (c21/h)*xk1;
xk2   = U\(L\rhs);
```

Note in particular how the LU factors are used again (they do not have to be recomputed at each stage).

7. Finally, the third stage is executed essentially in the same way as the first and second stages, but with $i = 3$ in Eq. (2.44) and using X_{k1} from the first stage and X_{k2} from the second stage

```
% Stage 3
xs    = x0 + a31*xk1 + a32*xk2;
[xt]  = feval(odefunction,t0+alpha(3)*h,xs);
rhs   = xt + (c31/h)*xk1 + (c32/h)*xk2;
xk3   = U\(L\rhs);
```

Again, the LU factors can be used as originally computed.

8. Equations (2.45a) and (2.45b) are then used to step the second- and third-order solutions from k to $k+1$

```
% Second-order step
x2 = x0 + mc(1)*xk1 + mc(2)*xk2 + mc(3)*xk3;
% Third-order step
x3 = x0 + m(1)*xk1 + m(2)*xk2 + m(3)*xk3;
```

9. The truncation error can be estimated as the difference between the second- and third-order solutions

```
% Estimated error and solution update
e = x3 - x2;
t = t0 + h;
x = x3;
```

The solution is taken to be the third-order result at the end of `ssros23p`.

In order to use function `ssros23p`, we require a calling function that will also adjust the integration step in accordance with user-specified tolerances and the estimated truncation error from `ssros23p`. A routine, analogous to `rkf45_solver`, is listed in `ros23p_solver`.

```
function [tout, xout, eout] = ros23p_solver(odefunction,...
                              jacobian,time_derivative,t0,tf,...
                              x0,hmin,nstepsmax,abstol,reltol,...
                              Dtplot)
% [tout, yout] = ros23p_solver('f','J','Ft',t0,tf,x0,hmin,...
%                               nstepsmax,abstol, reltol,Dtplot)
% Integrates a non-autonomous system of differential equations
% y'=f(t,x) from t0 to tf with initial conditions x0.
%
% Each row in solution array xout corresponds to a value
% returned in column vector tout.
% Each row in estimated error array eout corresponds to a
% value returned in column vector tout.
%
% ros23p_solver.m solves first-order differential equations
% using a variable-step implicit Rosenbrock method for a
% series of points along the solution by repeatedly calling
% function ssros23p for a single Rosenbrock step.
%
% The truncation error is estimated along the solution to
% adjust the integration step according to a specified error
% tolerance.
%
% Argument list
%
% f          - String containing name of user-supplied
%              problem description
%                     Call: xdot = fun(t,x) where f = 'fun'
%                     t      - independent variable (scalar)
%                     x      - Solution vector.
%                     xdot   - Returned derivative vector;
%                              xdot(i) = dx(i)/dt
%
% J          - String containing name of user-supplied Jacobian
```

```
%                    Call: Jac = fun(t,x) where J = 'fun'
%                    t      - independent variable (scalar)
%                    x      - Solution vector.
%                    Jac    - Returned Jacobian matrix;
%                               Jac(i,j) = df(i)/dx(j)
%
% Ft          - String containing nam of user-supplied
%               function time derivative
%                    Call: Ft = fun(t,x) where Ft = 'fun'
%                    t      - independent variable (scalar)
%                    x      - Solution vector.
%                    Ft     - Returned time derivative;
%                               Ft(i) = df(i)/dt
%
% t0          - Initial value of t
% tf          - Final value of t
% x0          - Initial value vector
% hmin        - minimum allowable time step
% nstepsmax - maximum number of steps
% abstol      - absolute error tolerance
% reltol      - relative error tolerance
% Dtplot      - Plot interval
%
% tout        - Returned integration points (column-vector).
% xout        - Returned solution, one solution row-vector per
%               tout-value.

% Initial integration step
h = 10*hmin;

% method parameters
gamma = 0.5+sqrt(3)/6;
a21 = 1.267949192431123;
a31 = 1.267949192431123;
a32 = 0.0;
c21 =-1.607695154586736;
c31 = -3.464101615137755;
c32 = -1.732050807567788;
d(1) = 7.886751345948129e-01;
d(2) = -2.113248654051871e-01;
d(3) = -1.077350269189626e+00;
alpha(1) = 0;
alpha(2) = 1.0;
alpha(3) = 1.0;
m(1) = 2.000000000000000e+00;
m(2) = 5.773502691896258e-01;
m(3) = 4.226497308103742e-01;
mc(1) = 2.113248654051871e+00;
mc(2) = 1.000000000000000e+00;
mc(3) = 4.226497308103742e-01;
% Start integration
t = t0;
tini = t0;
tout = t0;                          % initialize output value
xout = x0';                         % initialize output value
eout = zeros(size(xout));           % initialize output value
nsteps = 0;                         % initialize step counter
nplots = round((tf-t0)/Dtplot);     % number of outputs

% Initial integration step
h = 10*hmin;

% Step through nplots output points
for i = 1:nplots

    %   Final (output) value of the independent variable
```

```
    tplot = tini+i*Dtplot;
    % While independent variable is less than the final value,
    % continue the integration
    while t <= tplot*0.9999
        % If the next step along the solution will go past the
        % final value of the independent variable, set the step
        % to the remaining distance to the final value
        if t+h > tplot, h = tplot-t; end
        %  Single ros23p step
        [t,x,e] = ssros23p(odefunction,jacobian,...
            time_derivative,t0,x0,h,gamma,a21,a31,a32,...
            c21,c31,c32,d,alpha,m,mc);

        % Check if any of the ODEs have violated the error
        % criteria
        if max( abs(e) > (abs(x)*reltol + abstol) )
            % Error violation, so integration is not complete.
            % Reduce integration step because of error violation
            % and repeat integration from the base point. Set
            % logic variable for rejected integration step.
            h = h/2;

            % If the current step is less than the minimum
            % allowable step, set the step to the minimum
            % allowable value
            if h < hmin,  h = hmin; end

            % If there is no error violation, check if there is
            % enough "error margin" to increase the integration
            % step
        elseif max( abs(e) > (abs(x)*reltol + abstol)/8 )
            % The integration step cannot be increased, so leave
            % it unchanged and continue the integration from the
            % new base point.
            x0 = x; t0 = t;

            % There is no error violation and enough "security
            % margin"
        else

            % double integration step and continue integration
            % from new base point
            h = 2*h; x0 = x; t0 = t;
        end %if

        % Continue while and check total number of integration
        % steps taken
        nsteps=nsteps+1;
        if(nsteps > nstepsmax)
      fprintf('\n nstepsmax exceeded; integration terminated\n');
      break;
        end
    end %while

    % add latest result to the output arrays and continue for
    % loop
    tout = [tout ; t];
    xout = [xout ; x'];
    eout = [eout ; e'];
end % for
%
% End of ros23p_solver
```

Function ros23p_solver Routine for a variable step solution of an ODE system ros23p_solver closely parallels rkf45_solver, so we point out just a few differences:

1. After a set of comments explaining the arguments and operation of ros23p_solver, the parameters for the *ROS23P* algorithm are set in accordance with the values defined in Table 2.5.
2. The variable step algorithm is then implemented as discussed for rkf45_solver, based on the estimated truncation error vector from a call to ssros23p

```
function [t,x,e] = ssros23p(odefunction,jacobian,...
                            time_derivative,t0,x0,h,gamma,...
                            a21,a31,a32,c21,c31,c32,d,...
                            alpha,m,mc)
```

In order to apply this algorithm to an example we need function to define the ODE model. If the logistic equation is chosen as an example, such function is the same as in logistic_ode. The exact solution to the logistic equation is again programmed in logistic_exact.

Since *ROS23P* requires the Jacobian matrix in ssros23p, we provide function logistic_jacobian for this purpose.

```
function [Jac] = logistic_jacobian(t , N)
% Set global variables
global a b K N0

% Jacobian of logistic equation Nt=(a-b*N)*N
Jac = a - 2*b*N;
```

Function logistic_jacobian Routine for the computation of the Jacobian matrix of logistic equation (2.1)

Note that, in this case, the logistic equation has a 1×1 Jacobian matrix (single element), which is just the derivative of the RHS of Eq. (2.1) with respect to N. ROS23P also requires a function to compute the time derivative of the ODE function, which is only useful for nonautonomous problems (so not in this case).

We now have all of the programming elements for the *ROS23P* solution of the logistic equation. Execution of the main program gives the numerical output of Table 2.6 (the plotted output is essentially identical to Fig. 2.2 and therefore is not repeated here).

A comparison of Tables 2.3 and 2.6 clearly indicates that *RKF45* in this case was much more effective in controlling the integration error than *ROS23P* (all of the integration parameters such as error tolerances were identical for the two integrators, i.e., the parameters of Table 2.2). However, *ROS23P* did meet the relative error tolerance, relerr = 1e-03. As explained previously, the large absolute error is due to the dominance of the total error by the relative error, which produces a typical

Table 2.6 Numerical output obtained with `ros23p_solver` for the logistic equation (2.1)

t	x(t)	xex(t)	abserr	relerr
0.0	1000.00	1000.00	0.00000	0.00000
0.5	1596.67	1596.92	-0.25079	-0.00016
1.0	2501.33	2503.22	-1.88884	-0.00075
1.5	3814.04	3817.18	-3.13425	-0.00082
2.0	5595.63	5600.09	-4.46446	-0.00080
2.5	7808.18	7813.68	-5.49428	-0.00070
3.0	10271.87	10277.73	-5.86125	-0.00057
3.5	12703.07	12708.50	-5.42835	-0.00043
4.0	14832.45	14836.83	-4.37216	-0.00029
4.5	16511.65	16514.31	-2.66122	-0.00016
5.0	17728.93	17730.17	-1.23181	-0.00007
⋮	⋮	⋮	⋮	⋮
15.0	19999.89	19999.88	0.00746	0.00000

term in the error test of `ros23p_solver` at `t = 1.5` of `3814.04(0.001) = 3.814` whereas the absolute error (in absolute terms) was smaller, i.e., `3.13425`. Thus, although *ROS23P* did not control the integration error as tightly as *RKF45*, its performance can still probably be considered acceptable for most applications, i.e., 1 part in 1,000. This poorer error control can be explained by the lower order of the error estimator of *ROS23P* relative to *RKF45*.

In other words, the error tolerances for *ROS23P* solver of `abserr = 1e-03`, `relerr = 1e-03` are rather loose. If they are tightened to `abserr = 1e-05`, `relerr = 1e-05` (i.e., 1 part in 10^5) the resulting output from `ros23p_solver` is

```
t        x(t)     xex(t)     abserr    relerr
0.0   1000.00    1000.00    0.00000   0.00000
0.5   1596.92    1596.92   -0.00623  -0.00000
1.0   2503.20    2503.22   -0.01681  -0.00001
1.5   3817.15    3817.18   -0.03137  -0.00001
2.0   5600.04    5600.09   -0.04699  -0.00001
2.5   7813.62    7813.68   -0.05887  -0.00001
3.0  10277.67   10277.73   -0.06277  -0.00001
3.5  12708.44   12708.50   -0.05726  -0.00000
4.0  14836.78   14836.83   -0.04438  -0.00000
4.5  16514.29   16514.31   -0.02130  -0.00000
5.0  17730.19   17730.17    0.02374   0.00000
.                                  .
.                                  .
.                                  .
15.0  19999.89   19999.88   0.00639   0.00000
```

The enhanced performance of *ROS23P* using the tighter error tolerances is evident (*ROS23P* achieved the specified accuracy of 1 part in 10^5). This example illustrates the importance of error tolerance selection and the possible differences in error control between different integration algorithms. In other words, some experimentation with the error tolerances may be required to establish the accuracy (reliability) of the computer solutions.

Since *ROS23P* has a decided advantage over *RKF45* for stiff problems (because of superior stability), we now illustrate this advantage with the 2×2 ODE system of Eq. (2.21) with again evaluating the numerical solution using the analytical solution of Eq. (2.23). The coding of the 2×2 problem to run under *ROS23P* is essentially the same as for *RKF45* and can also be found in the companion library. However, a routine for the Jacobian matrix is required by *ROS23P* (see function `jacobian_stiff_odes`).

```
function Jac = jacobian_stiff_odes(t,x)

% Set global variables
global a b

% Jacobian matrix
Jac = [-a   b;
        b  -a];
```

Function jacobian_stiff_odes Jacobian matrix of 2×2 ODE Eq. (2.21)

Here we are evaluating the Jacobian matrix $f_x(x_k, t_k)$ in Eq. (2.44) as required in `ssros23p`. For example, the first row, first element of this matrix is $\frac{\partial f_1}{\partial x_1} = -a$. Note that since Eq. (2.21) are *linear constant coefficient ODEs, their Jacobian matrix is a constant matrix*, and therefore function `jacobian_stiff_odes` would only have to be called once. However, since `ros23p_solver` and `ssros23p` are general purpose routines (they can be applied to nonlinear ODEs for which the Jacobian matrix is not constant, but rather is a function of the dependent variable vector), `jacobian_stiff_odes` will be called at each point along the solution of Eq. (2.21) through `ros23p_solver`.

We should note the following points concerning the Jacobian matrix:

1. If we are integrating an nth order ODE system (n first-order ODEs in n unknowns or a $n \times n$ ODE system), the *Jacobian matrix is of size* $n \times n$. This size increases very quickly with n. For example, if $n = 100$ (a modest ODE problem), the Jacobian matrix is of size $100 \times 100 = 10{,}000$.
2. In other words, we need to compute the $n \times n$ partial derivatives of the Jacobian matrix, and for large n (e.g., $n > 100$), this becomes difficult if not essentially impossible (not only because there are so many partial derivatives, but also, because the actual analytical differentiation may be difficult depending on the complexity of the derivative functions in the RHS of Eq. (1.12) that are to be differentiated).
3. Since analytical differentiation to produce the Jacobian matrix is impractical for large n, we generally have to resort to a numerical procedure for computing the required partial derivatives. Thus, a *numerical Jacobian* is typically used in

Table 2.7 Numerical output from `ssros23p` for the stiff problem (2.21) with $a = 500000.5$ and $b = 499999.5$

a = 500000.500			b = 499999.500	
t	x(t)	xex(t)	abserr	relerr
0.00	0.0000000	0.0000000	0.0000000	0.0000000
	2.0000000	2.0000000	0.0000000	0.0000000
0.10	0.9048563	0.9048374	0.0000189	0.0000208
	0.9048181	0.9048374	-0.0000193	-0.0000213
0.20	0.8187392	0.8187308	0.0000084	0.0000103
	0.8187187	0.8187308	-0.0000120	-0.0000147
0.30	0.7408210	0.7408182	0.0000028	0.0000038
	0.7408101	0.7408182	-0.0000082	-0.0000110
0.40	0.6703187	0.6703200	-0.0000013	-0.0000020
	0.6703128	0.6703200	-0.0000072	-0.0000108
0.50	0.6065277	0.6065307	-0.0000029	-0.0000049
	0.6065246	0.6065307	-0.0000061	-0.0000100
⋮	⋮	⋮	⋮	⋮
10.00	0.0000454	0.0000454	-0.0000000	-0.0007765
	0.0000454	0.0000454	-0.0000000	-0.0007765

the solution of stiff ODE systems. The calculation of a numerical Jacobian for Eq. (2.21) is subsequently considered.

The numerical output from these functions is listed in abbreviated form in Table 2.7. As in the case of Table 2.4, two rows are printed at each time instant. The first row corresponds with the output for the first state variable (x_1) while the second row corresponds with the second state variable (x_2). This solution was computed with good accuracy and modest computation effort (the number of calls to IVP solver was `nsteps = 14`). Note how the two solutions, $x_1(t)$, $x_2(t)$, merged almost immediately and were almost identical by $t = 0.1$ (due to the large eigenvalue $\lambda_2 = -10^6$ in Eq. (2.24) so that the exponential $e^{-\lambda_2 t}$ decayed to insignificance almost immediately). This example clearly indicates the advantage of a stiff integrator (*RKF45* could not handle this problem with reasonable computational effort since it would take an extremely small integration step because of the stiffness). Clearly the solution of the simultaneous equations of Eq. (2.44) was worth the effort to maintain stability with an acceptable integration step (the integration step could be monitored by putting it in an output statement in `ros23p_solver`). Generally, this example illustrates the advantage of an implicit (stiff) integrator (so that the additional computation of solving simultaneous linear algebraic equations is worthwhile).

There is one detail that should be mentioned concerning the computation of the solution in Table 2.7. Initially `ros23p_solver` and `ssros23p` failed to compute a solution. Some investigation, primarily by putting output statements in ros23p_solver, indicated that the problem was the selection of an initial integration interval in ros23p_solver according to the statement

```
% Initial integration step
h = Dtplot/10;
```

In other words, since Dtplot = 0.1 (see the output interval of Table 2.7), the initial integration step is 0.01. Recall that ros23p_solver is trying to compute a solution according to Eq. (2.24) in which an exponential decays according to $e^{-10^6 t}$. If the initial step in the numerical integration is $h = 0.01$, the exponential would be $e^{-10^6(0.01)} = e^{-10^4}$ and it therefore has decayed to insignificance. The automatic adjustment of the integration step in ros23p_solver would have to reduce the integration step from 0.01 to approximately 10^{-7} so that the exponential is $e^{-10^6(10^{-7})} = e^{-0.1}$ and it apparently was unable to do this. By changing the initial integration programming to

```
% Initial integration step
h = Dtplot/1.0e+5;
```

the numerical integration proceeded without difficulty and produced the numerical solution of Table 2.7. This discussion illustrates that some experimentation with the initial integration step may be required, particularly for stiff problems. An alternative would be to use an available algorithm for the initial integration step, but this would increase the complexity of ros23p_solver. We therefore opted to use a manual adjustment of the initial integration interval to successfully start the numerical integration.

As indicated previously, the use of an analytical Jacobian with a stiff integrator is not practical for large ODE problems (i.e., $n > 100$), but rather, a numerical Jacobian should be used. To illustrate this approach, we consider the use of a numerical Jacobian for the solution of Eq. (2.21). For example, we can use the *finite difference approximations*.

$$f_x = \begin{bmatrix} \frac{\partial f_1}{\partial x_1} & \frac{\partial f_1}{\partial x_2} \\ \frac{\partial f_2}{\partial x_1} & \frac{\partial f_2}{\partial x_2} \end{bmatrix} \approx \begin{bmatrix} \frac{f_1(x_1+\Delta x_1,x_2)-f_1(x_1,x_2)}{\Delta x_1} & \frac{f_1(x_1,x_2+\Delta x_2)-f_1(x_1,x_2)}{\Delta x_2} \\ \frac{f_2(x_1+\Delta x_1,x_2)-f_2(x_1,x_2)}{\Delta x_1} & \frac{f_2(x_1,x_2+\Delta x_2)-f_2(x_1,x_2)}{\Delta x_2} \end{bmatrix} \tag{2.49}$$

Application of Eq. (2.49) to Eq. (2.21) gives

$$f_x \approx \begin{bmatrix} \frac{-a(x_1+\Delta x_1)+bx_2-(ax_1+bx_2)}{\Delta x_1} & \frac{-ax_1+b(x_2+\Delta x_2)-(-ax_1+bx_2)}{\Delta x_2} \\ \frac{f_2(x_1+\Delta x_1,x_2)-f_2(x_1,x_2)}{\Delta x_1} & \frac{f_2(x_1,x_2+\Delta x_2)-f_2(x_1,x_2)}{\Delta x_2} \end{bmatrix} = \begin{bmatrix} -a & b \\ b & -a \end{bmatrix}$$

Thus, Eq. (2.49) gives the exact Jacobian (see jacobian_stiff_odes) because the finite differences are exact for linear functions (i.e., the linear functions of Eq. (2.21)). However, generally this will not be the case (for nonlinear functions) and therefore the finite difference approximation of the Jacobian matrix will introduce errors in the numerical ODE solution. The challenge then is to compute the numerical Jacobian with sufficient accuracy to produce a numerical solution of acceptable accuracy. Generally, this has to do with the selection of the increments Δx_1, Δx_2. But the principal advantage in using finite differences to avoid analytical differentiation is generally well worth the additional effort of computing a sufficiently accurate numerical Jacobian. To illustrate the programming of a numerical Jacobian, we use routine `jacobian_stiff_odes_fd` instead of `jacobian_stiff_odes`.

```
function [Jac] = jacobian_stiff_odes_fd(t,x)
% Function jacobian_stiff_odes_fd computes the Jacobian matrix
% by finite differences

% Set global variables
global a b

% Jacobian of the 2 x 2 ODE system
% Derivative vector at base point
xb  = x;
xt0 = stiff_odes(t,x);

% Derivative vector with x1 incremented
dx1  = x(1)*0.001 + 0.001;
x(1) = x(1) + dx1;
x(2) = xb(2);
xt1  = stiff_odes(t,x);

% Derivative vector with x2 incremented
x(1) = xb(1);
dx2  = x(2)*0.001 + 0.001;
x(2) = x(2) + dx2;
xt2  = stiff_odes(t,x);

% Jacobian matrix computed by finite differences in place of
% Jac = [-a b; b -a]);
Jac(1,1) = (xt1(1)-xt0(1))/dx1;
Jac(1,2) = (xt2(1)-xt0(1))/dx2;
Jac(2,1) = (xt1(2)-xt0(2))/dx1;
Jac(2,2) = (xt2(2)-xt0(2))/dx2;
```

Function jacobian_stiff_odes_fd Routine for the computation of the Jacobian matrix of Eq. (2.21) using the finite differences of Eq. (2.49)

We can note the following points about this routine:

1. While the Jacobian routine `jacobian_stiff_odes_fd` appears to be considerably more complicated than the routine `jacobian_stiff_odes`, it has one important advantage: *analytical differentiation is not required.* Rather, the finite difference approximation of the Jacobian partial derivatives requires only calls to the ODE routine, `stiff_odes` (which, of course, is already available for the problem ODE system).
2. To briefly explain the finite differences as expressed by Eq. (2.49), we first need the derivative functions at the base (current) point along the solution (to give

the derivative functions $f_1(x_1, x_2), f_2(x_1, x_2)$ in Eq. (2.49)). These derivatives are computed by the first call to `stiff_odes`

3. Then we need the derivatives with x_1 incremented by Δx_1, that is $f_1(x_1+\Delta x_1, x_2)$, $f_2(x_1 + \Delta x_1, x_2)$ which are computed by

```
% Derivative vector with x1 incremented
dx1   = x(1)*0.001 + 0.001;
x(1)  = x(1) + dx1;
x(2)  = xb(2);
xt1   = stiff_odes(t,x);
```

   Note that in computing the increment `dx1` we use $0.001x_1 + 0.001$; the second 0.001 is used in case $x_1 = 0$ (which would result in no increment) as it does at the initial condition $x_1(0) = 0$.

4. Next, the derivatives with x_2 incremented by Δx_2 are computed

```
% Derivative vector with x2 incremented
x(1)  = xb(1);
dx2   = x(2)*0.001 + 0.001;
x(2)  = x(2) + dx2;
xt2   = stiff_odes(t,x);
```

5. Then the four partial derivatives of the Jacobian matrix are computed. For example, $\frac{\partial f_1}{\partial x_1} \approx \frac{f_1(x_1+\Delta x_1, x_2)-f_1(x_1, x_2)}{\Delta x_1}$ is computed as

```
% Jacobian matrix (computed by finite differences in
% place of Jac = [-a b; b -a]);
Jac(1,1)  =  (xt1(1)-xt0(1))/dx1;
```

The numerical solution with `jacobian_stiff_odes` (analytical Jacobian) replaced with `jacobian_stiff_odes_fd` (numerical Jacobian) gave the same solution as listed in Table 2.7. This is to be expected since the finite difference approximations are exact for the linear ODEs Eq. (2.21).

To summarize this discussion of *ROS23P*, we have found that this algorithm:

1. Has a reliable error estimate (the difference between the second- and third-order solutions), but not as accurate as *RKF45* (the difference between the fourth- and fifth-order solutions).
2. Requires only the solution of linear algebraic equations at each step along the solution (thus the name LIRK where "LI" denotes *linearly implicit*). We should also note that the accuracy of the numerical solution is directly dependent on the accuracy of the Jacobian matrix. This is in contrast with implicit ODE integrators that require the solution of nonlinear equations, but the Jacobian can often be approximate or inexact since all that is required is the convergence of the nonlinear equation solutions, usually by Newton's method or some variant, which can converge with an inexact Jacobian.
3. Has excellent stability properties (which are discussed in detail in [12]).

Table 2.8 MATLAB solvers for differential equations

Initial value problem solvers for ODEs (if unsure about stiffness, try `ode45` *first, then* `ode15s`)	
`ode45`	Solve nonstiff differential equations, medium-order method
`ode23`	Solve nonstiff differential equations, low-order method
`ode113`	Solve nonstiff differential equations, variable-order method
`ode23t`	Solve moderately stiff ODEs and DAEs index 1, trapezoidal rule
`ode15s`	Solve stiff ODEs and DAEs index 1, variable-order method
`ode23s`	Solve stiff differential equations, low-order method
`ode23tb`	Solve stiff differential equations, low-order method
Initial value problem solvers for fully implicit ODEs/DAEs $F(t, y, y') = 0$	
`decic`	Compute consistent intial conditions
`ode15i`	Solve implicit ODEs or DAEs i ndex 1
Initial value problem solver for delay differential equations (DDEs)	
`dde23`	Solve delay differential equations (DDEs) with constant delays
Boundary value problem solver for ODEs	
`bvp4c`	Solve two-point boundary value problems for ODEs by collocation
1D Partial differential equation solver	
`pdepe`	Solve initial-boundary value problems for parabolic-elliptic PDE

Thus, we now have a choice of high accuracy nonstiff (*RKF45*) and stiff (*ROS23P*) algorithms that are implemented in library routines for solution of the general $n \times n$ initial value ODE problem. The ambition of these solvers is not to compete with the high-quality integrators that are included in the MATLAB ODE Suite or within SCILAB or OCTAVE, but they are easy to use and to understand and will also be easily translated into various environments. Before testing them in additional applications, we introduce briefly the MATLAB ODE Suite.

2.4 MATLAB ODE Suite

We have previously discussed the use of advanced integrators, which are part of the MATLAB ODE suite or library. Examples include the use of `ode45` and `ode15s` in `Main_bacteria`, and `ode15s` in `Main_two_tanks`. The MATLAB ODE integrators have options beyond the basic integrators; furthermore, there is an extensive selection of integrators as given in Table 2.8. We cannot go into all of the features available because of limited space. The code for these integrators is relatively long and complex, so they are typically used without modification. Additional details about the MATLAB integrators are available in [1].

The application examples described in the next section are used to compare some of the time integrators developed in the previous sections, as well as several integrators from the MATLAB ODE Suite.

2.5 Some Additional ODE Applications

The ODE applications considered previously in this chapter were quite modest in complexity and could therefore be solved analytically or numerically; comparison of these two types of solutions was used to establish the validity of the numerical algorithms and associated codes. Now that the numerical integrators have been developed and tested, we consider some ODE applications that are sufficiently complex to preclude analytical solution; thus, we have only the numerical integrators as a viable approach to solutions, which is the usual case in realistic applications, i.e., numerical methods can be applied when analytical methods are not tractable.

2.5.1 Spruce Budworm Dynamics

The first application we consider is an ecological model of the interaction of spruce budworm populations with forest foliage [13]. The full model is a 3 × 3 system of logistic-type ODEs. However, before considering the full model, we consider a simplified version that has only one ODE, but which exhibits interesting dynamics. The idea of the simplified model is to consider that the slow variables (associated with foliage quantity and quality, $S(t)$) are held fixed, and to analyze the long-term behavior of the fast variable, i.e., the budworm density, $B(t)$.

The budworm's growth follows a logistic equation

$$\frac{\mathrm{d}B}{\mathrm{d}t} = r_B B\left(1 - \frac{B}{k_B}\right) - g \tag{2.50}$$

where the carrying capacity $k_B = kS$ is proportional to the amount of foliage available, i.e., proportional to S.

The effect of predation (consumption of budworms by birds) is represented by:

$$g = \frac{\beta B^2}{\alpha^2 + B^2} \tag{2.51}$$

which has the following properties:

- The consumption of prey (budworms) by individual predators (birds) is limited by saturation to the level β; note that $g \to \beta$ for large B.
- There is a decrease in the effectiveness of predation at low prey density (with the limit $g \to 0, B \to 0$, i.e., the birds have a variety of alternative foods).
- α determines the scale of budworm densities at which saturation takes place, i.e., α^2 relative to B^2.

Clearly g from Eq. (2.51) introduces a strong nonlinearity in ODE (2.50) which is a principal reason for using numerical integration in the solution of Eq. (2.50).

We can now use Eq. (2.50) to plot $\frac{\mathrm{d}B}{\mathrm{d}t}$ versus B as in Fig. 2.5, termed a *phase plane plot*. Note that there are three equilibrium points for which $\frac{\mathrm{d}B}{\mathrm{d}t} = 0$ (the fourth point

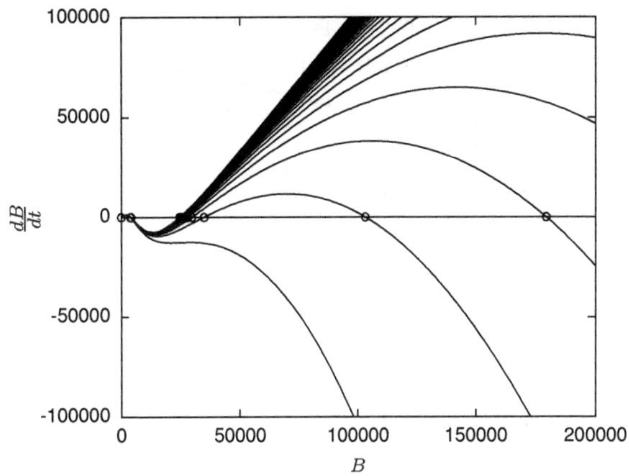

Fig. 2.5 $\frac{dB}{dt}$ as a function of B from Eq. (2.50), for increasing value of S (S evolves from 200 to 10,000 by steps of 200)

at the origin $B = 0, t = 0$ is not significant). These equilibrium points are the roots of Eq. (2.50) (combined with Eq. (2.51)) with $\frac{dB}{dt} = 0$ i.e.,

$$0 = r_B B\left(1 - \frac{B}{K_B}\right) - \frac{\beta B^2}{\alpha + B^2} \tag{2.52}$$

The middle equilibrium point (at approximately $B = 1.07 \times 10^5$) is not stable while the two outlying equilibrium points are stable. Figure 2.5 is parametrized with increasing values of S (from 200 to 10,000, by steps of 200) corresponding to increasing values of the carrying capacity $K_B = kS$.

Qualitatively, the evolution of the system can be represented as in Fig. 2.6. Consider the situation where there are three equilibrium points as in Fig. 2.5. In this case, the intermediate point is unstable, whereas the upper and lower equilibrium points are stable. Assume that the system is in a lower equilibrium point. As the forest foliage, $S(t)$, slowly increases, the system moves along the heavy lower equilibrium line. This happens because the budworm density, $B(t)$ is low (endemic population), and the forest can grow under this good condition (low $B(t)$). At the end of this slow process, the system moves quickly along a vertical arrow to reach the upper equilibrium branch, which corresponds to a much higher budworm density (outbreak population). Now, old trees are more susceptible to defoliation and die. As a consequence, $S(t)$ slowly decreases, and the system slowly moves along the heavy upper equilibrium line. This continues up to a certain stage, where the system jumps along a vertical line (budworm population collapses), and goes back to a status characterized by a low budworm density.

In summary, slow processes are depicted by the heavy equilibrium lines, whereas fast processes are represented by vertical arrows in Fig. 2.6. The interacting spruce budworm population and forest follow limit cycles characterized by two periods of

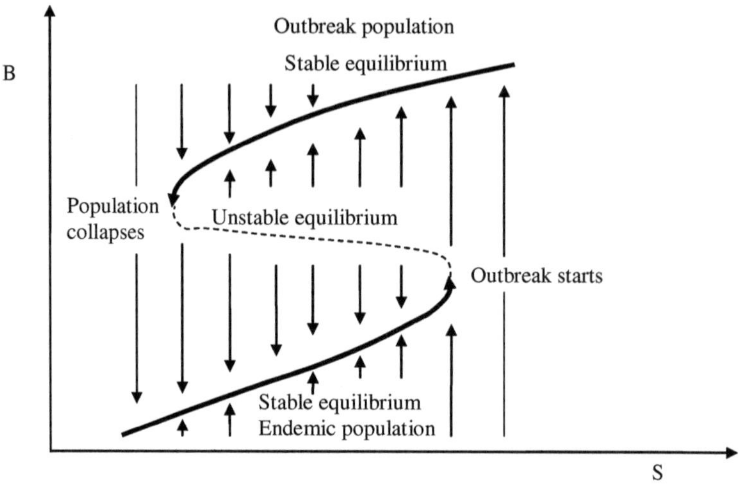

Fig. 2.6 Qualitative evolution of the budworm-forest system

slow change and two periods of fast change. An analysis of the system must include the fast processes along with the slow processes. Even though the system dynamics is dictated by the slow processes for most of the time (i.e., when the system moves slowly along heavy equilibrium lines), fast transitions explain budworm population outbreaks and collapses, which would be completely unforeseen if the fast dynamics is neglected.

We now consider the full model in which the carrying capacity is given by

$$K_B = \frac{kSE^2}{E^2 + T_E^2} \tag{2.53}$$

i.e., it is again proportional to the amount of foliage available, $S(t)$, but also depends on the physiological condition (energy) of the trees, $E(t)$; K_B declines sharply when E falls below a threshold T_E.

The effect of predation is still represented by Eq. (2.51), but, in addition, the half-saturation density α is proportional to the branch surface area S, i.e., $\alpha = aS$.

The total surface area of the branches in a stand then follows the ODE

$$\frac{\mathrm{d}S}{\mathrm{d}t} = r_S S\left(1 - \frac{S}{K_S}\frac{K_E}{E}\right) \tag{2.54}$$

that allows S to approach its upper limit K_S. An additional factor $\frac{K_E}{E}$ is inserted into the equation because S inevitably decreases under stress conditions (death of branches or even whole trees).

Table 2.9 Spruce budworm versus forest—parameter values

Parameter	Value	Units
r_B	1.52	year^{-1}
r_S	0.095	year^{-1}
r_E	0.92	year^{-1}
k	355	larvae/branch
a	1.11	larvae/branch
β	43,200	larvae/acre/year
K_S	25,440	branch/acre
K_E	1	–

The energy reserve also satisfies an equation of the logistic type

$$\frac{dE}{dt} = r_E E \left(1 - \frac{K_E}{E}\right) - P\frac{B}{S} \tag{2.55}$$

where the second term on the RHS describes the stress exerted on the trees by the budworm's consumption of foliage. In this expression $\frac{B}{S}$ represents the number of budworms per branch. The proportionality factor P is given by

$$P = \frac{pE^2}{E^2 + T_E^2} \tag{2.56}$$

as the stress on the trees is related to the amount of foliage consumed (P declines sharply when E falls below a threshold T_E).

The initial conditions (ICs) are taken as:

$$B(t = 0) = 10; \qquad S(t = 0) = 7,000; \qquad E(t = 0) = 1 \tag{2.57}$$

The model parameters are given in Table 2.9, [13]. The 3 × 3 ODE model—Eqs. (2.50)–(2.56)—are solved by the code in function `spruce_budworm_odes`.

```
function xt = spruce_budworm_odes(t,x)

% Set global variables
global rb k beta a rs Ks re Ke p Te

% Transfer dependent variables
B = x(1);
S = x(2);
E = x(3);

% Model Parameters
Kb    = k*S*E^2/(E^2+Te^2);
alpha = a*S;
g     = beta*B^2/(alpha^2+B^2);
P     = p*E^2/(Te^2+E^2);
```

```
% Temporal derivatives
Bt  = rb*B*(1-B/Kb) - g;
St  = rs*S*(1-(S/Ks)*(Ke/E));
Et  = re*E*(1-E/Ke) - P*B/S;

% Transfer temporal derivatives
xt = [Bt St Et]';
```

Function spruce_budworm_odes Function for the solution of Eqs. (2.50)–(2.56) and associated algebraic equations

We can note the following details about this function:

1. After defining the global variables which are shared with the main program to be discussed next, the dependent variable vector received from the integration function, x, is transferred to problem oriente variables to facilitate programming

```
% Global variables
global rb k beta a rs Ks re Ke p Te

% Transfer dependent variables
B = x(1);
S = x(2);
E = x(3);
```

2. The problem algebraic variables are computed from the dependent variable vector $(B, S, E)^T$

```
% Temporal derivatives

Kb    = k*S*E^2/(E^2+Te^2); alpha = a*S; g     =
beta*B^2/(alpha^2+B^2); P      = p*E^2/(Te^2+E^2);
```

Note the importance of ensuring that all variables and parameters on the RHS of these equations are set numerically before the calculation of the RHS variables.

3. The ODEs, Eqs. (2.50), (2.54), (2.55) are then programmed and the resulting temporal derivatives are transposed to a column vector as required by the integrator

```
% Temporal derivatives
Bt  = rb*B*(1-B/Kb)  - g;
St  = rs*S*(1-(S/Ks)*(Ke/E));
Et  = re*E*(1-E/Ke)  - P*B/S;
%
% Transfer temporal derivatives
xt = [Bt St Et]';
```

The main program that calls the ODE function spruce_budworm_odes is shown in Budworm_main

```
close all
clear all
```

```
% start a stopwatch timer
tic

% set global variables
global rb k beta a rs Ks re Ke p Te

% model parameters
rb   = 1.52;
k    = 355;
beta = 43200;
a    = 1.11;
rs   = 0.095;
Ks   = 25440;
re   = 0.92;
Ke   = 1.0;
p    = 0.00195;
Te   = 0.03;

% initial conditions
t0 = 0;
tf = 200;
B  = 10;
S  = 7000;
E  = 1;
x  = [B S E]';

% call to ODE solver (comment/decomment one of the methods
% to select a solver)

% method = 'Euler'
% method = 'rkf45'
% method = 'ode45'
 method = 'ode15s'
%
switch method
    % Euler
    case('Euler')
        Dt = 0.01;
        Dtplot = 0.5;
        [tout, yout] = euler_solver(@spruce_budworm_odes,...
                                    t0, tf, x, Dt, Dtplot);
    % rkf45
    case('rkf45')
        hmin = 1e-3;
        nstepsmax = 1000;
        abstol = 1e-3;
        reltol = 1e-3;
        Dtplot = 0.5;
        [tout,yout,eout] = rkf45_solver(@spruce_budworm_odes,...
                               t0,tf,x,hmin,nstepsmax,abstol,...
                               reltol,Dtplot);
        figure(5)
        plot(tout,eout)
        xlabel('t');
        ylabel('truncation error');
    % ode45
    case('ode45')
        options = odeset('RelTol',1e-6,'AbsTol',1e-6);
```

```
        t=[t0:0.5:tf];
        [tout, yout] = ode45(@spruce_budworm_odes,t,x,...
                                options);
    % ode15s
    case('ode15s')
        options = odeset('RelTol',1e-6,'AbsTol',1e-6);
        t=[t0:0.5:tf];
        [tout, yout] = ode15s(@spruce_budworm_odes,t,x,...
                                options);

end
% plot results
figure(1)
subplot(3,1,1)
plot(tout,yout(:,1),'k');
%     xlabel('t [years]');
ylabel('B(t)','FontName','Helvetica','FontSize',12);
%     title('Budworm density');
subplot(3,1,2)
plot(tout,yout(:,2),'k');
%     xlabel('t [years]');
ylabel('S(t)','FontName','Helvetica','FontSize',12);
%     title('Branch density')
subplot(3,1,3)
plot(tout,yout(:,3),'k');
xlabel('t [years]','FontName','Helvetica','FontSize',12);
ylabel('E(t)','FontName','Helvetica','FontSize',12);
set(gca,'FontName','Helvetica','FontSize',12);
%     title('Energy');
figure(2)
plot3(yout(:,1),yout(:,2),yout(:,3),'k')
xlabel('B(t)','FontName','Helvetica','FontSize',12);
ylabel('S(t)','FontName','Helvetica','FontSize',12);
zlabel('E(t)','FontName','Helvetica','FontSize',12);
grid
title('3D phase plane plot','FontName','Helvetica',...
      'FontSize',12);
set(gca,'FontName','Helvetica','FontSize',12);

% read the stopwatch timer
tcpu=toc;
```

Script Budworm_main Main program for the solution of Eqs. (2.50)–(2.56)

We can note the following points about this main program:

1. First, variables are defined as global so they can be shared with the ODE routine
2. The model parameters are then defined numerically
3. The time scale and the initial conditions of Eq. (2.57) are defined
4. The parameters of the integrator, e.g., error tolerances, are set and an integrator is selected among four possible choices, e.g., `euler_solver`, `rkf45_solver` (among the basic integrators introduced earlier in this chapter) or `ode45` and `ode15s` (among the integrators available in the MATLAB ODE suite). Table 2.10 shows a comparison of the performance of these and other IVP solvers applied to

Table 2.10 Performance of different IVP solvers applied to the spruce budworm problem

	Euler	Heun	rkf45	ros23p	ode45	ode15s Adams	BDF
N. steps	20,000	20,000	1,122	20,520	837	2,109	2,117
CPU time	2.5	4.9	1.0	4.24	1.1	2.2	2.2

Computational times are normalized with respect to the time spent by the most efficient solver, in this example, rkf45

Eqs. (2.50)–(2.56). Absolute and relative tolerances were fixed to 10^{-6}. For this particular example, the most efficient solver is rkf45 while Heun's method requires the highest computational cost. These results show that the system of ODEs is not stiff (rkf45 is an explicit time integrator) and that time step size adaptation is an important mechanism to ensure specified tolerances and to improve computational efficiency (we have used a conservatively small step size with fixed step integrators such as Euler's and Heun's method).

5. After calculating the solution, a series of plots displays the numerical solution.

The plotted solutions correspond with Fig. 2.7, for the time evolution of the three states, and Fig. 2.8, for the 3D phase plane. The oscillatory nature of the solution resulting from the combination of slow and fast dynamics depicted in Fig. 2.7 is clear. Also, the number of nonlinear ODEs (three) and associated algebraic equations demonstrates the utility of the numerical solution; in other words, analytical solution of this model is precluded because of its size and complexity.

2.5.2 Liming to Remediate Acid Rain

As a final ODE example application, we consider the modeling of an ecological system described by a 3×3 system of nonlinear ODEs. Specifically, the effect of acid rain on the fish population of a lake and the effect of remedial liming is investigated. This model is described in [14].

The fish population $N(t)$ is growing logistically, i.e.,

$$\frac{\mathrm{d}N}{\mathrm{d}t} = r(C)N - \frac{r_0 N^2}{K(C)} - H, \quad N(0) = N_0 > 0 \tag{2.58}$$

where $r(C)$ is the specific growth rate, which depends on the acid concentration $C(t)$ in the following way:

$$r(C) = \begin{cases} r_0 & \text{if} \quad C < C_{\text{lim}} \\ r_0 - \alpha(C - C_{\text{lim}}) & \text{if} \quad C_{\text{lim}} < C < C_{\text{death}} \\ 0 & \text{if} \quad C_{\text{death}} < C < Q/\delta \end{cases}, \tag{2.59}$$

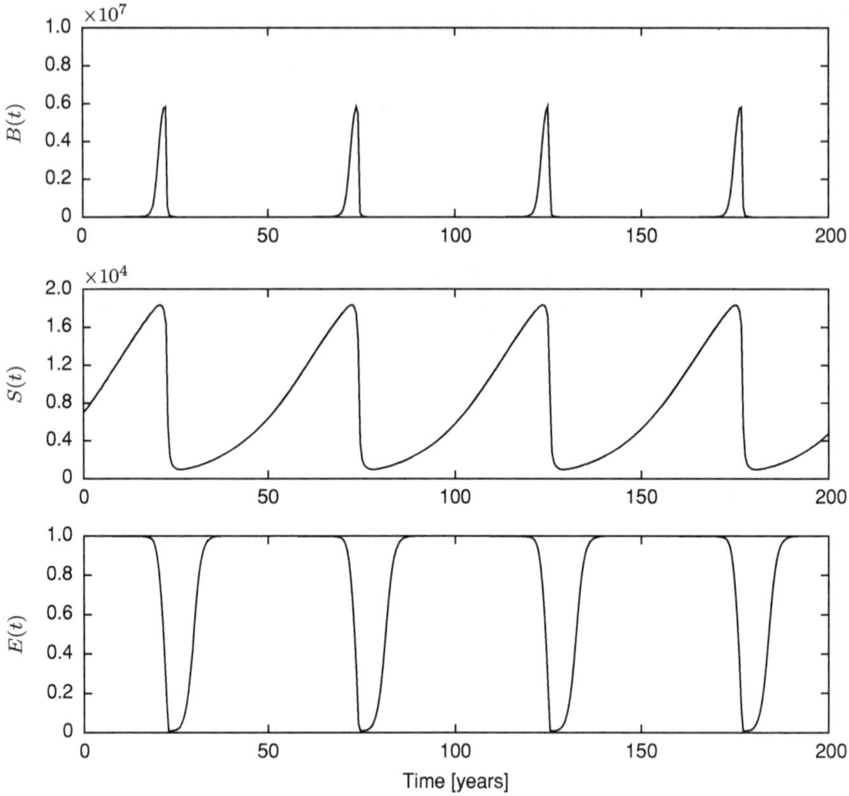

Fig. 2.7 Evolution of the three state variables ($B(t)$, $S(t)$ and $E(t)$) from the problem described by Eqs. (2.50), (2.54) and (2.55)

Fig. 2.8 Composite plot of the dependent variable vector from Eqs. (2.50), (2.54) and (2.55)

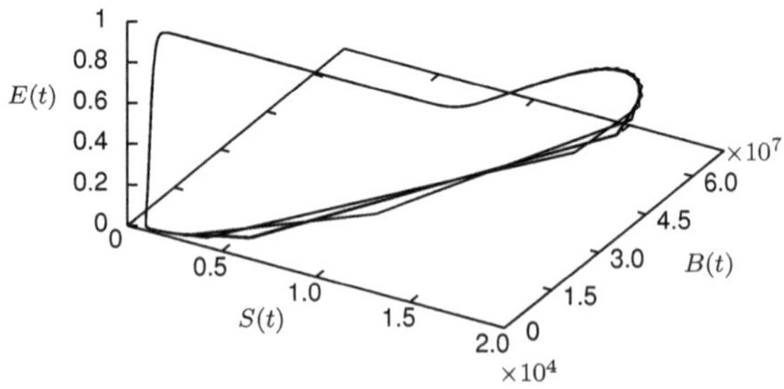

$K(C)$ is the carrying capacity (i.e., the maximum population density that the ecosystem can support), which also depends on $C(t)$

$$K(C) = \begin{cases} K_0 & \text{if} \quad C < C_{\text{lim}} \\ K_0 - \beta(C - C_{\text{lime}}) & \text{if} \quad C_{\text{lim}} < C < C_{\text{death}} \\ K_{\text{lim}} & \text{if} \quad C_{\text{death}} < C < Q/\delta \end{cases}, \tag{2.60}$$

Table 2.11 Table of parameter values for Eqs. (2.58)–(2.62)

$r0 = 0.02$	$Q = 2$
$C_{\text{lim}} = 50$	$\delta = 0.002$
$\alpha = 0.0001$	$\delta_0 = 0.005$
$\beta = 0.05$	$\eta = 0.04$
$K_0 = 100{,}000$	$\eta_0 = 0.004$
$K_{\text{lim}} = 100$	$H = 100$

and H is the harvesting rate. In Eqs. (2.59)–(2.60), C_{lim} denotes the critical value of the acid concentration (between 0 and C_{lim}, acid is harmless to the fish population) and $C_{\text{death}} = (r_0 - \alpha C_{\text{lime}})/\alpha$ is the concentration above with the fish population completely stops growing.

We suggest a careful study of the RHS of Eq. (2.58) and the switching functions of Eqs. (2.59)–(2.60) since these functions reflect several features of the model.

The acid concentration $C(t)$ is described by the ODE

$$\frac{\mathrm{d}C}{\mathrm{d}t} = Q - \delta C - \delta_0 E \tag{2.61}$$

where Q is the inlet acid flow rate (due to acid rain), δ is the natural depletion rate, while δ_0 is the depletion rate due to liming. E is the liming effort applied to maintain the lake at a permissible acid concentration C_{lim}, and is given by the ODE

$$\frac{\mathrm{d}E}{\mathrm{d}t} = \eta(C - C_{\text{lim}}) - \eta_0 E \tag{2.62}$$

where η represents the control action and η_0 is the natural depletion rate of E.

Again, we suggest a careful analysis of the RHS functions of Eqs. (2.61) and (2.62). The complexity of the model is clearly evident from Eqs. (2.58)–(2.62), e.g., the nonlinear switching functions of Eqs. (2.59) and (2.60). Thus, although some analytical analysis is possible as we demonstrate in the subsequent discussion, a numerical solution of Eq. (2.24) is the best approach to gain an overall understanding of the characteristics of the model. In particular, we will compute the state space vector (the solution of Eqs. (2.58), (2.61) and (2.62)), $N(t)$, $C(t)$, $E(t)$ by numerical ODE integration.

The initial conditions for Eqs. (2.58), (2.61) and (2.62) are taken as

$$N(t=0) = 72{,}500; \quad C(t=0) = 80; \quad E(t=0) = 190 \tag{2.63}$$

and the parameter values are given in Table 2.11.

If harvesting is below a certain threshold value, i.e., $H < \frac{K(C^*)\{r(C^*)\}^2}{4r_0}$, there exist two equilibrium points $P_i(N^*, C^*, E^*)$ with $i = 1, 2$ in the state space. The notation * indicates that the state is at equilibrium.

$$E^* = \frac{\eta\left(\frac{Q}{\delta} - C_{\text{lim}}\right)}{\eta_0 + \eta\frac{\delta_0}{\delta}} \tag{2.64}$$

$$C^* = \frac{Q - \delta_0 E^*}{\delta} \tag{2.65}$$

$$N_1^* = \frac{K(C^*)r(C^*)}{2r_0}\left(1 - \sqrt{1 - \frac{4r_0 H}{K(C^*)\{r(C^*)\}^2}}\right) \tag{2.66}$$

$$N_2^* = \frac{K(C^*)r(C^*)}{2r_0}\left(1 + \sqrt{1 - \frac{4r_0 H}{K(C^*)\{r(C^*)\}^2}}\right) \tag{2.67}$$

The equilibrium point P_1 corresponding to N_1^* of Eq. (2.66) is unstable (two eigenvalues of the Jacobian matrix have negative real parts, but the third one is real positive, so that P_1 is a saddle point) whereas P_2 corresponding to N_2^* of Eq. (2.67) is locally asymptotically stable (the three eigenvalues of the Jacobian matrix have negative real parts). In fact, when C and E tends to their steady-state values C^* and E^*, it is possible to rewrite the ODE for N, Eq. (2.58), as follows

$$\frac{dN}{dt} = -\frac{r_0}{K(C^*)}(N - N_1^*)(N - N_2^*) \tag{2.68}$$

Since $N_2^* > N_1^*$, the RHS of Eq. (2.68) expression shows that

$$\frac{dN}{dt} < 0 \quad \text{if} \quad N < N_1^* \quad \text{or} \quad N > N_2^* \tag{2.69}$$

$$\frac{dN}{dt} > 0 \quad \text{if} \quad N_1^* < N < N_2^* \tag{2.70}$$

In turn, the fish population tends to extinction if $N(0) < N_1^*$, and tends to N_2^* if $N(0) > N_1^*$. Therefore, the equilibrium point P_2 is globally asymptotically stable in the region

$$\left\{(N, C, E): \quad N_1^* \le N \le K(0), \quad 0 \le C \le \frac{Q}{\delta}, \quad 0 \le E \le \frac{\eta}{\eta_0}\left(\frac{Q}{\delta} - C_{\text{lim}}\right)\right\} \tag{2.71}$$

For the parameter value of Table 2.11, the equilibrium points are [14]

$$N_1^* = 6{,}671; \quad N_2^* = 75{,}069 \tag{2.72}$$

The code `fish_odes` implements a solution to Eqs. (2.58)–(2.62) including several variations. First, the function to define the model equations is listed.

```
function xt = fish_odes(t,x)
```

```
% Set global variables
global r0 Clim alpha Cdeath K0 Klim beta H Q delta delta0
global eta eta0

% x has three columns corresponding to three different
% column solution vectors (which can be used for
% numerical evalution of the Jacobian):
ncols = size(x,2);

for j=1:ncols

    % Transfer dependent variables
    N = x(1,j);
    C = x(2,j);
    E = x(3,j);

    % Temporal derivatives
    if C < Clim
        r = r0;
    elseif Clim <= C < Cdeath
        r = r0-alpha*(C-Clim);
    else
        r = 0;
    end
    %
    if C < Clim
        K = K0;
    elseif Clim <= C < Cdeath
        K = K0-beta*(C-Clim);
    else
        K = Klim;
    end

    Nt  = r*N - r0*N^2/K - H;
    Ct  = Q - delta*C - delta0*E;
    Et = eta*(C-Clim) - eta0*E;

    % Transfer temporal derivatives
    % (One column for each column of x)
    xt(:,j) = [Nt Ct Et]';
end;
```

Function fish_odes Function for Eqs. (2.58)–(2.62)

We can note the following points about `fish_odes`.

1. After defining a set of global variables, a $3\times$ `ncol` matrix, `x(3,1)`, is defined, where the first index defines a row dimension of three for the state vector $(N, C, E)^T$, and the second index defines a column dimension of one for one computed solution corresponding to a single set of parameters and initial conditions defined in the main program that calls `fish_odes` (discussed next)

```
% x has three columns corresponding to three different
% column solution vectors
ncols = size(x,2);
% Step through the ncol solutions
for j=1:ncols
    %
    % Transfer dependent variables
    N = x(1,j);
```

```
        C = x(2,j);
        E = x(3,j);
```

2. The expressions (2.59) to (2.60) are then programmed

```
    % Parameters r, K set
        %
        if C < Clim
            r = r0;
        elseif Clim <= C < Cdeath
            r = r0-alpha*(C-Clim);
        else
            r = 0;
        end
        %
        if C < Clim
            K = K0;
        elseif Clim <= C < Cdeath
            K = K0-beta*(C-Clim);
        else
            K = Klim;
        end
```

   Note in particular the switching to three possible values of the parameters r and k depending on the current value of C.
3. The temporal derivatives for the three solutions are then computed according to ODEs, (2.58), (2.61), and (2.62)

```
    % Temporal derivatives
    Nt  = r*N - r0*N^2/K - H;
    Ct  = Q - delta*C - delta0*E;
    Et = eta*(C-Clim) - eta0*E;
```

4. The temporal derivatives are then stored in a 3×1 matrix for return to the ODE integrator (a total of $3 \times 1 = 3$ derivatives)

```
    % Transfer temporal derivatives
    % (One column for each column of x)
    xt(:,j) = [Nt Ct Et]';
```

Note the index for the solutions, `j`, is incremented by the `for` statement at the beginning of `fish_odes` that is terminated by the `end` statement. In the present case, `ncol=1` (just one solution is computed). However, this approach to computing multiple solutions in parallel could be used to conveniently compare solutions, generated, for example, for multiple sets of initial conditions or model parameters. In other words, having the solutions available simultaneously would facilitate their comparison, e.g., by plotting them together. This feature of computing and plotting multiple solutions in a parametric study is illustrated subsequently in Fig. 2.11.

The main program that calls `fish_odes` is presented in `fish_main`.

```
close all
clear all

% start a stopwatch timer
tic

% set global variables
global r0 Clim alpha Cdeath K0 Klim beta H Q delta delta0
global eta eta0 fac thresh vectorized

% model parameters
r0     = 0.02;
Clim   = 50;
alpha  = 0.0001;
Cdeath = (r0+alpha*Clim)/alpha;
K0     = 100000;
Klim   = 100;
beta   = 0.05;
H      = 100;
Q      = 2;
delta  = 0.002;
delta0 = 0.005;
eta0   = 0.004;

% select the control action 'eta' (comment/decomment one
% of the actions)
 action = 'strong'
% action = 'moderate'
% action = 'weak'
switch action
    % strong
    case('strong')
        eta = 0.5;
    % moderate
    case('moderate')
        eta = 0.1;
    % weak
    case('weak')
        eta = 0.04;
end

% equilibrium points
[N1star,N2star] = equilibrium_fish(r0,Clim,alpha,...
                     Cdeath,K0,Klim,beta,H,Q,delta,...
                     delta0,eta,eta0)

% initial conditions
t0 = 0;
tf = 3560;
N  = 72500;
%      N = 5000;
C = 80;
E = 190;
x = [N C E]';

% call to ODE solver (comment/decomment one of the methods
% to select a solver)
```

```
%       method = 'euler'
%       method = 'midpoint'
%       method = 'heun'
%       method = 'heun12'
%       method = 'rk4'
%       method = 'rkf45'
%       method = 'ros3p'
method = 'ros23p'
%       method = 'ode45'
%       method = 'ode15s'
%       method = 'lsodes'
switch method
    % Euler
    case('euler')
        Dt     = 0.1;
        Dtplot = 20;
        [tout,xout] = euler_solver(@fish_odes,t0,tf,x,...
                                   Dt,Dtplot);
    % midpoint
    case('midpoint')
        Dt     = 5;
        Dtplot = 20;
        [tout,xout] = midpoint_solver(@fish_odes,t0,...
                                      tf,x,Dt,Dtplot);
    % Heun
    case('heun')
        Dt     = 5;
        Dtplot = 20;
        [tout,xout] = heun_solver(@fish_odes,t0,tf,x,...
                                  Dt, Dtplot);
    % Heun12
    case('heun12')
       hmin      = 0.0001;
       nstepsmax = 1e5;
       abstol    = 1e-3;
       reltol    = 1e-3;
       Dtplot    = 20;
       [tout,xout] = heun12_solver(@fish_odes,t0,tf,...
                                   x,hmin,nstepsmax,...
                                   abstol,reltol,Dtplot);
    % rk4
    case('rk4')
        Dt     = 5;
        Dtplot = 20;
        [tout, xout] = rk4_solver(@fish_odes,t0,tf,x,...
                                  Dt,Dtplot);
    % rkf45
    case('rkf45')
        hmin      = 0.0001;
        nstepsmax = 1e5;
        abstol    = 1e-3;
        reltol    = 1e-3;
        Dtplot    = 20;
        [tout,xout,eout] = rkf45_solver(@fish_odes,t0,...
                                  tf,x,hmin,nstepsmax,...
                                  abstol,reltol,Dtplot);
        figure(2)
        plot(tout/356,eout(:,1),':r');
        hold on
```

```
        plot(tout/356,eout(:,2),'--b');
        plot(tout/356,eout(:,3),'--g');
        xlabel('t');
        ylabel('e(t)');
    % ros3p
    case('ros3p')
        Dt     = 1;
        Dtplot = 20;
        fac    = [];
        thresh = 1e-12;
        [tout,xout] = ros3p_solver(@fish_odes,...
                                   @jacobian_num_fish,...
                                   t0,tf,x,Dt,Dtplot);
    % ros23p
    case('ros23p')
        hmin      = 0.0001;
        nstepsmax = 1e5;
        abstol    = 1e-3;
        reltol    = 1e-3;
        Dtplot    = 1;
        fac       = [];
        thresh    = 1e-12;
        [tout,xout,eout] = ros23p_solver(@fish_odes,...
                                   @jacobian_num_fish,...
                                   @ft_fish,t0,tf,x,...
                                   hmin,nstepsmax,...
                                   abstol,reltol,Dtplot);
        figure(2)
        plot(tout,eout(:,1),':r');
        hold on
        plot(tout,eout(:,2),'--b');
        plot(tout,eout(:,3),'--g');
        xlabel('t');
        ylabel('e(t)');
    % ode45
    case('ode45')
        options = odeset('RelTol',1e-3,'AbsTol',1e-3);
        Dtplot = 20;
        t = [t0:Dtplot:tf];
        [tout,xout] = ode45(@fish_odes,t,x,options);
    % ode15s
    case('ode15s')
        options = odeset('RelTol',1e-3,'AbsTol',1e-3);
        Dtplot=20;
        t=[t0:Dtplot:tf];
        [tout, xout] = ode15s(@fish_odes,t,x,options);
    % lsodes
    case('lsodes')
        resname = 'fish_odes';
        jacname = '[]';
        neq     = 3;
        Dtplot  = 20;
        tlist   = [t0+Dtplot:Dtplot:tf];
        itol    = 1;
        abstol  = 1e-3;
        reltol  = 1e-3;
        itask   = 1;
        istate  = 1;
        iopt    = 0;
```

```
        lrw    = 50000;
        rwork  = zeros(lrw,1);
        liw    = 50000;
        iwork  = zeros(liw,1);
        mf     = 222;
        [tout,xout] = lsodes(resname,jacname,neq,x,t0,...
                             tlist,itol,reltol,abstol,...
                             itask,istate,iopt,rwork,...
                             lrw,iwork,liw,mf);
end

% plot results
figure(1)
subplot(3,1,1)
plot(tout/356,xout(:,1),'-');
%     xlabel('t [years]');
ylabel('N(t)');
%     title('Fish population');
subplot(3,1,2)
plot(tout/356,xout(:,2),'-');
%     xlabel('t [years]');
ylabel('C(t)');
%     title('Acid concentration')
subplot(3,1,3)
plot(tout/356,xout(:,3),'-');
xlabel('t [years]');
ylabel('E(t)');
%     title('Liming effort');

% read the stopwatch timer
tcpu=toc;
```

Script fish_main Main program that calls subordinate routine `fish_odes`

We can note the following points about `fish_main`.

1. First, global variables are defined and the model parameters are set.
2. An integrator is then selected from a series of basic or advanced integrators (also including LSODES, which is a ODE solver from ODEPACK [2] transformed into a MEX-file - we will give more details on how to create MEX-files in a subsequent section) by uncommenting a line, in this case for `ros23p_solver` discussed previously.
3. The degree of liming is then set by a switch function which in this case selects a character string for "strong."
4. Calls to the various integrators are listed. The call to `ros23p` is similar to that of the main program `Budworm_main`. The details for calling the various integrators are illustrated with the coding of the calls. Table 2.12 shows a comparison of the performance of these solvers. Note that the performance of the solvers increases as the control law becomes weaker, in particular, for Euler's and Heun's methods.
5. The numerical solutions resulting from the calls to the various integrators are plotted. Note in the case of `ros23p`, the estimated errors (given by the sub-

Table 2.12 Performance of different IVP solvers applied to the liming to remediate acid rain problem

							ode15s	
Control action		Euler	Heun	rkf45	ros23p	ode45	Adams	BDF
Strong	N. steps	17,800	1,780	212	213	101	281	271
	CPU time	38.9	8.4	3.4	5.0	2.4	4.4	4.2
Moderate	N. steps	3560	356	193	203	44	123	127
	CPU time	8.0	1.8	3.2	4.9	1.4	2.8	2.6
Weak	N. steps	1780	178	193	193	29	79	82
	CPU time	4.0	1.0	3.2	4.6	1.2	2.2	2.1

Computational times are normalized with respect to the time spent by the most efficient solver, in this example, Heun's solver with weak control law

traction of Eq. (2.45a) and (2.45b) and implemented in ssros23p) are also plotted.

To complete the programming of Eqs. (2.58)–(2.62), we require a function for the calculation of the equilibrium points (called by fish_odes) and a function for the Jacobian matrix (called by ssros23p). Function equillibrium_fish is a straightforward implementation of Eqs. (2.64)–(2.67).

```
function [N1star,N2star] = equilibrium_fish(r0,Clim,...
                          alpha,Cdeath,K0,Klim,beta,...
                          H,Q,delta,delta0,eta,eta0)
%
Estar = eta*(Q/delta-Clim)/(eta0+eta*(delta0/delta));
Cstar = (Q-delta0*Estar)/delta;

%
if Cstar < Clim
    r = r0;
elseif Clim <= Cstar < Cdeath
    r = r0-alpha*(Cstar-Clim);
else
    r = 0;
end

%
if Cstar < Clim
    K = K0;
elseif Clim <= Cstar < Cdeath
    K = K0-beta*(Cstar-Clim);
else
    K = Klim;
end

%
N1star = (K*r)/(2*r0)*(1-sqrt(1-(4*r0*H)/(K*r^2)));
N2star = (K*r)/(2*r0)*(1+sqrt(1-(4*r0*H)/(K*r^2)));
```

Function equillibrium_fish Function for the calculation of the equilibrium points from Eq. (2.64) to (2.67)

Function `jacobian_fish` computes the analytical Jacobian of the ODE system, Eqs. (2.58), (2.61) and (2.62)

```
function Jac = jacobian_fish(t,x)
%
% Set global variables
global r0 Clim alpha Cdeath K0 Klim beta H Q delta delta0
global eta eta0
%
% Transfer dependent variables
N = x(1);
C = x(2);
E = x(3);
%
% Jacobian matrix
%
%       Nt  = r*N - r0*N^2/K - H;
%       Ct  = Q - delta*C - delta0*E;
%       Et = eta*(C-Clim) - eta0*E;
%
%
if C < Clim
    r = r0;
    K = K0;
    Jac = [r-2*r0*N/K     0           0          ;
              0              -delta     -delta0 ;
              0              eta        -eta0   ];
elseif Clim <= C < Cdeath
    r = r0-alpha*(C-Clim);
    K = K0-beta*(C-Clim);
    Jac = [r-2*r0*N/K   -alpha*N+beta*r0*N^2/K^2   0        ;
             0             -delta                       -delta0 ;
             0             eta                          -eta0   ];
else
    r = 0;
    K = Klim;
    Jac = [r-2*r0*N/K     0           0          ;
             0              -delta     -delta0 ;
             0              eta        -eta0   ];
end
```

Function jacobian_fish Function for the calculation of the Jacobian matrix of Eqs. (2.58), (2.61) and (2.62)

Note that the Jacobian matrix programmed in `jacobian_fish` also has a set of three switching functions, which give different elements of the Jacobian matrix depending on the current value of C. To illustrate the origin of the elements of the Jacobian matrix, consider the first case for $C < C_{\text{lim}}$.

```
if C < Clim
    r = r0;
    K = K0;
Jac = [r-2*r0*N/K    0          0  ;
         0         -delta     -delta0 ;
         0           eta        -eta0   ];
```

The first row has the three elements for the Jacobian matrix from Eq. (2.58). The RHS function of Eq. (2.58) is:

$$r(C)N - \frac{r_0 N^2}{K(C)} - H$$

with $r = r_0$, $K = K_0$, the RHS function becomes

$$r_0 N - \frac{r_0 N^2}{K_0} - H$$

We now have the following expressions for the derivative of this function with respect to each of the state variables

$$\frac{\partial\left(r_0 N - \frac{r_0 N^2}{K_0} - H\right)}{\partial N} = r_0 - \frac{2r_0 N}{K_0}; \qquad \frac{\partial\left(r_0 N - \frac{r_0 N^2}{K_0} - H\right)}{\partial C} = 0$$

$$\frac{\partial\left(r_0 N - \frac{r_0 N^2}{K_0} - H\right)}{\partial E} = 0$$

which are programmed as

```
Jac = [r-2*r0*N/K     0            0            ;
```

The remaining six elements of the Jacobian matrix follow in the same way from the RHS functions of Eqs. (2.61) and (2.62).

We can note three important features of the programming of the Jacobian matrix:

- If the state vector is of length n (i.e., n ODEs), the Jacobian matrix is of size $n \times n$. This size grows very quickly with increasing n. Thus for large systems of ODEs, e.g., $n > 100$, the Jacobian matrix is difficult to evaluate analytically (we must derive $n \times n$ derivatives).
- The Jacobian matrix can have a large number of zeros. Even for the small 3×3 ODE problem of Eqs. (2.58), (2.61), and (2.62), the Jacobian matrix has four zeros in a total of nine elements. Generally, for physical problems, the fraction of zeros increases rapidly with n and is typically 0.9 or greater. In other words, Jacobian matrices tend to be *sparse*, and algorithms that take advantage of the sparsity by not storing and processing the zeros can be very efficient. Therefore, scientific computation often depends on the use of sparse matrix algorithms, and scientific software systems such as MATLAB have sparse matrix utilities.
- Because the differentiation required for an analytical Jacobian is often difficult to develop, numerical methods for producing the required $n \times n$ partial derivatives are frequently used, as illustrated with Eq. (2.49) and in function `jacobian_stiff_odes_fd`. A routine for calculating the numerical Jacobian for Eqs. (2.58), (2.61), and (2.62) is presented in function `jacobian_num_fish`.

```
function Jac = jacobian_num_fish(t,x)
%
% Set global variables
global fac thresh vectorized
%
% numerical Jacobian matrix
tstar = t;
xstar = x;
xtstar = fish_odes(tstar,xstar);
threshv = [thresh;thresh;thresh];
vectorized = 1;
[Jac, fac] = numjac(@fish_odes,tstar,xstar,xtstar,threshv,...
                    fac,vectorized);
```

Function jacobian_num_fish Function for the numerical calculation of the Jacobian of Eqs. (2.58), (2.61) and (2.62)

We will not consider the details of `jacobian_num_fish`, but rather, just point out the use of `fish_odes` for ODEs (2.58), (2.61) and (2.62), and the MATLAB routines `threshv` and `numjac` for the calculation of a numerical Jacobian. Clearly calls to the MATLAB routines for a numerical Jacobian will be more compact than the programming of the finite difference approximations for large ODE systems (as illustrated by the small 2×2 ODE system in function jacobian_stiff_odes_fd).

Coding to call `jacobian_num_fish` is illustrated in the use of `ros3p` (fixed step LIRK integrator) in `fish_main`.

This completes the coding of Eqs. (2.58), (2.61) and (2.62). We conclude this example by considering the plotted output from the MATLAB code `fish_main` given in Figs. 2.9 and 2.10. We note in Fig. 2.9 that the solution $N(t)$ reaches a stable equilibrium point $N_2^* = 93{,}115$, which is the value for strong liming ($\eta = 0.5$) as set in `fish_main`. This constrasts with the stable equilibrium point of Eq. (2.72) for weak liming ($\eta = 0.04$). As might be expected, the equilibrium point for weak liming is below that for strong liming (increased liming results in a higher equilibrium fish population).

Figure 2.10 indicates that the maximum estimated error for $N(t)$ is about 3.6. Since $N(t)$ varies between 72,500 (the initial condition set in fish_main) and 93,115 (the final equilibrium value of $N(t)$), the maximum fractional error is 3.6/72500 = 0.0000496, which is below the specified relative error of `reltol=1e-3` set in `fish_main`. Of course, this error of 3.6 is just estimated, and may not be the actual integration error, but this result suggests that the numerical solution has the specified accuracy.

Also, the estimated errors for the other two state variables, $C(t)$ and $E(t)$, are so small they are indiscernible in Fig. 2.10. This is to be expected since the magnitudes of these variables are considerably smaller than for $N(t)$ (see Fig. 2.9). In conclusion, these results imply that the error estimate of the embedded LIRK algorithm performed as expected in this application.

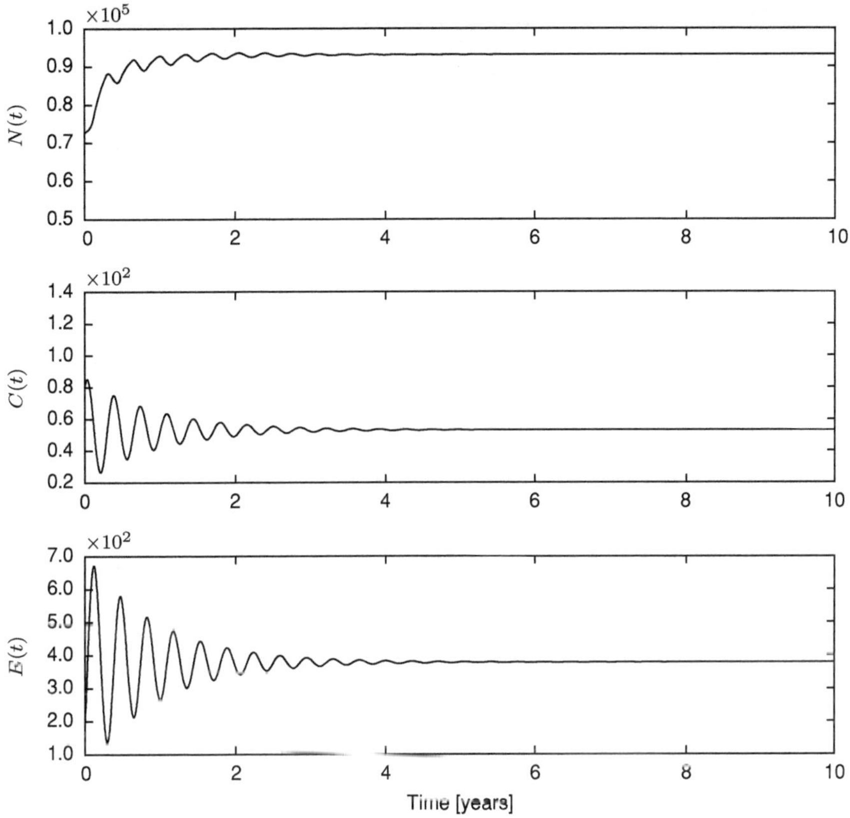

Fig. 2.9 Solution of Eqs. (2.58), (2.61) and (2.62) from `fish_main` for a strong control action $\eta = 0.5$

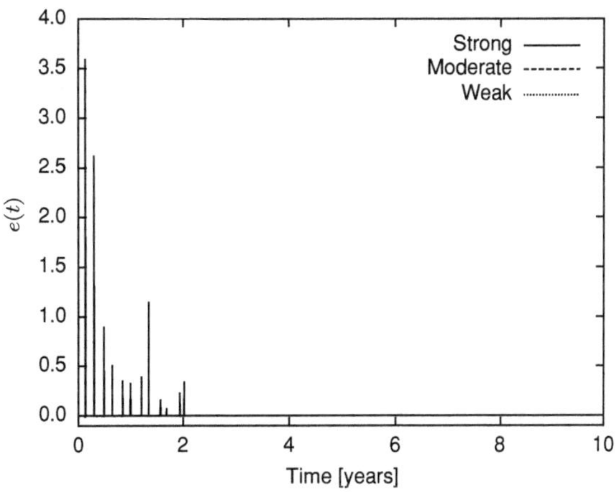

Fig. 2.10 Estimated error from `ros23p` for Eqs. (2.58), (2.61) and (2.62) from fish_main

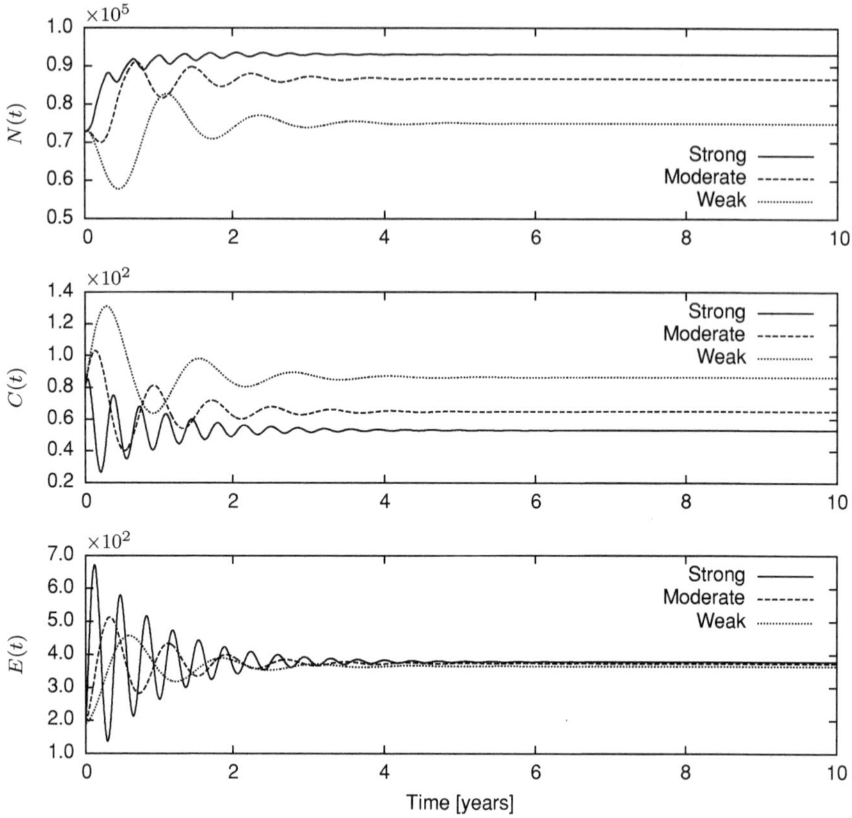

Fig. 2.11 State variable plots for Eqs. (2.58), (2.61) and (2.62) for the three liming conditions corresponding to $\eta = 0.04, 0.1, 0.5$

We conclude this discussion by including a few plots—see Fig. 2.11—with the solutions for the three liming conditions ($\eta = 0.04, 0.1, 0.5$) superimposed so that the effect of the liming is readily apparent.

As another result, we can mention that for the weak liming condition ($\eta = 0.04$), if the initial fish population is lower than the first unstable equilibrium point $N_1^* = 6{,}661$, e.g. $N_1^* = 5{,}000$, the fish population decreases and eventually vanishes as predicted by the theoretical analysis. Also, for this case, an event detection is necessary to prevent the fish population from becoming negative (a function `events.m` monitors the value of $N(t)$). In this respect, the model equations are not well formulated (they allow $N(t)$ to become negative).

2.6 On the Use of SCILAB and OCTAVE

As mentioned before, MATLAB has been selected as the main programming environment because of its convenient features for vector/matrix operations that are central to the solution of AE/ODE/PDE systems. In addition, MATLAB provides a very complete library of numerical algorithms (for numerical integration, matrix operations, eigenvalue computation,...), e.g., the MATLAB ODE SUITE [1], that can be used advantageously in combination with the proposed MOL toolbox.

However, there exist very powerful open source alternatives, such as SCILAB (for Scientific Laboratory) or OCTAVE, that can be used for the same purposes. SCILAB and OCTAVE provide high-level, interpreted programming environments, with matrices as the main data type, similar to MATLAB.

Initially named Ψlab (Psilab), the first software environment was created in 1990 by researchers from INRIA and *École nationale des ponts et chaussées (ENPC)* in France. The SCILAB Consortium was then formed in 2003 [15] to broaden contributions and promote SCILAB in academia and industry. In 2010, the Consortium announced the creation of SCILAB Enterprises, which develops and maintain SCILAB as an open source software but proposes commercial services to industry (professional software and project development). With regard to the solution of ordinary differential equations, SCILAB has an interface, called *ode*, to several solvers, especially those from the FORTRAN library ODEPACK originally developed by Alan Hindmarsh [2].

OCTAVE [16] was initially developed by John Eaton in the early 1990s, and then developed further by many contributors following the terms of the GNU General Public License (GPL) as published by the Free Software Foundation. The name OCTAVE is inspired by Octave Levenspiel, former Chemical Engineering Professor of John Eaton, who was known for his ability to solve numerical problems. OCTAVE has a built-in ODE solver based on LSODE [3], and a DAE solver, DASSL, originally developed by Linda Petzold [4]. Additional contributed packages are also available, e.g., OdePkg, which has a collection of explicit and implicit ODE solvers and DAE solvers.

An interesting feature of SCILAB and OCTAVE, as we will see, is the degree of compatibility with MATLAB codes. In many cases, slight modifications of the MATLAB codes will allow us to use them in SCILAB or OCTAVE.

Let us now show the programming of the spruce budworm dynamics—see Eqs. (2.50)–(2.55)—in SCILAB highlighting the main differences with MATLAB.

The SCILAB script `spruce_budworm_main.sci` is the main program, which now has an extension .sci instead of .m. The main features of this program are the following:

- Commented lines (these lines are not read by the interpreter but are very useful to explain the code) in SCILAB start with a double slash (//) as in C++, which is the equivalent to the percentage symbol (%) in MATLAB.

- The information to be displayed during the execution of the program can be changed in SCILAB with the `mode(k)` command. With `k=-1` the code runs silently (no information is displayed).
- The stopwatch timer commands coincide with the MATLAB ones (`tic`, `toc`)
- Setting the global variables in SCILAB is slightly different than in MATLAB. In SCILAB, each variable is inside quotation marks and it is separated from the other variables by commas.
- One important difference with respect to the MATLAB codes is that the functions must be loaded with the `exec` command. For instance, the ODEs are programmed in function `spruce_budworm_odes.sci` and it is loaded with the command `exec('spruce_budworm_odes.sci')`. In MATLAB any function defined in the path can be called.
- The definition of model parameters and initial conditions is exactly the same in MATLAB and SCILAB.
- After the definition of the initial conditions, the ODE solver is chosen. In this case two possibilities are offered, namely, *Euler's* method and a *Runge–Kutta–Fehlberg* implementation. Note that all our basic time integrators (presented in this chapter) can easily be translated to SCILAB, and are provided in the companion software. In the application example, *Euler's* method is chosen by commenting out the line `method = 'rkf45'`. Then the `select` command with two different `cases` is used. This command is equivalent to the `switch` command in MATLAB. As `'Euler'` is the method of choice, the code will run the lines corresponding to this selection while the lines corresponding to `'rkf45'` will be blind to the execution. An example of use of the SCILAB ODE library is included in Sect. 3.12.
- In order to use the Euler solver, it is required to load the function that implements it: `exec('euler_solver.sci')`. Then the function where the ODEs are defined is called (`spruce_budworm_odes.sci`) practically in the same way as in MATLAB. In this sense, in SCILAB we have

```
[tout,yout] = euler_solver(''spruce_budworm_odes(t,x)'',...
                             t0,tf,x,Dt,Dtplot);
```

while in MATLAB this line is written as:

```
[tout,yout] = euler_solver(@spruce_budworm_odes,...
                             t0, tf, x, Dt, Dtplot);
```

- Finally, the solution is plotted using the same commands as in MATLAB

```
// Display mode
mode(-1);

// Clear previous workspace variables
clear

// start a stopwatch timer
```

```
tic

// set global variables
global("rb","k","pbeta","a","rs","Ks","re","Ke","p","Te")

// Load the subroutines
exec('spruce_budworm_odes.sci')

// model parameters
rb    = 1.52;
k     = 355;
pbeta = 43200;
a     = 1.11;
rs    = 0.095;
Ks    = 25440;
re    = 0.92;
Ke    = 1;
p     = 0.00195;
Te    = 0.03;

// initial conditions
t0 = 0;
tf = 200;
B  = 10;
S  = 7000;
E  = 1;
x  = [B,S,E]';

// call to ODE solver (comment/decomment one of the methods
// to select a solver)
method = 'Euler'
//method = 'rkf45'

select method,
    case 'Euler' then
        // Load the Euler solver subroutine
        exec('euler_solver.sci')
        Dt           = 0.01;
        Dtplot       = 0.5;
        [tout,yout] = euler_solver("spruce_budworm_odes(t,x)",...
                                    t0,tf,x,Dt,Dtplot);
    case 'rkf45' then
        // Load the rkf45 solver subroutines
        exec('rkf45_solver.sci')
        exec('ssrkf45.sci')
        hmin       = 0.001;
        nstepsmax = 1000;
        abstol     = 0.001;
        reltol     = 0.001;
        Dtplot     = 0.5;
        [tout,yout,eout] =...
            rkf45_solver("spruce_budworm_odes(t,x)",t0,tf,x,...
                          hmin,nstepsmax,abstol,reltol,Dtplot);
end

// Plot the solution
subplot(3,1,1)
plot(tout,yout(:,1))
ylabel('B(t)','FontSize',2);
```

```
subplot(3,1,2)
plot(tout,yout(:,2))
ylabel('S(t)','FontSize',2);
subplot(3,1,3)
plot(tout,yout(:,3))
xlabel('Time [years]','FontSize',2);
ylabel('E(t)','FontSize',2);

// read the stopwatch timer
tcpu = toc();
```

Script spruce_budworm_main.sci Main program that calls functions `Euler_solver.sci` and `spruce_budworm_odes.sci`

The other two SCILAB functions required to solve the problem are: `spruce_budworm_odes.sci` (where the RHS of the ODEs are defined) and `Euler_solver.sci` (containing the implementation of the Euler solver). These codes are practically the same as in MATLAB, the main differences are those already mentioned in the main script, i.e., the symbol used to comment the lines (//), the way of setting the global variables and the `mode` command to select the information printed during execution.

```
// Display mode
mode(-1);

function [xt] = spruce_budworm_odes(t,x)

// Output variables initialisation (not found in input
// variables)
xt=[];

// Set global variables
global("rb","k","pbeta","a","rs","Ks","re","Ke","p","Te")

// Transfer dependent variables
B = x(1);
S = x(2);
E = x(3);

// Model Parameters
Kb    = ((k*S)*(E^2))/(E^2+Te^2);
alpha = a*S;
g     = (pbeta*(B^2))/(alpha^2+B^2);
P     = (p*(E^2))/(Te^2+E^2);

// Temporal derivatives
Bt = (rb*B)*(1-B/Kb) - g;
St = (rs*S)*(1-(S/Ks)*(Ke/E));
Et = ((re*E)*(1-E/Ke)-(P*B)/S);

// Transfer temporal derivatives
xt = [Bt,St,Et]';
endfunction
```

Function spruce_budworm_odes.sci Right hand side of the ODE system (2.50)–(2.55).

Finally, the SCILAB function `Euler_solver.sci` contains another difference with respect to MATLAB: the `feval` command

```
xnew = x + feval(odefunction,t,x)*Dt;
```

is substituted by

```
xnew = x+evstr(odefunction)*Dt;
```

where `odefunction` is an input parameter of type *string*.

```
// Display mode
mode(-1);

function [tout,xout] = euler_solver(odefunction,t0,tf,...
                                    x0,Dt,Dtplot)

// Output variables initialisation (not found in input
// variables)
tout=[];
xout=[];

// Initialization
plotgap = round(Dtplot/Dt);// number of computation
// steps within a plot interval
Dt      = Dtplot/plotgap;
nplots = round((tf-t0)/Dtplot);// number of plots
t       = t0;// initialize t
x       = x0;// initialize x
tout    = t0;// initialize output value
xout    = x0';// initialize output value

// Implement Euler''s method
for i = 1:nplots
  for j = 1:plotgap
    // Use MATLAB''s feval function to access the
    // function file, then take Euler step
    xnew = x+evstr(odefunction)*Dt;
    t = t+Dt;
    x = xnew;
  end;
  //   Add latest result to the output arrays
  tout = [tout;t];
  xout = [xout;x'];
  //
end;
endfunction
```

Function Euler_solver.sci SCILAB version of the basic Euler ODE integrator

As mentioned before, SCILAB has an interface, called *ode*, to several solvers. The use of function `ode` in SCILAB will be illustrated with a bioreactor example (see Sect. 3.12). The different options for the solvers include: `lsoda` (that automatically selects between Adams and BDF methods), adaptive RK of order 4, and Runge–Kutta–Fehlberg.

On the other hand, OCTAVE comes with LSODE (for solving ODEs) and DASSL, DASPK and DASRT (designed for DAEs). However, a whole collection of around 15 solvers is included in the package OdePkg http://octave.sourceforge.net/odepkg/overview.html. This package include explicit RK solvers of different orders for ODE problems, backward Euler method and versions of different FORTRAN solvers for DAEs (as RADAU5, SEULEX, RODAS, etc.), as well as more sophisticated solvers for implicit differential equations and delay differential equations which are out of the scope of this book.

It must be mentioned that OCTAVE is even more compatible with MATLAB than SCILAB. The main difference between MATLAB and OCTAVE is the call to the ODE solvers when built-in functions are used. If time integration is carried out using our solvers (Euler, rkf45, ros23p, etc.), the MATLAB codes developed in this chapter can be directly used in OCTAVE. However, when built-in solvers are used, slight modifications are required. For instance, the call to ode15s in MATLAB for the spruce budworm problem is of the form:

```
options = odeset('RelTol',1e-6,'AbsTol',1e-6);
t       = [t0:0.5:tf];
[tout, yout] = ode15s(@spruce_budworm_odes,t,x,options);
```

where `spruce_budworm_odes` is the name of the function where the RHS of the ODE equations is implemented, `t` is the time span for the integration, x are the initial conditions and `options` is an optional parameter for setting integration parameters as the tolerances or the maximum step size.

The call to the lsode solver in OCTAVE is carried out as follows:

```
lsode_options('absolute tolerance',1e-6);
lsode_options('relative tolerance',1e-6);
tout                 = [t0:0.5:tf];
[yout, istat, msg] = lsode(@spruce_budworm_odes, x, tout);
```

The most important difference with respect to MATLAB is the order of the dependent and independent problem variables x and t. Note that in OCTAVE x is the second input parameter and t the third. This also affects function `spruce_budworm_odes`. In MATLAB the first line of this code reads as:

```
function xt = spruce_budworm_odes(t,x)
```

while in OCTAVE we have

```
function xt = spruce_budworm_odes(x,t)
```

Note that if we want to reuse this code for integration with our solvers, they must be slightly modified accordingly. For instance, in MATLAB, Euler solver make calls to function `spruce_budworm_odes`

```
xnew = x + feval(odefunction,t,x)*Dt;
```

Table 2.13 Performance of different IVP solvers in several environments for the spruce budworm and liming to remediate acid rain problems

		Spruce budworm	Acid rain
MATLAB	Euler	23.85	2.96
	rkf45	10.00	2.53
	ode45	10.77	1.77
	ode15s	21.92	3.27
SCILAB	Euler	117.31	224.61
	rkf45	24.23	9.61
	lsode (Adams)	10.38	5.00
	lsode (BDF)	13.46	7.69
OCTAVE	Euler	43.85	52.30
	rkf45	19.23	5.00
	lsode	6.15	1.00
	dassl	13.46	1.77

In the acid rain problem a *strong* control law, which is the most challenging, is used. Times have been normalized with respect to the most efficient case, i.e., LSODE in OCTAVE for the acid rain problem

In OCTAVE this must be modified to

```
xnew = x + feval(odefunction,x,t)*Dt;
```

To conclude this section, a comparison of the performance of different solvers in MATLAB, SCILAB, and OCTAVE is included in Table 2.13. When our simple IVP solvers are used, i.e., euler and rkf45, MATLAB is by far the most efficient environment for both problems. The computational cost in OCTAVE is clearly the largest one. On the other hand, when built-in solvers are used, OCTAVE is the most efficient alternative (especially with LSODE) while SCILAB and MATLAB computational costs are of comparable magnitude (SCILAB is more efficient than MATLAB for solving the spruce budworm problem and the reverse is true when solving the acid rain problem). However, it should be noted that in general, when the size and complexity of the problem increases, MATLAB will usually appear as the most efficient alternative even when using built-in solvers.

2.7 How to Use Your Favorite Solvers in MATLAB?

This last section is intended for the reader with some background knowledge in programming, and we suggest for the reader who is not interested in implementation details to skip this material, and possibly to come back to it later on.

MATLAB EXecutable files (MEX-files) are dynamically linked subroutines produced from C or FORTRAN source code that, when compiled, can be run within MATLAB in the same way as MATLAB M-files or built-in functions www.mathworks.com/support/tech-notes/1600/1605.html.

The main reasons for using MEX-files are listed below:

1. During the last few decades, a huge collection of C and FORTRAN numerical algorithms have been created and largely tested by different research and industrial organizations. Such codes cover a large variety of fields such as linear algebra, optimization, times series analysis among many others. Particularly interesting for this chapter is the collection of efficient and reliable ODE solvers. Using the MEX-files we can call these subroutines without the need of rewriting them in MATLAB.
2. MATLAB is a high-level language and as such the programming is easier than in low-level languages as C/C++ or FORTRAN. As a drawback, MATLAB codes are, in general, slower than C and FORTRAN. MEX-files allow us to increase the efficiency of MATLAB M-files or built-in functions.

In order to create a MEX-file, first a gateway routine is required. The gateway is the routine through which MATLAB accesses the rest of the routines in MEX-files. In other words, it is the connection bridge between MATLAB and the FORTRAN or C subroutines. The standard procedure for creating gateways is described in http://www.mathworks.com/support/tech-notes/1600/1605.html?BB=1. This procedure is, however, tiresome, prone to mistakes and with a complex debugging only recommendable for those with good programming skills.

There is however a tool (OPENFGG) for generating semiautomatically the gateways. This tool can be downloaded from the url http://www8.cs.umu.se/~dv02jht/exjobb/download.html and includes a graphical user interphase. There are four steps to create the gateway with this tool:

- Create a new project *File* → *New*
- Add the source files to parse. In this step we include the FORTRAN code for which we want to generate the mex file (e.g. `mycode.f`)
- Select the inputs and the outputs of `mycode.f`
- Press the *Compile* buttom. The gateway `mycodegw.f` is generated

One of the main drawbacks of OPENFGG is that it does not allow to create reverse gateways, this is, codes for calling MATLAB files from FORTRAN.

It must be noted that for creating a MEX file an adequate FORTRAN or C/C++ compiler must be installed.

Now we are ready to create the MEX-file. To that purpose we open a MATLAB session and type:

```
mex -setup
```

And choose one of the installed compilers. This step is only required to be done once. After this, type

```
mex mycode.f mycodegw.f
```

Which creates the MEX-file that can be used at the MATLAB command prompt in the same way as any M-file or built-in function.

From the MATLAB version 7.2 (R2006a), it is possible to use open source FORTRAN and C compilers (g95, gfortran, gcc) through the tool GNUMEX. GNUMEX allows to set up the Windows versions of gcc (also gfortran and g95) to compile mex files. The procedure is described in detail at http://gnumex.sourceforge.net, However, the main steps are summarized below:

- Install the packages for the gcc compiler. The easiest way is to Download Mingw from http://www.mingw.org/ and install it. The installation path `C:\mingw` is recommended.
- Download the g95 compiler from http://www.g95.org/downloads.shtml and install it. The option *Self-extracting Windows x86 (gcc 4.1, experimental)* is recommended.
- Download GNUMEX http://sourceforge.net/projects/gnumex/ unzip the files and place the folder in a directory (e.g `C:\gnumex`).
- Open a MATLAB session, go to the path where gnumex was unzipped and type `gnumex`. A window will open. Through this window we will modify the `mexopts.bat` which contains the information about the compilers and options used for creating the MEX-file.
- In the new window
  - Indicate the place where mingw and g95 binaries are located
  - In `language for compilation` chose C/C++ or g95 depending on if the MEX-file will be created from a C or a FORTRAN subroutine.
  - Press `Make options file`

Now we should be able to create the MEX-file using the command `mex` in MATLAB.

A simple example of how to create a MEX-file is described in Sect. 2.7.1.

2.7.1 A Simple Example: Matrix Multiplication

In order to illustrate the procedure for constructing a MEX-file and how such MEX-file can speed-up our code, we consider in this section the simple example of matrix multiplication.

We first list the MATLAB algorithm for matrix multiplication `matrix_mult.m`

```
function C = matrix_mult(A,B,nrA,ncA,ncB)

% Code for matrix multiplication
C = zeros(nrA,ncB);
for ii = 1:nrA
    for jj = 1:ncB
        for kk = 1:ncA
            C(ii,jj) = C(ii,jj) + A(ii,kk)*B(kk,jj);
        end
    end
end
```

Function matrix_mult.m Simple MATLAB algorithm for matrix multiplication

It should be noted that the MATLAB symbol * contains a much more efficient algorithm than `matrix_mult.m` thus `matrix_mult.m` is only used for illustration purposes.

The FORTRAN code for matrix multiplication is written in `matmult.f`

```
C matmult subroutine in FORTRAN

      SUBROUTINE matmult(A,B,C,nrA,ncA,ncB)

      IMPLICIT NONE
      INTEGER nrA,ncA,ncB
      REAL*8 A(nrA,ncA)
      REAL*8 B(ncA,ncB)
      REAL*8 C(nrA,ncB)

      INTEGER I,J,K

      DO I=1,nrA
          DO J=1,ncB
              C(I,J) = 0.0
                  DO K=1,ncA
                      C(I,J) = C(I,J) + A(I,K)*B(K,J)
                  END DO
          END DO
      END DO
      RETURN
      END
```

Function matmult.f Simple FORTRAN algorithm for matrix multiplication

The next step is to create the gateway for the FORTRAN subroutine. For that purpose, we will employ the tool `OpenFGG`. In this tool, we must define the input and output variables of the subroutine, so we define `A`, `B`, `nrA`, `ncA`, and `ncB` as input variables and `C` as the output variable. After this step the gateway `matmultgw.f` is generated.[1] It must be noted here that the gateway `matmultgw.f` contains almost 700 lines of code which gives us an idea of the complexity of creating FORTRAN gateways by hand.

The next step is to create the MEX-file using the following command in MATLAB:

```
mex -O matmult.f matmultgw.f
```

where the `-O` option is to optimize the code. This step creates the MEX-file whose extension will depend on the platform and the version of MATLAB. In our case, we obtain `matmult.mexglx` since we are runing MATLAB 2009b under a Linux 32 bits platform.

[1] The code for the gateway `matmultgw.f` is not included in this document because of its length.

Another option is to create the MEX file for a C function. The C code for matrix multiplication is written in `mmcsubroutine.c`

```
void mmcsubroutine(
                        double    C[],
                        double    A[],
                        double    B[],
                        int       nrA,
                        int       ncA,
                        int       ncB
          )
{
    int i,j,k,cont;
    cont = 0;
    for (j=0; j<ncB; j++) {
        for (i=0; i<nrA; i++) {
            for (k=0; k<ncA; k++) {
                C[cont] += A[nrA*k+i]*B[k+j*ncA];
            }
            cont++;
        }
    }
    return;
}
```

Function mmcsubroutine.c Simple C algorithm for matrix multiplication

There is no software for the automatic generation of the gateway thus we have to create it by hand. In this particular case, this task is not too difficult but some programming skills are required. The gateway is written in `matmultc.c`

```
/* Gateway MATMULTC.C for matrix multiplication in C   */
#include <math.h>
#include "mex.h"

/* Input Arguments */
#define A_IN    prhs[0]
#define B_IN    prhs[1]
#define NRA_IN  prhs[2]
#define NCA_IN  prhs[3]
#define NCB_IN  prhs[4]

/* Output Arguments */
#define C_OUT   plhs[0]

/* Mex function */
void mexFunction( int nlhs, mxArray *plhs[],
          int nrhs, const mxArray*prhs[] )
{
    double *C;
    double *A,*B;
    int nrA,ncA,ncB;
    mwSize m,n,nA,mB;

    /* Check for proper number of arguments */
```

```
    if (nrhs != 5) {
    mexErrMsgTxt("Five input arguments required.");
    } else if (nlhs > 1) {
    mexErrMsgTxt("Too many output arguments.");
    }

    /* Check the dimensions of A and B */
    nA = mxGetN(A_IN);
    mB = mxGetM(B_IN);
    if ((nA != mB)) {
    mexErrMsgTxt("N cols. in A different from N rows in B.");
    }

    /* Create a matrix for the return argument */
    m = mxGetM(A_IN);
    n = mxGetN(B_IN);
    C_OUT = mxCreateDoubleMatrix(m, n, mxREAL);

    /* Assign pointers to the various parameters */
    C   = mxGetPr(C_OUT);
    A   = mxGetPr(A_IN);
    B   = mxGetPr(B_IN);
    nrA = (int) mxGetScalar(NRA_IN);
    ncA = (int) mxGetScalar(NCA_IN);
    ncB = (int) mxGetScalar(NCB_IN);

    /* Do the actual computations in a subroutine */
    mmcsubroutine(C,A,B,nrA,ncA,ncB);
    return;
}
```

Function matmultc.c C gateway for the subroutine `mmcsubroutine.c`

We can note the following details about `matmultc.c`:

1. The code starts by defining the input and output arguments of the subroutine `mmcsubroutine.c`.

```
/* Input Arguments */
#define A_IN    prhs[0]
#define B_IN    prhs[1]
#define NRA_IN  prhs[2]
#define NCA_IN  prhs[3]
#define NCB_IN  prhs[4]
/* Output Arguments */
#define C_OUT   plhs[0]
```

   There are five inputs and one output. The part `rhs` in `prhs[*]` calls for the right hand side while `lhs` in `plhs[*]` calls for the left hand side. The order of the parameters when calling the subroutine is indicated by the number between brackets.
2. After defining the inputs and outputs, we start by constructing the MEX function.

```
/* Mex function */
void mexFunction( int nlhs, mxArray *plhs[],
          int nrhs, const mxArray*prhs[] )
```

The name of the MEX function must always be mexFunction. This part is common to all MEX-files.

3. Next the parameters used in the MEX function are defined

```
double *C;
double *A,*B;
int nrA,ncA,ncB;
mwSize m,n,nA,mB;
```

4. The matrix output is created

```
/* Create a matrix for the return argument */
 m = mxGetM(A_IN);
 n = mxGetN(B_IN);
 C_OUT = mxCreateDoubleMatrix(m, n, mxREAL);
```

5. Then the pointers are assigned to the different input/output parameters

```
/* Assign pointers to the various parameters */
C   = mxGetPr(C_OUT);
A   = mxGetPr(A_IN);
B   = mxGetPr(B_IN);
nrA = (int) mxGetScalar(NRA_IN);
ncA = (int) mxGetScalar(NCA_IN);
ncB = (int) mxGetScalar(NCB_IN);
```

6. At this point, the subroutine which performs the matrix multiplication computation is called

```
/* Do the actual computations in a subroutine */
mmcsubroutine(C,A,B,nrA,ncA,ncB);
```

7. Alternatively some code for checking the input/output arguments can be included

```
/* Check for proper number of arguments */
    if (nrhs != 5) {
    mexErrMsgTxt(''Five input arguments required.'');
    } else if (nlhs > 1){
    mexErrMsgTxt(''Too many output arguments.'');
    }
    /* Check the dimensions of A and B */
    nA = mxGetN(A_IN); mB = mxGetM(B_IN);
    if ((nA != mB)) {
mexErrMsgTxt(''N cols. in A different from N rows in B.'');
    }
```

Finally, the MEX-file is created as in the previous case:

```
mex -O matmultc.c mmcsubroutine.c
```

Now we are ready to call the mex-file from MATLAB. In order to compare the computational times obtained with the MATLAB function and with the MEX-file, the MATLAB script main_matrix_mult will be used.

```
% Main program calling the matrix_mult code and the mex file
clear all
clc

% Dimension of matrices
nrA = 200;
ncA = 1000;
ncB = 500;

% Create the matrices
A = rand(nrA,ncA);
B = rand(ncA,ncB);

% Call the MATLAB function
tt = cputime;
C1 = matrix_mult(A,B,nrA,ncA,ncB);
fprintf('Matlab function time  = %2.2f s\n',cputime-tt);


% Call the FORTRAN MEX-file
tt = cputime;
C2 = matmult(A,B,nrA,ncA,ncB);
fprintf('Fortran Mex-file time = %2.2f s\n',cputime-tt);

% Call the C MEX-file
tt = cputime;
C3 = matmultc(A,B,nrA,ncA,ncB);
fprintf('C Mex-file time       = %2.2f s\n',cputime-tt);
```

Script main_matrix_mult Main program that calls MATLAB function `matrix_mult.m` and mex-file `matmult.mexglx`

As a result, we obtain the following MATLAB screen-print:

```
MATLAB function time  = 3.60 s
Fortran Mex-file time = 0.56 s
C Mex-file time       = 0.45 s
```

showing that for this particular case, the MEX-file obtained from the FORTRAN code is more than six times faster than the MATLAB function while the MEX-file obtained from the C code is even faster.

2.7.2 MEX-Files for ODE Solvers

Efficient FORTRAN or C time integrators can also be exploited within MATLAB using the concept of MEX-files. The creation of these MEX-files is too complex to be detailed in this introductory text and we content ourselves with an application example. A comparison against the IVP solver `RKF45` is shown in Table 2.14, which lists computational times normalized with respect to the smallest cost, i.e., that corresponding to the simulation of the logisticequation using the FORTRAN version.

Table 2.14 Computation times required for solving different ODE systems considered in this chapter with versions of the RKF45 solver implemented in FORTRAN and MATLAB

	FORTRAN	MATLAB
Logistic equation (Sect. 2.1)	1	1,500
Stiff ODEs (Sect. 2.2)	14.71	5882.4
Spruce budworm (Sect. 2.5)	229.4	56,971

As shown in the table, the FORTRAN MEX-file is orders of magnitude faster than the MATLAB version. Several MEX-files are available in the companion software. The details of their construction are omitted due to space limitation.

2.8 Summary

In this chapter, a few time integrators are detailed and coded so as to show the main ingredients of a good ODE solver. First, fixed-step integrators are introduced, followed by variable-step integrators that allow to achieve a prescribed level of accuracy. However, stability appears as an even more important issue than accuracy, limiting the time-step size in problems involving different time scales. A Rosenbrock's method is then considered as an example to solve efficiently this class of problems. After the presentation of these several basic time integrators, we turn our attention to the MATLAB ODE suite, a powerful library of time integrators which allow to solve a wide range of problems. We then apply several of these integrators to two more challenging application examples, i.e., the study of spruce budworm dynamics and the study of liming to remediate the effect of acid rain on a lake. On the other hand, we introduce the use of SCILAB and OCTAVE, two attractive open-source alternatives to MATLAB, to solve ODE and DAE problems, and we highlight the main syntaxic differences. As MATLAB is an interpreted language, the computational cost can however be an issue when computing the solution of large system of ODEs, or when solving repeatedly the same problem (as for instance when optimizing a cost function, involving the solution of a ODE model). The use of compiled functions can be advantageous in this case, and this is why we end the presentation of ODE integrators by the use of MATLAB MEX-files.

References

1. Shampine LF, Gladwell I, Thompson S (2003) Solving ODEs with MATLAB. Cambridge University Press, Cambridge
2. Hindmarsh AC (1983) ODEPACK, a systematized collection of ODE solvers, in scientific computing. In: Stepleman RS (ed) IMACS transactions on scientific computation, vol. 1. Elsevier, Amsterdam

3. Hindmarsh AC (1980) Lsode and lsodi, two new initial value ordinary differential equation solvers. ACM Signum Newslett 15(4):10–11
4. Breman KE, Campbell SL, Petzold LR (1989) Numerical solution of initial-value problems in differential-algebraic equations. Elsevier, New York
5. Verhulst PF (1938) Notice sur la loi que la population poursuit dans son accroissement. Correspondance mathmatique et physique 10:113–121
6. Butcher JC (1987) The numerical analysis of ordinary differential equations. Wiley, New York
7. Lambert JD (1991) Numerical methods for ordinary differential systems. Wiley, London
8. Hairer E, Wanner G (1996) Solving ordinary differential equations II: stiff and differential algebraic problems. Springer, New York
9. Hairer E, Norsett SP, Wanner G (1993) Solving ordinary differential equations I: nonstiff problems. Springer, New York
10. Rosenbrock HH (1963) Some general implicit processes for the numerical solution of differential equations. Comput J 5:329–330
11. Verwer JG, Spee EJ, Blom JG, Hundsdorfer W (1999) A second order rosenbrock method applied to photochemical dispersion problems. J Sci Comput 20:1456–1480
12. Lang J, Verwer JG (2001) Ros3p—an accurate third-order rosenbrock solver designed for parabolic problems. BIT 21:731–738
13. Ludwig D, Jones DD, Holling CS (1995) Qualitative analysis of insect outbreak systems: the spruce budworm and forest. J Anim Ecol 47:315–332
14. Ghosh M (2002) Effect of liming on a fish population in an acidified lake: a simple mathematical model. Appl Math Comput 135(2–3):553–560
15. Gomez C (2003) SCILAB consortium launched. Technical report 54, ERCIM News
16. Eaton JW, Bateman D, Hauberg S, Wehbring R (2013) GNU Octave: a high-level interactive language for numerical computations, 3rd edn. Network Theory Ltd, Bristol

Chapter 3
Finite Differences and the Method of Lines

As we recall from Chap. 1, the method of lines (MOL) is a convenient procedure for solving time-dependent PDEs, which proceeds in two separate steps:

- approximation of the spatial derivatives using finite differences, finite elements or finite volume methods (or any other techniques), and
- time integration of the resulting semi-discrete (discrete in space, but continuous in time) ODEs.

William Schiesser has significantly contributed to the development and popularization of the method and has written one of the first books on the subject [1].

In this chapter, we discuss this method in more details, considering mostly finite difference approximations, and we successively address the following questions:

- Is the numerical scheme stable?
- How accurate is the numerical scheme?
- How can we implement the finite difference approximations efficiently?
- How can we translate mixed-type (possibly complex) boundary conditions?

The first question is of fundamental importance since an unstable numerical scheme would be completely useless. However, stability analysis is a difficult subject that will only be sketched here, as a full investigation is out of the scope of this introductory text. To support our analysis, we will consider a simple linear convection-diffusion-reaction equation, which allows important properties related to stability and dynamics of the semi-discrete ODE system to be introduced, and use basic finite difference schemes (FDs) and time integrators. Then, more attention will be paid to accuracy, and higher-order finite FDs will be derived and formulated using the concept of a differentiation matrix. These higher-order FDs can be used in conjunction with the higher-order time integrators reviewed in Chaps. 1 and 2. In addition, alternative ways to take boundary conditions into account will be presented and the computation of the Jacobian matrix of the ODE system will be revisited. Finally, we discuss the use of SCILAB and OCTAVE to solve PDE problems using the method of lines.

A. Vande Wouwer et al., *Simulation of ODE/PDE Models with MATLAB®, OCTAVE and SCILAB*, DOI: 10.1007/978-3-319-06790-2_3,

3.1 Basic Finite Differences

FDs can be used to evaluate the derivatives of a function $x(z)$ which is known only at a set of $N+1$ discrete points z_i:

z	$x(z)$
z_0	x_0
z_1	x_1
$\vdots$	$\vdots$
z_i	x_i
$\vdots$	$\vdots$
z_N	x_N

To simplify the presentation, we first consider that the points z_i are regularly distributed (uniformly spaced) between z_0 and z_N:

$$z_i = z_0 + i\Delta z; \qquad \Delta z = \frac{z_N - z_0}{N}; \qquad i = 0, \ldots, N \tag{3.1}$$

The FD approximation of the nth-order derivative of $x(z)$ in z_i, i.e., $\frac{\mathrm{d}^n x(z_i)}{\mathrm{d}z^n}$, is based on Taylor series expansions of $x(z)$ at points z_k close to z_i. As an example, consider the evaluation of the first and second derivatives of $x(z)$ at z_i, and write the Taylor series expansions at z_{i+1} and z_{i-1}.

$$x_{i+1} = x_i + \frac{\Delta z}{1!}\left.\frac{\mathrm{d}x}{\mathrm{d}z}\right|_{z_i} + \frac{\Delta z^2}{2!}\left.\frac{\mathrm{d}^2x}{\mathrm{d}z^2}\right|_{z_i} + \frac{\Delta z^3}{3!}\left.\frac{\mathrm{d}^3x}{\mathrm{d}z^3}\right|_{z_i} + \frac{\Delta z^4}{4!}\left.\frac{\mathrm{d}^4x}{\mathrm{d}z^4}\right|_{z_i} + \cdots \tag{3.2}$$

$$x_{i-1} = x_i - \frac{\Delta z}{1!}\left.\frac{\mathrm{d}x}{\mathrm{d}z}\right|_{z_i} + \frac{\Delta z^2}{2!}\left.\frac{\mathrm{d}^2x}{\mathrm{d}z^2}\right|_{z_i} - \frac{\Delta z^3}{3!}\left.\frac{\mathrm{d}^3x}{\mathrm{d}z^3}\right|_{z_i} + \frac{\Delta z^4}{4!}\left.\frac{\mathrm{d}^4x}{\mathrm{d}z^4}\right|_{z_i} - \cdots \tag{3.3}$$

Subtracting Eq. (3.3) from Eq. (3.2), we obtain:

$$x_{i+1} - x_{i-1} = 2\frac{\Delta z}{1!}\left.\frac{\mathrm{d}x}{\mathrm{d}z}\right|_{z_i} + 2\frac{\Delta z^3}{3!}\left.\frac{\mathrm{d}^3x}{\mathrm{d}z^3}\right|_{z_i} + \cdots$$

which can be rewritten as

$$\left.\frac{\mathrm{d}x}{\mathrm{d}z}\right|_{z_i} = \frac{x_{i+1} - x_{i-1}}{2\Delta z} - \frac{\Delta z^2}{3!}\left.\frac{\mathrm{d}^3x}{\mathrm{d}z^3}\right|_{z_i} - \cdots \tag{3.4}$$

so that for $\Delta z \to 0$,

$$\left.\frac{\mathrm{d}x}{\mathrm{d}z}\right|_{z_i} = \frac{x_{i+1} - x_{i-1}}{2\Delta z} + O(\Delta z^2) \tag{3.5}$$

since the higher order terms in Δz become negligibly small. $O(\Delta z^2)$ denotes a term "of order" or proportional to Δz^2.

Other useful approximations of the first-order derivative can be derived from Eqs. (3.2)–(3.3), e.g.,

$$\left.\frac{dx}{dz}\right|_{z_i} = \frac{x_{i+1} - x_i}{\Delta z} - \frac{\Delta z}{2!}\left.\frac{d^2x}{dz^2}\right|_{z_i} + \cdots = \frac{x_{i+1} - x_i}{\Delta z} + O(\Delta z) \tag{3.6}$$

and

$$\left.\frac{dx}{dz}\right|_{z_i} = \frac{x_i - x_{i-1}}{\Delta z} + \frac{\Delta z}{2!}\left.\frac{d^2x}{dz^2}\right|_{z_i} + \cdots = \frac{x_i - x_{i-1}}{\Delta z} + O(\Delta z) \tag{3.7}$$

Note that Eqs. (3.6)–(3.7) are less accurate than Eq. (3.5), since the error decreases linearly with Δz (rather than quadratically as in Eq. (3.5)). Note also that Eq. (3.5) is a centered approximation (i.e., z_i is between z_{i-1} and z_{i+1}) whereas Eqs. (3.6)–(3.7) are non-centered approximations. Depending on the direction of the "flow of information", informally called the "direction of the wind", these latter formulas are termed "upwind" or "downwind" .

To obtain an approximation of the second-order derivative, we add Eqs. (3.2) and (3.3) to obtain

$$x_{i+1} + x_{i-1} = 2x_i + 2\frac{\Delta z^2}{2!}\left.\frac{d^2x}{dz^2}\right|_{z_i} + 2\frac{\Delta z^4}{4!}\left.\frac{d^4x}{dz^4}\right|_{z_i} + \cdots$$

which can be rewritten as:

$$\left.\frac{d^2x}{dz^2}\right|_{z_i} = \frac{x_{i+1} - 2x_i + x_{i-1}}{\Delta z^2} - \frac{\Delta z^2}{12}\left.\frac{d^4x}{dz^4}\right|_{z_i} - \cdots \tag{3.8}$$

When $\Delta z \to 0$,

$$\left.\frac{d^2x}{dz^2}\right|_{z_i} = \frac{x_{i+1} - 2x_i + x_{i-1}}{\Delta z^2} + O(\Delta z^2) \tag{3.9}$$

since the higher order terms in Δz become negligibly small. As we shall see later on in this chapter, this reasoning can be pursued in order to construct higher-order approximations.

3.2 Basic MOL

To introduce the method of lines (MOL), we consider a simple example, i.e., the linear advection equation

$$x_t = -vx_z \tag{3.10}$$

where $x(z,t)$ is a function of space and time (for example, the concentration of a component flowing through a pipe, or the density of cars along a highway); $x_t = \frac{\partial x}{\partial t}$ and $x_z = \frac{\partial x}{\partial z}$ (these short, subscript notations will also be used in the following computer codes). v is a real positive constant, representing the velocity (e.g., fluid or car velocity) and $z_0 < z \leq z_N$ is the spatial domain over which this equation is defined.

This PDE problem is supplemented with an initial condition (IC), i.e., the initial distribution of x over space at time $t = 0$

$$x(z,0) = x^0(z) \tag{3.11}$$

and a boundary condition (BC), e.g., the value of x (inlet concentration or number of cars entering the highway) over time at $z = z_0$

$$x(z_0,t) = x_0(t) \tag{3.12}$$

This BC, that specifies the value of the field in the boundary, is of *Dirichlet* type. Other types of boundary conditions will be considered in the sequel.

The MOL proceeds as follows:

1. A uniform spatial grid is defined over N intervals

$$z_i = z_0 + i\Delta z; \qquad \Delta z = \frac{z_N - z_0}{N} = \frac{L}{N}; \qquad i = 0, \ldots, N \tag{3.13}$$

2. The PDE (3.10) is expressed at each of the grid points, with the exception of the point z_0 where BC (3.12) is imposed

$$(x_i)_t = -v(x_i)_z \qquad i = 1, \ldots, N \tag{3.14}$$

3. The spatial derivative is replaced by a finite difference formula, e.g., Eq. (3.7)

$$(x_i)_z = \frac{x_i - x_{i-1}}{\Delta z} \tag{3.15}$$

This approximation transforms Eq. (3.14) into a system of ODEs of the form

$$\begin{bmatrix} x_1 \\ x_2 \\ \vdots \\ x_i \\ \vdots \\ x_N \end{bmatrix}_t = \begin{bmatrix} -\frac{v}{\Delta z}(x_1) \\ -\frac{v}{\Delta z}(x_2 - x_1) \\ \vdots \\ -\frac{v}{\Delta z}(x_i - x_{i-1}) \\ \vdots \\ -\frac{v}{\Delta z}(x_N - x_{N-1}) \end{bmatrix} + \begin{bmatrix} \frac{v}{\Delta z}(x_0) \\ 0 \\ \vdots \\ 0 \\ \vdots \\ 0 \end{bmatrix} \tag{3.16}$$

Note that the information provided by BC (3.12) (i.e., the value of x_0, which is imposed) appears explicitly in the second term of the RHS, i.e., as an external input to the system.

4. The IC (3.11) provides the value of the unknown functions $x_1(t), \ldots, x_N(t)$ at $t = 0$, and the system of Eq. (3.16) can therefore be integrated using one of the methods reviewed in the previous chapters (e.g., Euler, Runge-Kutta, etc.).

Before actually solving Eq. (3.16), we ask the following question: Is the numerical integration of Eq. (3.16) numerically stable?

3.3 Numerical Stability: Von Neumann and the Matrix Methods

Von Neumann stability analysis is a useful method for understanding the propagation of errors in linear difference equations. These difference equations describe the whole computational process required to calculate the values of the functions $x_1(t), \ldots, x_N(t)$ at each integration time step. This scheme therefore depends on the choice of a particular time integrator. For simplicity, we consider a fixed-step, explicit Euler method. This method replaces $(x_i)_t$ in Eq. (3.16) by

$$(x_i)_t = \frac{x_i(t + \Delta t) - x_i(t)}{\Delta t} = \frac{x_i^{k+1} - x_i^k}{\Delta t} \tag{3.17}$$

In this latter expression, we explicitly consider the discrete time steps of Euler's method $t_k = k\Delta t$ and we note $x_i(k\Delta t) = x_i^k$. In Chap. 1, we used h for the time-step size. Here, we prefer the notation Δt (as opposed to Δz which is usually reserved for a spatial increment), but these notations, i.e., h or Δt , will be used interchangeably in the following chapters.

Finally, Eqs. (3.14)–(3.15) and Eq. (3.17) give the following numerical scheme

$$\frac{x_i^{k+1} - x_i^k}{\Delta t} = -v\frac{x_i^k - x_{i-1}^k}{\Delta z} \tag{3.18}$$

which is now purely algebraic, i.e., it does not contain any of the original derivatives of Eq. (3.10).

This scheme provides explicitly the value of x_i^{k+1} as a function of x_i^k and x_{i-1}^k

$$x_i^{k+1} = x_i^k - v\frac{\Delta t}{\Delta z}(x_i^k - x_{i-1}^k) \tag{3.19}$$

that is, we can explicitly move forward in t from k to $k + 1$.

Von Neumann's stability analysis can be used to verify the stability of Eq. (3.19), prior to the application of problem-specific BCs and ICs. This method is based on spatial Fourier transforms (FTs), which convert the time-space difference Eq. (3.19) into a time recursion in terms of the spatial FTs. This procedure can be introduced in several ways. Here, we use the classical Fourier series expansion.

Assume that the solution at time $t_k = k\Delta t$, i.e., $x^k = (x_0^k, x_1^k, \ldots, x_N^k)^T$, represents a (spatial) period of a periodic signal extending from $-\infty$ to $+\infty$. For the purpose of the stability analysis, the solution $x^k = (x_0^k, x_1^k, \ldots, x_N^k)^T$ defined on the spatial interval $[z_0, z_N]$, where $z_N = z_0 + L$, is therefore *repeated* on the intervals:

$$\ldots, [z_0 - 2L, z_0 - L], [z_0 - L, z_0], [z_0, z_N], [z_N, z_N + L], [z_N + L, z_N + 2L], \ldots$$

This implicitly assumes that *periodic boundary conditions* apply, i.e., $x(z_0, t) = x(z_N, t)$.

This *imaginary* periodic signal, which has the solution we are examining over a period L, can be expanded into a *finite Fourier series* (a finite Fourier series is the discrete analog of the Fourier series, which is used for sampled signals—recall that we consider spatial sampling at this stage)

$$x_i^k = \sum_{n=-N}^{+N} A_n^k e^{jn\omega_0(i\Delta z)} \tag{3.20}$$

where j is the imaginary number such that $j^2 = -1$, and

$$\omega_0 = \frac{\pi}{L} \tag{3.21}$$

If we note the angles $\phi_n = n\omega_0(\Delta z) = n\frac{\pi}{L}\Delta z = n\frac{\pi}{N}$, then

$$x_i^k = \sum_{n=-N}^{+N} A_n^k e^{j(\phi_n)} \tag{3.22}$$

Substitution of Eq. (3.22) into Eq. (3.19) gives the evolution of the single Fourier mode of wave number n

$$A_n^{k+1} e^{j(i\phi_n)} = \left(1 - \frac{v\Delta t}{\Delta z}\right) A_n^k e^{j(i\phi_n)} + \frac{v\Delta t}{\Delta z} A_n^k e^{j((i-1)\phi_n)} \tag{3.23}$$

At each time step, each Fourier mode is therefore amplified by a factor

$$G_n = \frac{A_n^{k+1}}{A_n^k} \tag{3.24}$$

which, if we note $\sigma = \frac{v\Delta t}{\Delta z}$, can be written as

$$G_n = (1 - \sigma) + \sigma e^{-j\phi_n} \tag{3.25}$$

For the numerical scheme to be stable, the modulus of this amplification factor must be less than unity for all n, i.e.,

$$|G_n| \leq 1 \tag{3.26}$$

otherwise some of the modes could grow up to infinity as k goes to infinity. G_n represents a complex number $G_n = \xi_n + j\eta_n$ where

$$\begin{aligned} \xi_n &= 1 - \sigma(1 - \cos\phi_n) \\ \eta_n &= -\sigma \sin\phi_n \end{aligned} \tag{3.27}$$

and ϕ_n varying in the interval $[-\pi, \pi]$ (note that ϕ_n takes only $2N$ discrete values in this interval), the geometric locus of G_n

$$[\xi_n - (1-\sigma)]^2 + \eta_n^2 = \sigma^2 \cos^2(\phi_n) + \sigma^2 \sin^2(\phi_n) = \sigma^2 \tag{3.28}$$

is a circle of radius σ centered in $(1-\sigma, 0)$. Thus, condition (3.26) is satisfied if the radius of this circle is smaller or equal to 1, i.e.,

$$0 \leq \sigma = \frac{v\Delta t}{\Delta z} < 1 \tag{3.29}$$

This condition is the famous *Courant-Friedrichs-Lewy condition* (CFL condition).

At the limit $\sigma = 1$, the amplification factors are $G_n = e^{-j\phi_n}$, i.e., they represent $2N$ points regularly distributed along the unit circle in the complex plane, given by the angles ϕ_n.

In summary, von Neumann analysis verifies that no spatial Fourier component in the system is growing exponentially with respect to time (i.e., verifies that no amplification factor is larger than one).

We now turn our attention to an alternative stability analysis, sometimes referred to as the *matrix method*. Consider again Eq. (3.19)

$$x_i^{k+1} = x_i^k - v\frac{\Delta t}{\Delta z}\left(x_i^k - x_{i-1}^k\right) = (1-\sigma)x_i^k + \sigma x_{i-1}^k \tag{3.30}$$

or in matrix form

$$\begin{bmatrix} x_1 \\ \cdots \\ \cdots \\ x_N \end{bmatrix}^{k+1} = \begin{bmatrix} 1-\sigma & 0 & \cdots & 0 \\ \sigma & \ddots & \ddots & \vdots \\ 0 & \ddots & \ddots & 0 \\ 0 & 0 & \sigma & 1-\sigma \end{bmatrix} \begin{bmatrix} x_1 \\ \cdots \\ \cdots \\ x_N \end{bmatrix}^{k} + \begin{bmatrix} \sigma x_0 \\ 0 \\ \cdots \\ 0 \end{bmatrix}^{k} \tag{3.31}$$

where x_0^k is defined by boundary condition (3.12). In compact format, Eq. (3.31) can also be written as:

$$\mathbf{x}^{k+1} = \mathbf{A}\mathbf{x}^k + \mathbf{u}_{CL}^k \tag{3.32}$$

One way to analyze the stability of (3.32) is to check that the truncation errors $\mathbf{e}^k$ do not grow exponentially. If $\mathbf{x}_e^k$ is the exact solution at time t_k

$$\mathbf{x}^k = \mathbf{x}_e^k + \mathbf{e}^k \tag{3.33}$$

and

$$\mathbf{x}_e^{k+1} = \mathbf{A}\mathbf{x}_e^k + \mathbf{u}_{CL}^k \tag{3.34}$$

By subtracting Eq. (3.34) from (3.32), and using Eq. (3.33), the following expression for the propagation of the error is obtained

$$\mathbf{x}_e^{k+1} = \mathbf{A}\mathbf{x}_e^k \tag{3.35}$$

This latter operation cancels the second term on the RHS, $\mathbf{u}_{CL}^k$, which is related to the boundary conditions (so that we can conclude that the BCs have no influence on the dynamics of $\mathbf{e}^k$). The homogeneous system (3.35) is asymptotically stable if

$$\lim_{k\to\infty} \|\mathbf{x}_e^k\| = 0 \tag{3.36}$$

A necessary and sufficient condition of stability of the discrete-time system (3.36) is given by Owens [2]:

$$\rho(\mathbf{A}) = \max_k |\lambda_k| < 1 \tag{3.37}$$

where λ_k are the eigenvalues of the matrix $\mathbf{A}$ (i.e., the eigenvalues must lie inside the familiar unit circle in the complex plane).

In Eq. (3.31), the N eigenvalues of the matrix $\mathbf{A}$ are equal to the elements of the main diagonal (and are therefore equal)

$$\lambda_k = 1 - \sigma; \qquad \forall k \tag{3.38}$$

and condition (3.37) becomes

$$0 \le |1 - \sigma| < 1 \tag{3.39}$$

or

$$0 \le \sigma < 2 \tag{3.40}$$

which is less restrictive than (3.29), that restricts σ to be in the interval [0, 1].

This apparent mismatch between the two analysis has a simple explanation: condition (3.37) ensures the asymptotic stability of Eq. (3.35), i.e., that the truncation error eventually vanishes as expressed in Eq. (3.36), but do not impose conditions on the transient behaviour of $\mathbf{e}^k$. In fact, the values taken by $\mathbf{e}^k$ are influenced by the initial error $\mathbf{e}^0$ (which could be due to rounding errors when manipulating the initial condition)

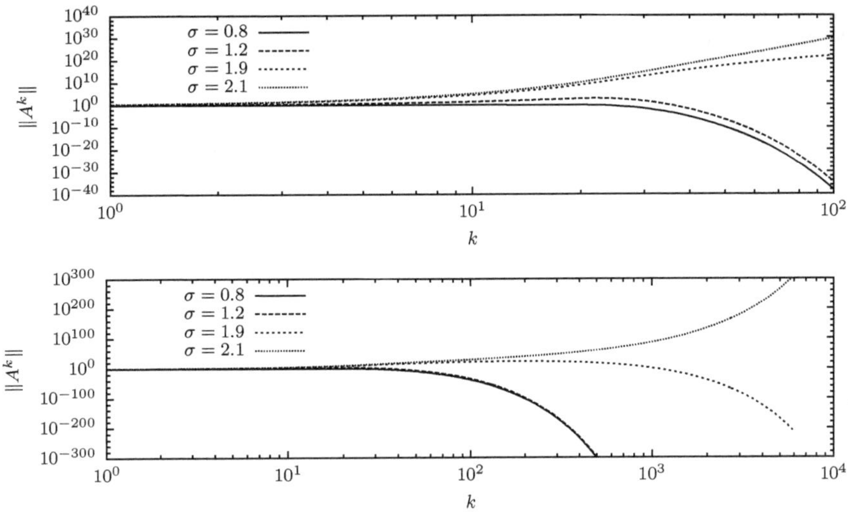

Fig. 3.1 Evolution of $\|(\mathbf{A})^k\|$ for different values of σ

$$\mathbf{e}^k = (\mathbf{A})^k \mathbf{e}^0 \tag{3.41}$$

where $(\mathbf{A})^k$ is the matrix $\mathbf{A}$ to the power k. The evolution of $\|(\mathbf{A})^k\|$ is illustrated in Fig. 3.1 on two time scales (upper and lower plots) and for four different values of σ.

- If $\sigma \leq 1$ (Von Neumann's condition (3.29)), then $\|\mathbf{e}^k\|$ do not grow at any time.
- If $1 < \sigma < 2$, $\|\mathbf{e}^k\|$ first grows and then decreases toward zero (as condition (3.37) is satisfied).
- If $2 \leq \sigma$, then the errors grow unbounded.

Figure 3.1 was obtained for $N = 20$ (the graphs have the same qualitative behavior for any value of N). The (spectral) norm of the matrix $\mathbf{A}$ is computed using the MATLAB function `norm(A)`.

One of the basic assumptions of Von Neumann's analysis is the existence of periodic boundary conditions $x(z_0, t) = x(z_N, t)$ or

$$x_0^k = x_N^k \tag{3.42}$$

If we introduce this condition into Eq. (3.31), we have

$$\begin{bmatrix} x_1 \\ \cdots \\ \cdots \\ x_N \end{bmatrix}^{k+1} = \begin{bmatrix} 1-\sigma & 0 & \cdots & \sigma \\ \sigma & \ddots & \ddots & \vdots \\ 0 & \ddots & \ddots & 0 \\ 0 & 0 & \sigma & 1-\sigma \end{bmatrix} \begin{bmatrix} x_1 \\ \cdots \\ \cdots \\ x_N \end{bmatrix}^{k} = \mathbf{A}'\mathbf{x} \tag{3.43}$$

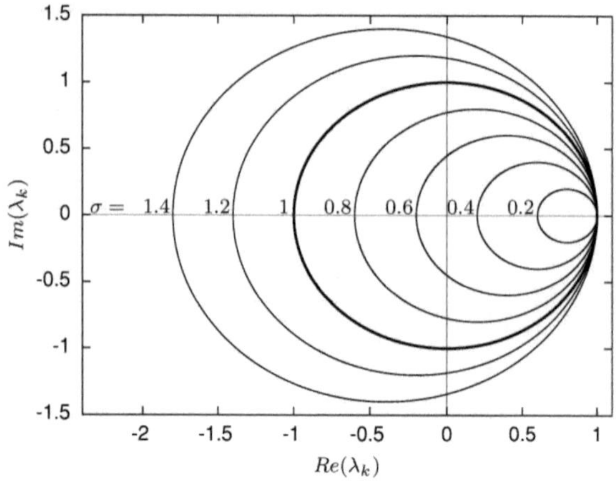

Fig. 3.2 Eigenvalue loci for different values of σ

In this case, the stability condition (3.37) involves the computation of the eigenvalues of the matrix $\mathbf{A}'$, which can be expressed in the general the form

$$\mathbf{A}' = \begin{bmatrix} c_1 & c_2 & c_3 & \cdot & c_{N-1} & c_N \\ c_N & c_1 & c_2 & c_3 & \cdot & c_{N-1} \\ c_{N-1} & c_N & c_1 & c_2 & c_3 & \cdot \\ \cdot & \cdot & \cdot & \cdot & \cdot & \cdot \\ c_3 & \cdot & c_{N-1} & c_N & c_1 & c_2 \\ c_2 & c_3 & \cdot & c_{N-1} & c_N & c_1 \end{bmatrix} \tag{3.44}$$

These eigenvalues are given by Varga [3]

$$\lambda_k = \sum_{i=0}^{N-1} c_{i+1} \exp\left(j(k-1)\frac{i2\pi}{N}\right); \qquad k = 1, \ldots, N \tag{3.45}$$

Applying this formula to Eq. (3.43), where $c_1 = 1 - \sigma$, $c_i = 0$, $i = 2, \ldots, N-1$, $c_N = \sigma$, we have

$$\begin{aligned}\lambda_k &= (1-\sigma) + \sigma e^{j(k-1)\frac{(N-1)2\pi}{N}} = (1-\sigma) + \sigma e^{j(k-1)2\pi} e^{-j(k-1)\frac{2\pi}{N}} \\ &= (1-\sigma) + \sigma e^{-j(k-1)\frac{2\pi}{N}}\end{aligned} \tag{3.46}$$

The eigenvalues are located in the complex plane on a circle of radius σ centered in $(1-\sigma, 0)$. Figure 3.2 shows these circles for $\sigma = 0.2, 0.4, \ldots, 1.4$. Clearly, the stability limit is

$$\sigma \leq 1 \tag{3.47}$$

since the eigenvalues must have a modulus strictly less than unity, in accordance with the necessary and sufficient condition of stability (3.37). Condition (3.47) is now equivalent to the results of Von Neumann's analysis.

The advantage of the matrix method is that it is possible to decouple the influences of spatial discretization and time integration. Returning to Eq. (3.43), where the time derivative is not yet discretized

$$\frac{\mathrm{d}\mathbf{x}}{\mathrm{d}t} = \mathbf{A}'\mathbf{x} \tag{3.48}$$

with

$$\mathbf{A}' = -\frac{v}{\Delta z}\begin{bmatrix} 1 & 0 & & & 1 \\ -1 & 1 & & & \\ 0 & & & & \\ & & & 1 & 0 \\ 0 & & 0 & -1 & 1 \end{bmatrix} \tag{3.49}$$

The stability of the semi-discrete system (3.48), which is continuous in time, is determined by the location of the eigenvalues λ_k of $\mathbf{A}'$, which should be such that [2]

$$Re(\lambda_k) \leq 0; \qquad \forall k \tag{3.50}$$

i.e., the eigenvalues are in the left half of the complex plane.

If $rank(\mathbf{A}') = N$, then the N eigenvectors

$$\mathbf{A}\mathbf{v}_k = \lambda_k \mathbf{v}_k \tag{3.51}$$

are linearly independent and form a basis in which the solution can be expressed as

$$\mathbf{x}(t) = \sum_{k=1}^{N} x_k(t)\mathbf{v}_k \tag{3.52}$$

where $x_k(t)$ are functions of time (yet to be determined).

If we introduce Eq. (3.52) into Eq. (3.48), we have

$$\frac{d}{\mathrm{d}t}\left[\sum_{k=1}^{N} x_k(t)\mathbf{v}_k\right] = \mathbf{A}'\sum_{k=1}^{N} x_k(t)\mathbf{v}_k = \sum_{k=1}^{N} x_k(t)\mathbf{A}'\mathbf{v}_k = \sum_{k=1}^{N} x_k(t)\lambda_k\mathbf{v}_k \tag{3.53}$$

Each mode k is therefore governed by

$$\frac{\mathrm{d}x_k}{\mathrm{d}t} = \lambda_k x_k \tag{3.54}$$

or in matrix form

$$\frac{d\mathbf{x}}{dt} = \Gamma \mathbf{x} \tag{3.55}$$

where $\mathbf{x}(t) = [x_1(t), x_2(t), \ldots, x_N(t)]^T$ and $\Gamma = diag(\lambda_1 \lambda_2 \ldots \lambda_N)$ is the diagonal matrix of eigenvalues.

The ODE system (3.48) can be integrated in time using a wide range of solvers, some of which have been reviewed in Chaps. 1 and 2. These solvers can be classified as single step (e.g., the Runge-Kutta methods) or multistep integrators (e.g., the Backward Differentiation Formulas integrators). Figure 1.6 shows the stability diagrams for the RK methods of orders 1–4, whereas Fig. 1.10 shows the stability diagrams for BDF methods of orders 1–6. These diagrams constrain the values of $\lambda_k \Delta t$ to lie in specific regions of the complex plane.

This latter analysis method is quite flexible, as it allows various space discretization schemes and time integration methods to be considered. Such method is now applied to the linear advection equation, for which we consider several alternative spatial discretization and time integration methods.

3.4 Numerical Study of the Advection Equation

We consider again Eq. (3.10):

$$x_t = -v x_z \tag{3.56}$$

with $0 \leq z \leq 1$, $t \geq 0$ and $v = 1$. The Dirichlet BC is given by:

$$x(z_0, t) = 0 \tag{3.57}$$

and the IC is taken as a wave of the form:

$$x(z, 0) = \frac{5.7}{\exp(\theta) + \exp(-\theta)}; \qquad \theta = 100(z - 0.25) \tag{3.58}$$

We first use a fixed-step explicit Euler integrator in combination with the alternative FD schemes:

$$\text{3-point cenetred:} \qquad (x_z)_t = \frac{x_{i+1} - x_{i-1}}{2\Delta z} \tag{3.59}$$

$$\text{2-point downwind:} \qquad (x_z)_t = \frac{x_{i+1} - x_i}{\Delta z} \tag{3.60}$$

$$\text{2-point upwind:} \qquad (x_z)_t = \frac{x_i - x_{i-1}}{\Delta z} \tag{3.61}$$

Note that Eq. (3.60) is a *downwind* scheme since x is flowing from left to right ($v = 1$ is positive) and the point z_{i+1} is situated to the right of the point z_i, where the derivative is evaluated. In contrast, Eq. (3.61) is an *upwind* scheme since it makes use of information at $z = z_{i-1}$ (i.e. x_{i-1}), to the left of the point z_i.

The application of these schemes leads to the following semi-discrete systems (where periodic boundary conditions are considered):

$$\text{3-point centered:} \quad \begin{bmatrix} x_1 \\ x_2 \\ x_3 \\ \cdots \\ x_N \end{bmatrix}_t = \frac{v}{2\Delta z} \begin{bmatrix} 0 & 1 & & & -1 \\ -1 & 0 & 1 & & \\ & -1 & 0 & 1 & \\ & & & \cdots & \\ 1 & & & -1 & 0 \end{bmatrix} \begin{bmatrix} x_1 \\ x_2 \\ x_3 \\ \cdots \\ x_N \end{bmatrix} \tag{3.62}$$

$$\text{2-point downwind:} \quad \begin{bmatrix} x_1 \\ x_2 \\ x_3 \\ \cdots \\ x_N \end{bmatrix}_t = \frac{v}{2\Delta z} \begin{bmatrix} -1 & 1 & & & \\ & -1 & 1 & & \\ & & -1 & 1 & \\ & & & \cdots & \\ 1 & & & & -1 \end{bmatrix} \begin{bmatrix} x_1 \\ x_2 \\ x_3 \\ \cdots \\ x_N \end{bmatrix} \tag{3.63}$$

$$\text{2-point upwind:} \quad \begin{bmatrix} x_1 \\ x_2 \\ x_3 \\ \cdots \\ x_N \end{bmatrix}_t = \frac{v}{2\Delta z} \begin{bmatrix} 1 & & & & -1 \\ -1 & 1 & & & \\ & -1 & 1 & & \\ & & & \cdots & \\ & & & 1 & 1 \end{bmatrix} \begin{bmatrix} x_1 \\ x_2 \\ x_3 \\ \cdots \\ x_N \end{bmatrix} \tag{3.64}$$

These matrices are in the form (3.44) and their eigenvalues can be computed using (3.45), i.e.,

- 3-point centered.

$$\lambda_k = -j\frac{v}{\Delta z}\sin\left[(k-1)\frac{2\pi}{N}\right] \tag{3.65}$$

- 2-point downwind:

$$\lambda_k = \frac{v}{\Delta z}\left(1 - \cos\left[(k-1)\frac{2\pi}{N}\right]\right) - j\frac{v}{\Delta z}\sin\left[(k-1)\frac{2\pi}{N}\right] \tag{3.66}$$

- 2-point upwind:

$$\lambda_k = \frac{v}{\Delta z}\left(\cos\left[(k-1)\frac{2\pi}{N}\right] - 1\right) - j\frac{v}{\Delta z}\sin\left[(k-1)\frac{2\pi}{N}\right] \tag{3.67}$$

with $k = 1, \ldots, N$. Figure 3.3 shows the location of these eigenvalues in the complex plane. On inspection of this figure, it is apparent that a 2-point downwind scheme leads to an unstable semi-discrete system (indeed the eigenvalues of Eq. (3.66) are in the right half-plane), whereas a centered scheme yields a purely oscillatory system (the eigenvalues are on the imaginary axis). Only a 2-point upwind scheme yields a (marginally) stable system (the system is said to be "marginally" stable as

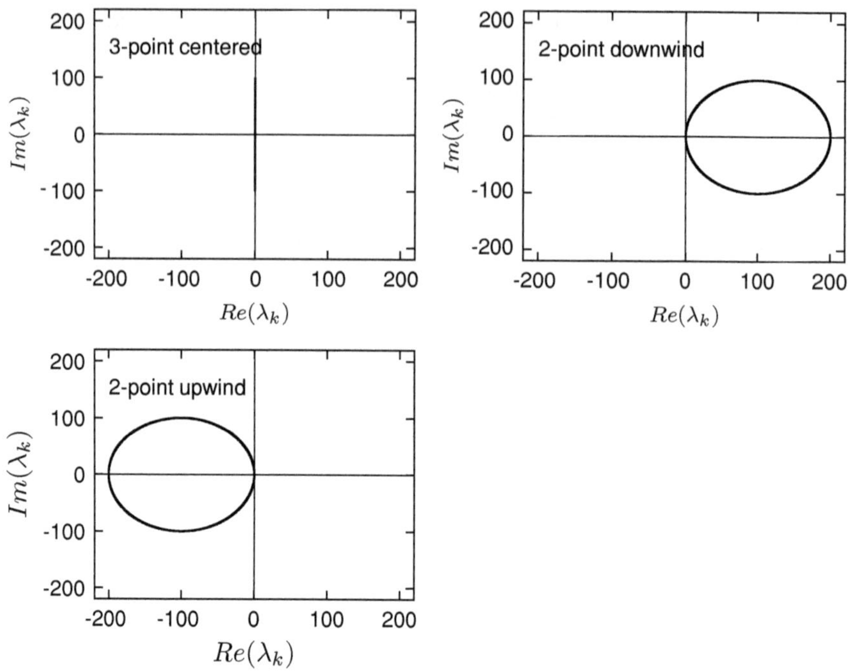

Fig. 3.3 Advection equation—Eigenvalue loci for alternative FD schemes

it has a zero-eigenvalue). In order to the scheme to be asymptotically stable all the eigenvalues must have negative real parts.

How does an explicit Euler algorithm perform on these semi-discrete systems? We know from Chap. 1 that $\lambda_k \Delta t$ must lie within a unit circle centered in $(-1, 0)$ (see Eq. (1.24) and Fig. 1.6). Obviously, this will work only for the 2-point upwind scheme, for which we have to make sure that the points $\lambda_k \Delta t$ (which are located on a circle as shown in Fig. 3.3) are inside the stability region of the explicit Euler algorithm.

To show that this is feasible, we have to manipulate Eq. (3.67). Let us define

$$\alpha = (k-1)\frac{2\pi}{N}; \qquad \sigma = \frac{v\Delta t}{\Delta z} \tag{3.68}$$

so that

$$\lambda_k \Delta t = \sigma(\cos\alpha - 1) - j\sigma \sin\alpha = x + jy \tag{3.69}$$

If we eliminate α from the two equations

$$\begin{aligned} x &= \sigma(\cos\alpha - 1) \\ y &= -\sigma \sin\alpha \end{aligned} \tag{3.70}$$

we get

$$\left(\frac{x+\sigma}{\sigma}\right)^2 + \left(\frac{y}{\sigma}\right)^2 = \cos^2\alpha + \sin^2\alpha = 1 \tag{3.71}$$

which is a circle of radius σ centered at $(-\sigma, 0)$. This circle is included inside the stability region of the explicit Euler algorithm if

$$\sigma = \frac{v\Delta t}{\Delta z} \leq 1 \tag{3.72}$$

which is the CFL condition that we derived previously.

Now that we know the conditions under which we have a stable combination of spatial discretization and time integration, we are in the situation where we can actually solve Eq. (3.64). Figure 3.4 shows some results for $\sigma = \{0.95, 1, 1.05\}$. Several interesting observations can be made:

- For $\sigma \leq 1$, the solution is stable, whereas for $\sigma > 1$, the solution becomes unstable (this is a numerical check of (3.72)).
- For $\sigma = 1$, the numerical results are excellent. The solution is traveling from left to right at a velocity $v = 1$, preserving the shape of the initial condition. This is the expected result since the advection equation has a known analytic solution in the form

$$x(z, t) = x^0(z - vt) \tag{3.73}$$

  starting from the initial condition $x^0(z) = x(z, 0)$ (i.e., the convection term $-vx_z$ moves the initial profile at a constant velocity v).
- For $\sigma < 1$, the solution is stable but not accurate since we observe an important *attenuation* of the wave as it travels along the spatial domain.

Therefore, $\sigma = 1$ appears as a critical value. Where does this come from? The numerical scheme is given by

$$\frac{x_i^{k+1} - x_i^k}{\Delta t} = -v\frac{x_i^k - x_{i-1}^k}{\Delta z}$$

or

$$x_i^{k+1} = x_i^k - v\frac{\Delta t}{\Delta z}(x_i^k - x_{i-1}^k) = (1-\sigma)x_i^k + \sigma x_{i-1}^k \tag{3.74}$$

which, with $\sigma = 1$, corresponds to

$$x_i^{k+1} = x_{i-1}^k \tag{3.75}$$

This latter expression is called the *shift condition*, which exactly satisfies Eq. (3.73).

In summary, $\sigma = 1$ is a very fortuitous situation where the numerical solution is stable and exact. Any other value of $\sigma < 1$ yields a stable but relatively inaccurate

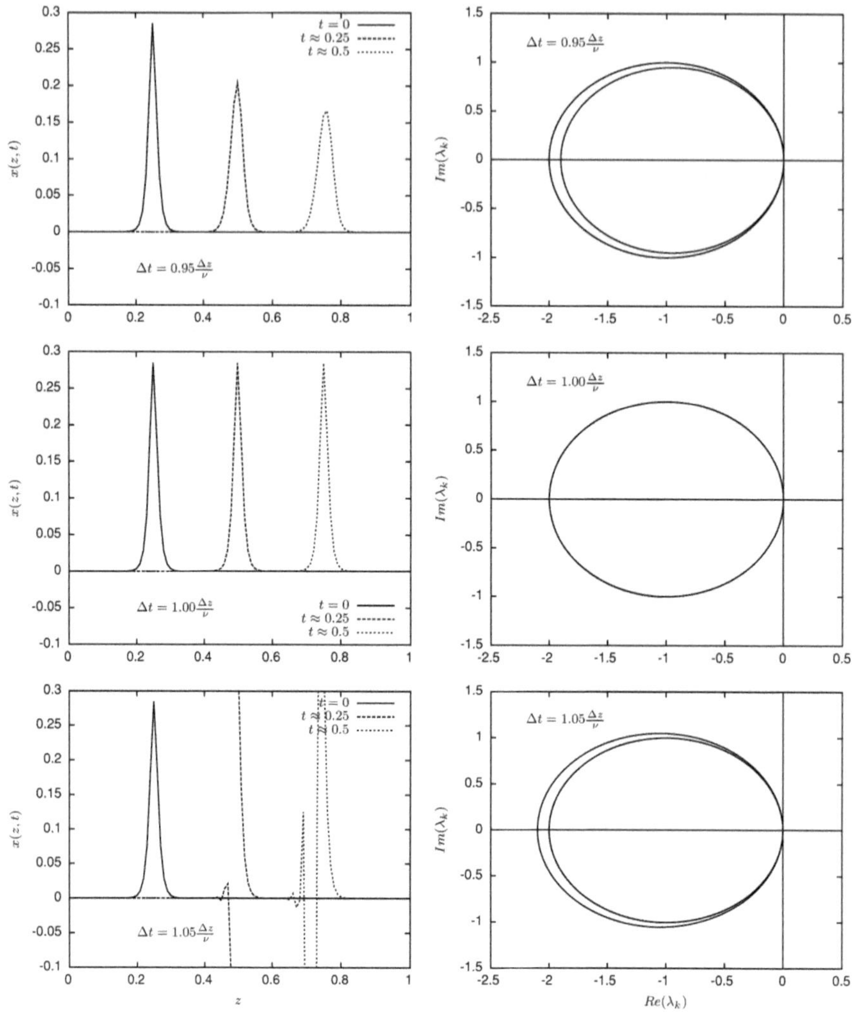

Fig. 3.4 Numerical solution of the initial boundary value problem (3.56)–(3.58) using 2-point upwind FDs and an explicit Euler method. The plotted times are $t = 0, 0.25, 0.5$

solution. Also, the use of higher-order FDs (which are in principle more accurate and will be detailed in the continuation of this chapter) will be less effective because the shift condition will generally not be satisfied for these schemes.

We now consider another time integrator, e.g. the leap-frog method introduced in Chap. 1 (see Eq. (1.19)). We know that the stability region of this method is a portion of the imaginary axis between the points $\pm j$ (see Fig. 1.7). The leap-frog integrator is therefore a good candidate to solve the semi-discrete system (3.62) which has purely imaginary eigenvalues (see Fig. 3.3). The eigenvalues (3.65) will be such that

$$-j \leq \lambda_k \Delta t = -j\sigma \sin\alpha \leq j \tag{3.76}$$

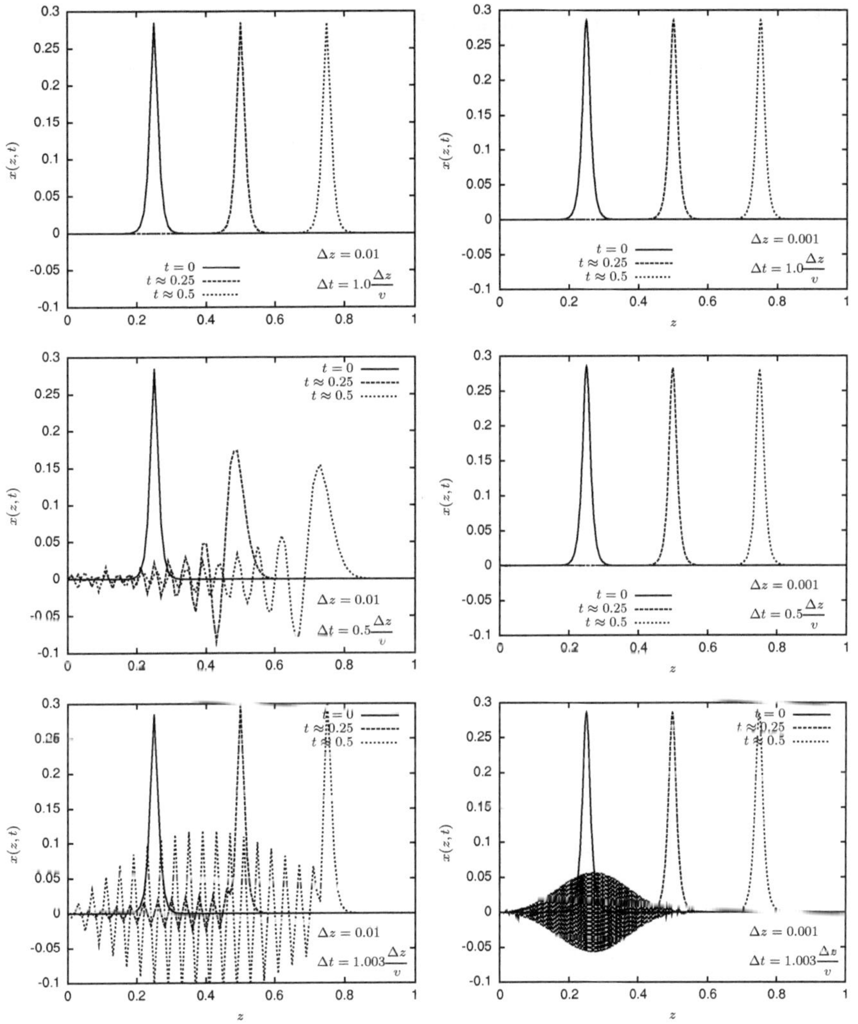

Fig. 3.5 Numerical solution of the initial boundary value problem (3.56–3.58) using 3-point centered FDs and a leap-frog method

if

$$\sigma = \frac{v\Delta t}{\Delta z} \leq 1 \tag{3.77}$$

which is again the CFL condition.

Figure 3.5 shows various numerical results obtained with the combination: “3-point centered FDs + leap-frog integrator”.

The following observations can be made:

- For $\sigma = 1$, the numerical results are excellent, even though the *shift condition* is not satisfied in this case.
- For $\sigma < 1$, the solution is stable but not accurate when using $\Delta z = 0.01$ (spurious oscillations appear). However, accuracy can be significantly improved by reducing Δz to 0.001. Logically, the accuracy depends on the fineness of the spatial grid: with a 3-point centered scheme, the truncation error is proportional to Δz^2, i.e. it is reduced by a factor 100 when Δz decreases from 0.01 to 0.001.
- When $\sigma > 1$, the solution very quickly becomes unstable.

These examples show that a combination "spatial discretization + time integrator" must be carefully selected. This combination of course depends on the equation to be solved, so that it is difficult to draw general conclusions and to suggest universal methods at this stage. In the following, we consider additional examples, and try however to highlight "easy" choices.

3.5 Numerical Study of the Advection-Diffusion Equation

We now consider a PDE combining advection (plug-flow motion) and a diffusion term, involving a second-order spatial derivative

$$x_t = -vx_z + Dx_{zz} \tag{3.78}$$

with $0 \leq z \leq 1, t \geq 0$.

This PDE is supplemented with two BCs of the Dirichlet-type (two BCs are required as the PDE is second-order in space)

$$x(z_0 = 0, t) = 0; \qquad x(z_N = 1, t) = 0 \tag{3.79}$$

Again, the IC is a wave in the form

$$x(z, 0) = \frac{5.7}{\exp(\theta) + \exp(-\theta)}; \qquad \theta = 164.14(z - 0.25) \tag{3.80}$$

We consider an approximation of the first-order derivative (advection term) using a 2-point upwind scheme and an approximation of the second-order derivative (diffusion term) using a 3-point centered scheme, i.e.

$$(x_z)_i = \frac{x_i - x_{i-1}}{\Delta z} \tag{3.81}$$

$$(x_{zz})_i = \frac{x_{i+1} - 2x_i + x_{i-1}}{\Delta z^2} \tag{3.82}$$

The semi-discrete ODE system then takes the form

$$\begin{bmatrix} x_1 \\ \cdots \\ \cdots \\ x_N \end{bmatrix}_t = -\frac{v}{\Delta z}\begin{bmatrix} 1 & & & & -1 \\ -1 & 1 & & & \\ & -1 & 1 & & \\ & & & \cdots & \\ & & & -1 & 1 \end{bmatrix}\begin{bmatrix} x_1 \\ x_2 \\ x_3 \\ \cdots \\ x_N \end{bmatrix} + \frac{D}{\Delta z^2}\begin{bmatrix} -2 & 1 & & & 1 \\ 1 & -2 & 1 & & \\ & 1 & -2 & 1 & \\ & & & \cdots & \\ 1 & & & 1 & -2 \end{bmatrix}\begin{bmatrix} x_1 \\ x_2 \\ x_3 \\ \cdots \\ x_N \end{bmatrix} \tag{3.83}$$

The eigenvalue spectrum of the resulting discretization matrix (sum of the two matrices on the RHS) can be computed using formula (3.45), which gives

$$\lambda_k = -\left(\frac{v}{\Delta z} + \frac{2D}{\Delta z^2}\right) + \frac{D}{\Delta z^2}\exp\left(j(k-1)\frac{2\pi}{N}\right) + \left(\frac{v}{\Delta z} + \frac{D}{\Delta z^2}\right)\exp\left(j(k-1)\frac{(N-1)2\pi}{N}\right) \tag{3.84}$$

or

$$\lambda_k = \left(\frac{v}{\Delta z} + \frac{2D}{\Delta z^2}\right)\left(\cos\left(\frac{k-1}{N}2\pi\right) - 1\right) - j\frac{v}{\Delta z}\sin\left(\frac{k-1}{N}2\pi\right) \tag{3.85}$$

These eigenvalues have negative real parts, for all positive values of v, D and Δz. In order to select an appropriate time integrator for solving (3.83), we first determine the $\lambda_k \Delta t$-locus in the complex plane. To this end, we use the same notation as in (3.68)

$$\alpha = (k-1)\frac{2\pi}{N}; \qquad \sigma = \frac{v\Delta t}{\Delta z} \tag{3.86}$$

and, in addition,

$$\beta = \frac{D\Delta t}{\Delta z^2} \tag{3.87}$$

so that

$$\lambda_k \Delta t = (\sigma + 2\beta)(\cos\alpha - 1) - j\sigma\sin\alpha = x + jy \tag{3.88}$$

The $\lambda_k \Delta t$-locus is therefore an ellipsis centered at $(-\sigma - 2\beta, 0)$ given by

$$\left(\frac{x + \sigma + 2\beta}{\sigma + 2\beta}\right)^2 + \left(\frac{y}{\sigma}\right)^2 = 1 \tag{3.89}$$

If we decide to use an explicit Euler method, the ellipsis (3.89) must be completely included inside the circle centered at

$$(x+1)^2 + y^2 = 1 \tag{3.90}$$

for the solution to be stable. This implies that the half axis of the ellipsis (3.89) must be smaller than the radius of the circle (3.90)

$$\sigma + 2\beta < 1; \qquad \sigma < 1 \tag{3.91}$$

Since σ and β are positive, the more severe condition is the first, which in terms of Δt and Δz, can be written as

$$\Delta t \leq \Delta t_{\max} = \frac{1}{\dfrac{v}{\Delta z} + \dfrac{2D}{\Delta z^2}} \tag{3.92}$$

Rigorously, we should also check that the ellipsis and the circle have only one intersection at (0, 0). In fact, this is the case if (3.92) is satisfied, and we skip the details of these algebraic manipulations.

We now check the validity of the stability condition (3.92) in solving the IBVP (3.78)–(3.80), using

$$v = 1; \qquad D = 0.005 \tag{3.93}$$

and different values of Δz, i.e., 0.01, 0.001, and 0.0001. On examination of the numerical solutions, $\Delta z = 0.001$ seems to be a good choice, and Fig. 3.6 shows three solutions obtained respectively with $\Delta t = \Delta t_{\max}$, $\Delta t = 0.1\Delta t_{\max}$ and $\Delta t = 1.003\Delta t_{\max}$. From this figure, it is apparent that:

- Δt may vary largely within the stability bounds, without much effect on the solution.
- As soon as Δt exceeds the predicted stability limit, the solution becomes unstable.

Using an upwind scheme for the first-order derivative in combination with the explicit Euler method is probably the most natural choice (and we definitely recommend it). However, we would like to stress that there are no general rules, and that a solution is also possible using a 2-point downwind scheme

$$(x_z)_t = \frac{x_{i+1} - x_i}{\Delta z} \tag{3.94}$$

For the pure advection equation (studied in the previous section), this latter scheme yields a semi-discrete equation system whose solution is unstable since the eigenvalues of the matrix of the semi-discrete system are in the right half-plane. Here, the influence of the diffusion term will be beneficial, as the following reasoning shows.

The semi-discrete equation system can be written as

$$\begin{bmatrix} x_1 \\ \cdots \\ \cdots \\ x_N \end{bmatrix}_t = -\frac{v}{\Delta z} \begin{bmatrix} -1 & 1 & & & \\ & -1 & 1 & & \\ & & -1 & 1 & \\ & & & \cdots & \\ 1 & & & & -1 \end{bmatrix} \begin{bmatrix} x_1 \\ x_2 \\ x_3 \\ \cdots \\ x_N \end{bmatrix}$$

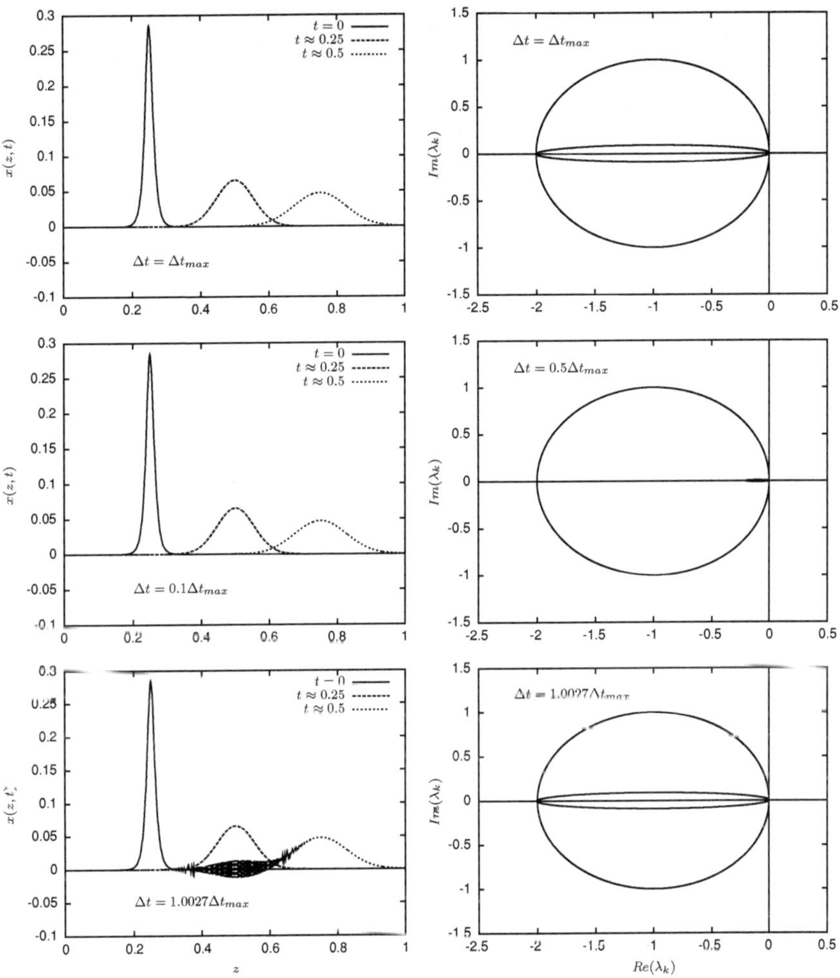

Fig. 3.6 Numerical solution of the initial boundary value problem (3.78–3.80) using 2-point upwind FDs for the first-order derivative (advection term), 3-point centered FDs for the second order derivative (diffusion term) and an explicit Euler method

$$+\frac{D}{\Delta z^2}\begin{bmatrix} -2 & 1 & & & 1 \\ 1 & -2 & 1 & & \\ & 1 & -2 & 1 & \\ & & & \cdots & \\ 1 & & & 1 & -2 \end{bmatrix}\begin{bmatrix} x_1 \\ x_2 \\ x_3 \\ \cdots \\ x_N \end{bmatrix} \tag{3.95}$$

Again, formula (3.45) can be used to compute the eigenvalues of the spatial discretization matrix (sum of the two matrices on the RHS)

$$\lambda_k = \left(\frac{v}{\Delta z} - \frac{2D}{\Delta z^2}\right) + \left(-\frac{v}{\Delta z} + \frac{D}{\Delta z^2}\right) \exp\left(j(k-1)\frac{2\pi}{N}\right)$$
$$+ \frac{D}{\Delta z^2} \exp\left(j(k-1)\frac{(N-1)2\pi}{N}\right) \tag{3.96}$$

or, with the notation (3.86)

$$\lambda_k = \left(\frac{2D}{\Delta z^2} - \frac{v}{\Delta z}\right)(\cos\alpha - 1) - j\frac{v}{\Delta z}\sin\alpha \tag{3.97}$$

For stability, it is therefore required that (eigenvalues must have negative real parts)

$$\frac{2D}{\Delta z^2} - \frac{v}{\Delta z} \geq 0 \tag{3.98}$$

This imposes a condition on spatial discretization

$$\Delta z \leq \frac{2D}{v} \tag{3.99}$$

Note that, when using a 2-point upwind scheme, the eigenvalues (3.85) have negative real parts, for all positive values of v, D and Δz. Here, on the contrary, the spatial grid must be fine enough to retain stability.

Assuming that condition (3.68) is satisfied, the $\lambda_k \Delta t$-locus must lie inside the circle (3.90). Based on the same algebraic manipulations as before, it is possible to show that this locus is again an ellipsis of the form

$$\left(\frac{x + 2\beta - \sigma}{2\beta - \sigma}\right)^2 + \left(\frac{y}{\sigma}\right)^2 = 1 \tag{3.100}$$

For this ellipsis to be included in the circle (3.90), several conditions have to be achieved:

- The centre of ellipsis $(-2\beta + \sigma, 0)$ must lie in the left half-plane

$$2\beta - \sigma > 0 \tag{3.101}$$

- The two half axis must be less than 1

$$2\beta - \sigma < 1; \qquad \sigma < 1 \tag{3.102}$$

  or (combining with condition (3.101))

$$0 < 2\beta - \sigma < 1; \qquad 0 < \sigma < 1 \tag{3.103}$$

or

$$\begin{aligned} 0 &< \sigma < 1 \\ \frac{\sigma}{2} &< \beta < \frac{1+\sigma}{2} \end{aligned} \tag{3.104}$$

- The only common point between the ellipsis and the circle must be (0, 0). Some algebraic manipulations show that conditions (3.104) are then restricted to

$$\begin{aligned} 0 &< \sigma < 1 \\ \frac{\sigma+\sigma^2}{2} &< \beta < \frac{1+\sigma}{2} \end{aligned} \tag{3.105}$$

These conditions cannot be directly expressed in terms of admissible values of Δz and Δt, but we can illustrate a particular situation. Assume arbitrarily that $\sigma = 0.5$. The second condition then gives $\frac{3}{8} < \beta < \frac{3}{4}$. In terms of Δz and Δt, this can be translated into

$$\Delta t = \frac{\Delta z}{2v}; \qquad \frac{2D}{3v} < \Delta z < \frac{4D}{3v}. \tag{3.106}$$

Figure 3.7 shows the $\lambda_k \Delta t$-loci for $\frac{2D}{3v}, \frac{D}{v}, \frac{4D}{3v}$ and the corresponding simulation results. These three cases are stable, even though the quality (accuracy) is questionable. Accuracy is in fact related to the value of Δz, whose smallest value corresponds to case 1, i.e. $\Delta z = \frac{D}{3v}$. With $v = 1$ and $D = 0.005$, the grid spacing is $\Delta z = 0.00333$. However, we have seen previously that the largest possible value (for achieving satisfactory accuracy) is $\Delta z = 0.001$.

To ensure satisfactory accuracy, we will impose Δz and use conditions (3.105) to determine the admissible values of Δt. The first condition can be written as

$$\Delta t < \frac{v}{\Delta z} \tag{3.107}$$

the second as

$$\frac{\sigma+\sigma^2}{2} < \beta \Longleftrightarrow \frac{v\Delta t}{\Delta z} + \left(\frac{v\Delta t}{\Delta z}\right)^2 < 2\frac{D\Delta t}{\Delta z^2} \Longleftrightarrow \Delta t < \frac{2D}{v^2} - \frac{\Delta z}{v} \tag{3.108}$$

and the third as

$$\beta < \frac{1+\sigma}{2} \Longleftrightarrow \frac{D\Delta t}{\Delta z^2} < \frac{1}{2}\left(1 + \frac{v\Delta t}{\Delta z}\right) \Longleftrightarrow \Delta t < \frac{1}{\frac{2D}{v^2} - \frac{v}{\Delta z}} \tag{3.109}$$

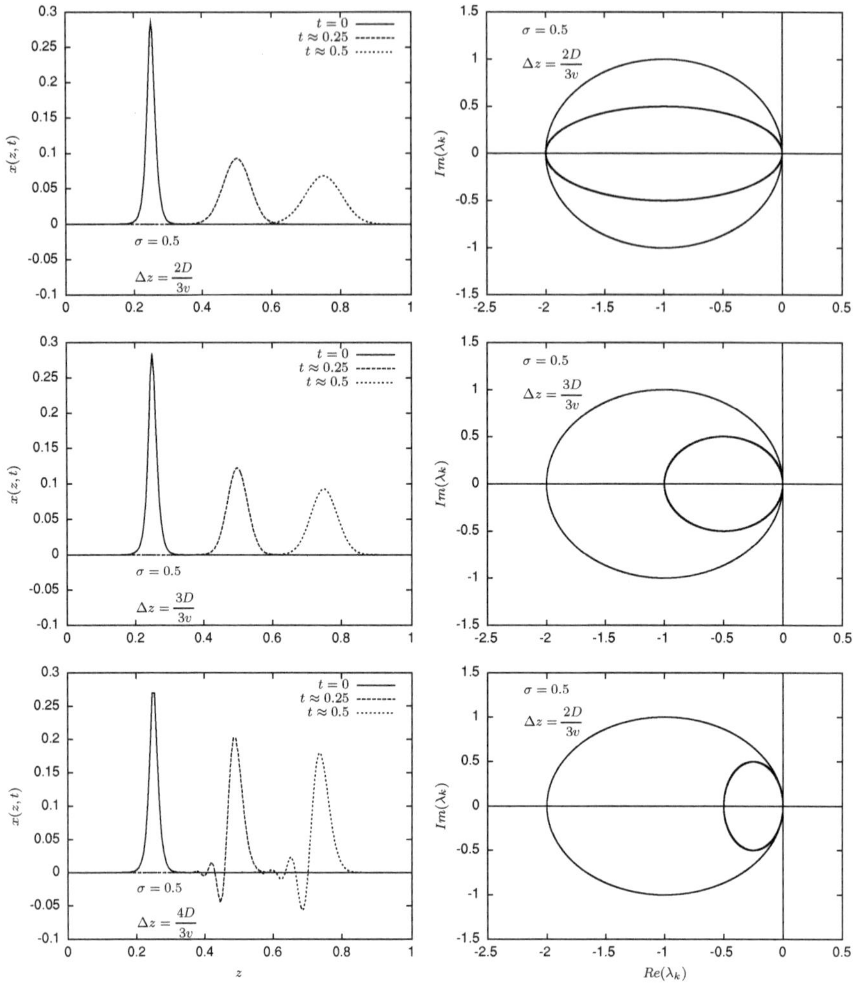

Fig. 3.7 Numerical solution of the initial boundary value problem (3.78–3.80) using 2-point downwind FDs for the first-order derivative (advection term), 3-point centered FDs for the second order derivative (diffusion term) and an explicit Euler method

The maximum admissible time-step size Δt_{max} is therefore given by

$$\Delta t_{\text{max}} = \min\left(\frac{v}{\Delta z}, \frac{2D}{v^2} - \frac{\Delta z}{v}, \frac{1}{\dfrac{2D}{v^2} - \dfrac{v}{\Delta z}}\right) \tag{3.110}$$

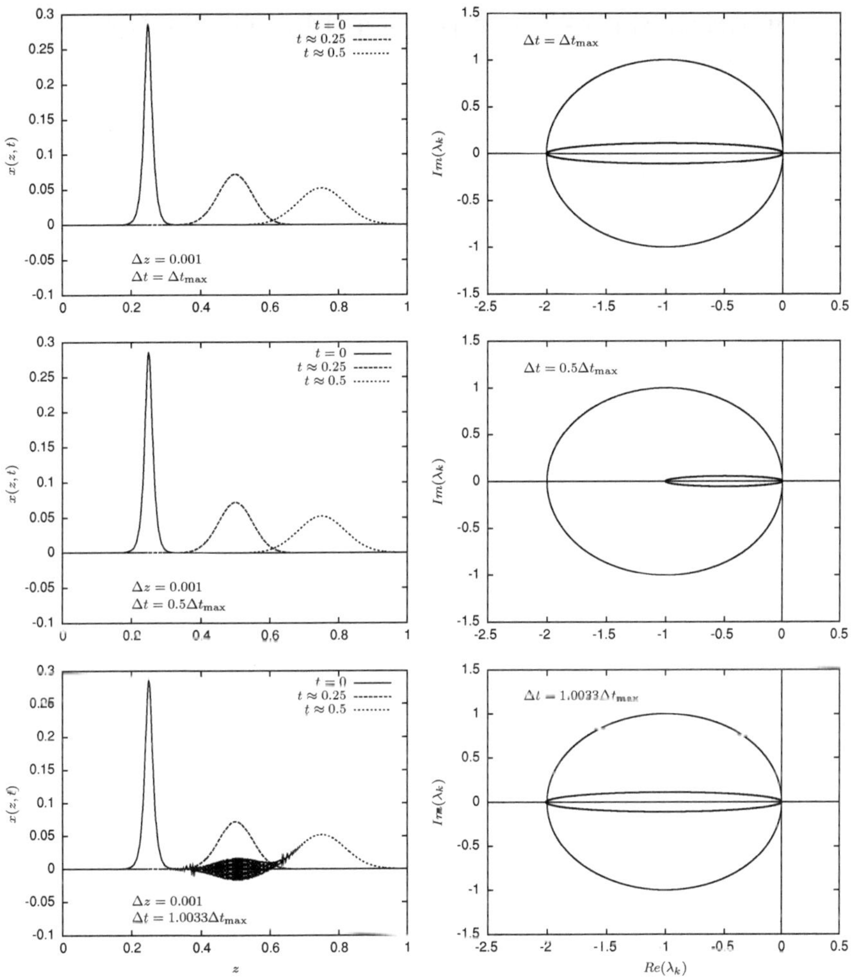

Fig. 3.8 Numerical solution of the initial boundary value problem (3.78–3.80) using 2-point downwind FDs for the first-order derivative (advection term), 3-point centered FDs for the second order derivative (diffusion term) and an explicit Euler method

For the numerical values considered in the example ($v = 1$, $D = 0.005$ and $\Delta z = 0.001$), $\Delta t_{max} = \dfrac{10^{-3}}{9}$. Figure 3.8 shows three different situations, e.g. $\Delta t = \Delta t_{max}$, $\Delta t = 0.5\Delta t_{max}$ and $\Delta t = 1.0035\Delta t_{max}$. We can note the following:

- For $\Delta t \leq \Delta t_{max}$, the numerical simulation is very satisfactory (and is comparable in quality with the results obtained with an upwind FD scheme, as presented in Fig. 3.6).
- The numerical scheme becomes unstable as soon as Δt exceeds Δt_{max}.

3.6 Numerical Study of the Advection-Diffusion-Reaction Equation

In addition to advection and diffusion, we now consider a source term $r(x)$, which could for instance model a chemical reaction rate

$$x_t = -vx_z + Dx_{zz} + r(x) \tag{3.111}$$

with $0 \leq z \leq 1, t \geq 0$.

The simplest such model would be a linear reaction rate of the form $r(x) = \alpha x$, yielding the PDE

$$x_t = -vx_z + Dx_{zz} + \alpha x \tag{3.112}$$

If we use the same spatial discretization schemes as for the advection-diffusion equation, i.e., Eqs. (3.81)–(3.82), the eigenvalues of the semi-discrete system are given by

$$\lambda_k = \alpha + \left(\frac{v}{\Delta z} + \frac{2D}{\Delta z^2}\right)\left(\cos\left(\frac{k-1}{N}2\pi\right) - 1\right) - j\frac{v}{\Delta z}\sin\left(\frac{k-1}{N}2\pi\right) \tag{3.113}$$

Compared to Eq. (3.85), this expression shows that the eigenvalues are shifted to the left or to the right by α (depending on the sign of α). If α is positive (i.e., if the quantity x is produced in the chemical reaction), then part of the eigenvalues have positive real parts so that the system is unstable (whatever the value of Δt). On the other hand, if α is negative (i.e., if the quantity x is consumed in the chemical reaction), the eigenvalue spectrum is shifted to left in the left half plane so that the system is stable. However, the constraints on Δt might be more stringent than for the advection-diffusion problem as the following reasoning shows.

In the previous section, the study of the stability conditions for the advection-diffusion equation showed that the $\lambda_k \Delta t$-locus is an ellipsis given by Eq. (3.100). Specifically, the stability condition (3.103) says that the half horizontal axis $2\beta - \sigma$ of the ellipsis must be smaller than the radius of the unit circle (so as to ensure that the ellipsis lies inside the unit circle). The consideration of an additional reaction term leads to essentially the same conclusions. After some (easy but technical) manipulations, it is possible to show that the $\lambda_k \Delta t$-locus is again an ellipsis

$$\left(\frac{x + \sigma + 2\beta - \alpha \Delta t}{\sigma + 2\beta}\right)^2 + \left(\frac{y}{\sigma}\right)^2 = 1 \tag{3.114}$$

which intersects the real axis in two points

$$\begin{aligned} p_1 &= \alpha \Delta t \\ p_2 &= \alpha \Delta t - 2\sigma - 4\beta \end{aligned} \tag{3.115}$$

These two abscissa are negative when α is negative and $p_2 < p_1$, so that the stability condition is

$$p_2 = \alpha \Delta t - 2\sigma - 4\beta \geq -2 \tag{3.116}$$

to ensure that the ellipsis lies in the unit circle. This can be rewritten as

$$\Delta t \leq \Delta t_{\max} = \frac{1}{\dfrac{v}{\Delta z} + \dfrac{2D}{\Delta z^2} - \dfrac{\alpha}{2}} \tag{3.117}$$

which is indeed smaller than the maximum time step size predicted by Eq. (3.92) when α is negative. This result is confirmed by simulation in Fig. 3.9, which repeats the tests of Fig. 3.6 with an additional reaction term corresponding to $\alpha = -5$.

3.7 Is it Possible to Enhance Stability?

As the previous examples show, both spatial discretization and time integration influence the stability of the numerical computation scheme.

The selection of a spatial discretization scheme influences the stability of the semi-discrete ODE system. When considering the advection-diffusion example, we saw that it is possible to use an upwind or a downwind FD scheme for approximating the first-order (advection) term. However, the use of an upwind scheme leads to an unconditionally stable semi-discrete ODE system, whereas a downwind scheme leads to a more restrictive stability condition (3.99). As mentioned before, it is difficult to draw general conclusions from a few examples concerning the choice of a spatial discretization scheme, but a few *practical guidelines* can be proposed based on experience:

- First-order spatial derivatives (representing advection or convection) are usually best approximated using upwind FD schemes (physically, useful information on the solution is "sensed" in the upwind direction, from where the solution "flows").
- Second-order derivatives (representing diffusion or dispersion) are usually best approximated using centered FD schemes (physically, diffusion has no preferential direction, and information is "sensed" in both directions).

Once spatial discretization has been carried out and a stable semi-discrete ODE system has been obtained, a sufficiently small time-step size Δt must be selected in order to remain inside the stability region imposed by the time integrator. This is particularly true when using explicit ODE integrators, as in the previous examples where we used an explicit Euler method. However, we have seen in Chap. 1 that implicit time integrators have much larger stability regions, and it can be very advantageous to use one.

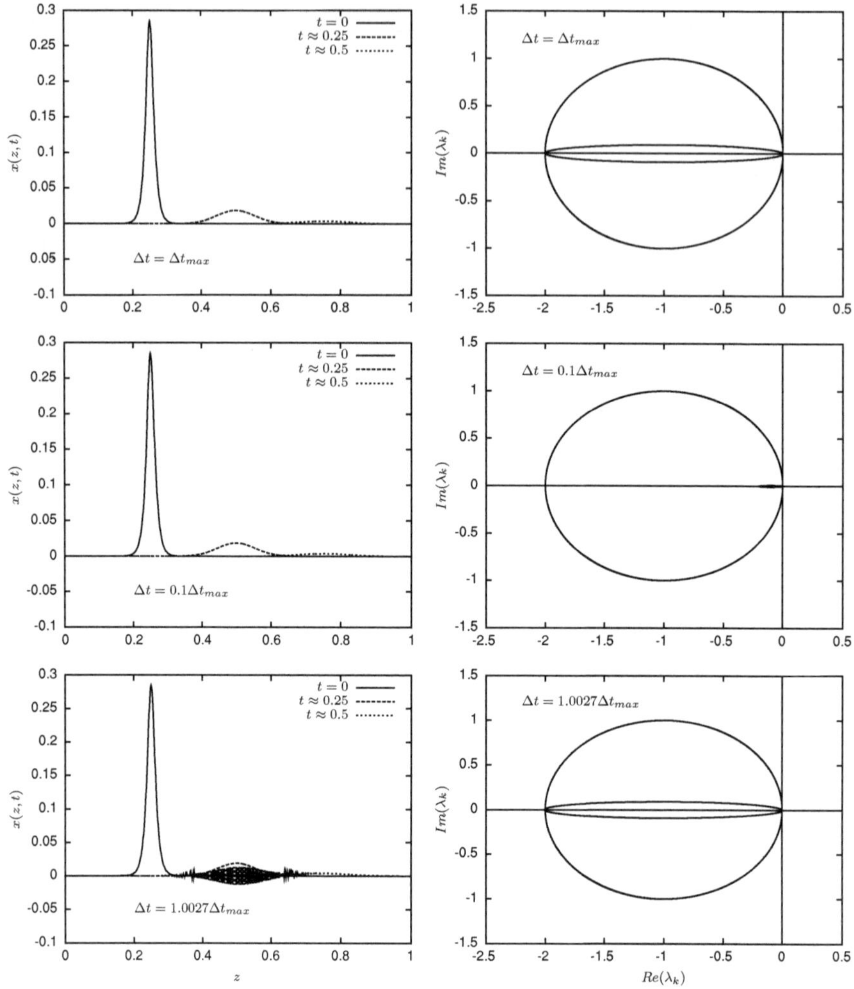

Fig. 3.9 Numerical solution of the initial boundary value problem (3.111, 3.79–3.80) using 2-point upwind FDs for the first-order derivative (advection term), 3-point centered FDs for the second order derivative (diffusion term) and an explicit Euler method

To illustrate this latter statement, we return to the advection-diffusion problem (3.78)–(3.80), that we approximate using the FD schemes (3.81–3.82)

$$(x_z)_i = \frac{x_i - x_{i-1}}{\Delta z} \tag{3.118}$$

$$(x_{zz})_i = \frac{x_{i+1} - 2x_i + x_{i-1}}{\Delta z^2} \tag{3.119}$$

i.e., we use an upwind FD scheme for the advection term.

However, instead of using an explicit Euler method, we now apply an implicit Euler method to the semi-discrete ODE system

$$\frac{x_i^{k+1} - x_i^k}{\Delta t} = -v(x_z)_i^{k+1} + D(x_{zz})_i^{k+1} \tag{3.120}$$

As discussed in Chap. 1, the implicit Euler method is unconditionally stable, i.e., there is no constraint on Δt due to stability. In fact, the only constraint on Δt is due to accuracy, i.e., the quality of the approximation of the temporal derivative is determined by $\dfrac{x_i^{k+1} - x_i^k}{\Delta t}$.

These conclusions are illustrated in Fig. 3.10, which shows that satisfactory numerical results are obtained with $\Delta t = \Delta t_{\text{exp, max}}$ (where $\Delta t_{\text{exp, max}}$ is the maximum time step size of the explicit Euler method), $\Delta t = 10\Delta t_{\text{exp, max}}$, and $\Delta t = 100\Delta t_{\text{exp, max}}$.

In this latter case, the numerical accuracy slightly deteriorates due to the very large time step size.

The use of higher-order implicit time integrators, such as the BDF methods of order 1–6 reviewed in Chap. 1, can be beneficial to accuracy but can lead to stability problems if the eigenvalues of the semi-discrete system lie close to the imaginary axis. For example, the semi-discrete system (3.62) has purely imaginary eigenvalues (see Fig. 3.3), and BDF methods of order larger than 2 will be unsuitable to solve this problem as they are not unconditionally stable in the entire left half plane (see Fig. 1.10 and the comments thereafter).

3.8 Stiffness

Up to now, we have mostly considered stability as a "yes" or "no" answer. There is an additional system characteristic, called *stiffness*, related to the presence of very different time scales, which can make time integration particularly delicate. For instance, a system response to an external perturbation can display small-scales features occurring on periods of a few seconds, and overall transients dominated by large time constants of a few hours. In this case, the limited stability regions of explicit time integrators will impose important restrictions on the time step size. In fact, large eigenvalues (corresponding to small time constants) will require small time steps, whereas numerical efficiency would require larger steps to compute the complete solution with reasonable effort. Again, the use of an implicit ODE solver will be the way around this problem.

To illustrate the concept of stiffness, we consider a classical example: the Brusselator, which was proposed by Prigogine [4, 5], as he was working in Brussels (explaining the origin of the name)

$$\begin{aligned} u_t &= A - (B+1)u + u^2 v + \alpha u_{zz} \\ v_t &= Bu - u^2 v + \alpha v_{zz} \end{aligned}; \qquad 0 \le z \le 1, \quad t \ge 0 \tag{3.121}$$

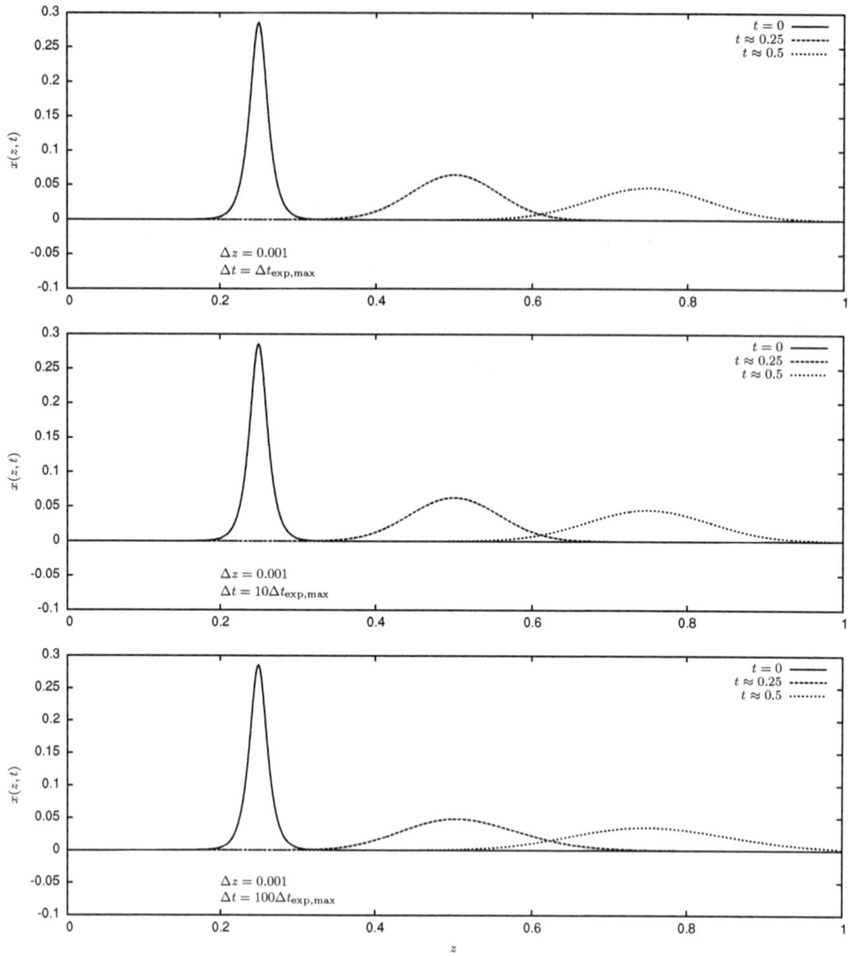

Fig. 3.10 Numerical solution of the initial boundary value problem (3.78–3.80) using 2-point upwind FDs for the first-order derivative (advection term), 3-point centered FDs for the second order derivative (diffusion term) and an implicit Euler method

supplemented by Dirichlet BCs

$$u(0, t) = u(1, t) = 1 \tag{3.122}$$

$$v(0, t) = v(1, t) = 3 \tag{3.123}$$

and ICs

$$u(z, 0) = 1 + \sin(2\pi z) \tag{3.124}$$

$$v(z, 0) = 3 \tag{3.125}$$

with the numerical parameter values: $A = 1$, $B = 3$, $\alpha = 0.02$.

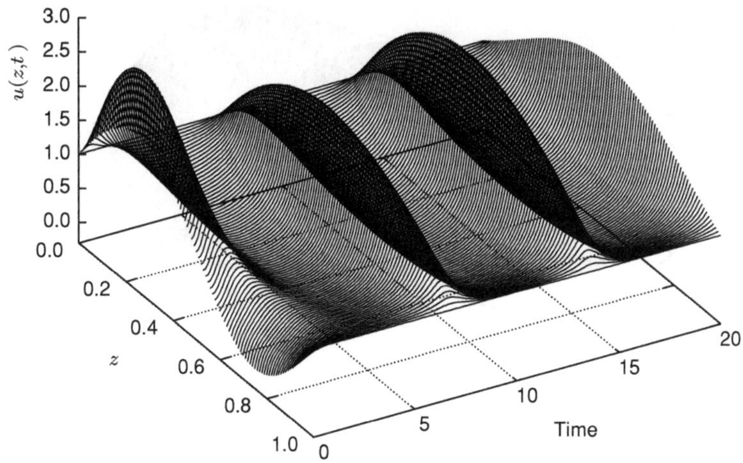

Fig. 3.11 Numerical solution (u) of the Brusselator problem (3.121)–(3.125) using 3-point centered FDs for the second-order derivative and a spatial grid with 101 points

In contrast with the previous examples, the PDEs (3.121) are nonlinear. If we use a 3-point centered FD scheme, as in (3.119), to approximate the second-order derivatives, we obtain a semi-discrete system in the form

$$\begin{bmatrix} u_1 \\ \vdots \\ u_i \\ \vdots \\ u_{N-1} \\ v_1 \\ \vdots \\ v_i \\ \vdots \\ v_{N-1} \end{bmatrix}_t = \begin{bmatrix} A-(B+1)u_1+u_1^2v_1+\alpha(-2u_1+u_2) \\ \vdots \\ A-(B+1)u_i+u_i^2v_i+\alpha(u_{i-1}-2u_i+u_{i+1}) \\ \vdots \\ A-(B+1)u_{N-1}+u_{N-1}^2v_{N-1}+\alpha(u_{N-2}-2u_{N-1}) \\ Bu_1-u_1^2v_1+\alpha\alpha(-2v_1+v_2) \\ \vdots \\ Bu_i-u_i^2v_i+\alpha(v_{i-1}-2v_i+v_{i+1}) \\ \vdots \\ Bu_{N-1}-u_{N-1}^2v_{N-1}+\alpha(v_{N-2}-2v_{N-1}) \end{bmatrix} + \begin{bmatrix} \alpha u_0 \\ 0 \\ \vdots \\ 0 \\ \alpha u_N \\ \alpha v_0 \\ 0 \\ \vdots \\ 0 \\ \alpha v_N \end{bmatrix} \tag{3.126}$$

where the BCs give the values of u_0, u_N, v_0 and v_N.

Using the methods presented in the previous sections, stability could only be studied locally, i.e., around a particular point in the (u, v) space. Indeed, as the semi-discrete system (3.126) is nonlinear, it would be necessary to evaluate the Jacobian matrix, and to compute the eigenvalues. We will not detail these computations here, but instead, we will compare the performance of two integrators of the MATLAB ODE Suite on this problem: (a) `ode45`, an explicit Runge-Kutta Fehlberg method and (b) `ode15s`, an implicit method based on backward differentiation formulas.

Both solvers give satisfactory results, which are represented in Figs. 3.11 and 3.12 for $0 \le t \le 20$. These solutions are computed on a spatial grid with $N = 100$ points,

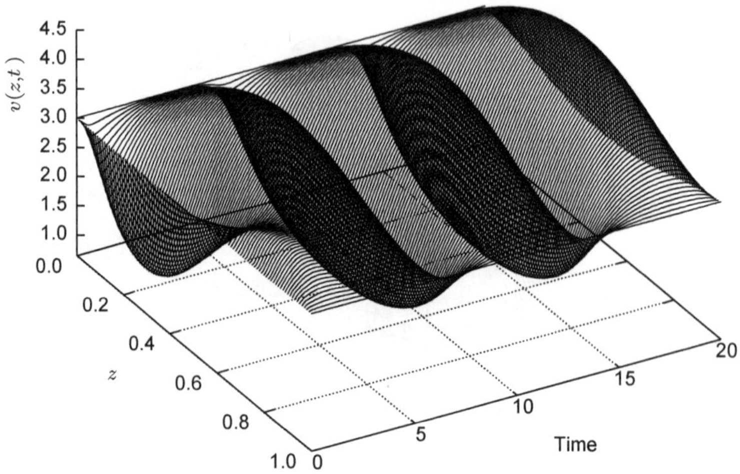

Fig. 3.12 Numerical solution (v) of the Brusselator problem (3.121)–(3.125) using 3-point centered FDs for the second-order derivative and a spatial grid with 101 points

Table 3.1 Computational statistics of `ode45` and `ode15s` (spatial grid with 101 points)

	`ode45`	`ode15s`
Sucessful steps	4821	137
Failed attemps	322	16
Function evaluations	30859	923
CPU (s)	2.2	0.4

and tolerances on time integration defined by `AbsTol = 10-3`, and `RelTol = 10-6`.

Even though the two solvers give a satisfactory solution, the computational load is quite different as apparent on inspection of Table 3.1

Clearly, `ode15s` performs much more efficiently on this problem, reducing the CPU by a factor of almost 6. This is a clear sign that the limited stability region of the explicit ODE integrator `ode45` restricts the time step size to (much) smaller values. In this example, stiffness mostly depends on the numerical values of two parameters: the number of grid points and the selection of a FD scheme for the second-order derivative. Table 3.2 shows the computational statistics when the number of grid points is increased to 201 (instead of 101). While the number of steps taken by `ode15s` remains unchanged, the number of steps taken by `ode45` is multiplied by 4. As a result, the CPU required by `ode15s` is just multiplied by 1.5 (this is logical since we are solving twice as much equations), whereas the CPU required by `ode45` is multiplied by about 4.

Table 3.3 shows the influence of the FD scheme used to approximate the second-order derivative. For instance, a 5-point centered scheme is used instead of a 3-point centered scheme. Again, the performance of `ode15s` is unaffected, whereas the CPU required by `ode45` increases.

Table 3.2 Computational statistics of ode45 and ode15s (spatial grid with 201 points)

	ode45	ode15s
Sucessful steps	19282	137
Failed attemps	1265	16
Function evaluations	123283	1523
CPU (s)	9.3	0.6

Table 3.3 Computational statistics of ode45 and ode15s (spatial grid with 101 points and 5-point centered FD scheme for the second-order derivative)

	ode45	ode15s
Sucessful steps	6425	137
Failed attemps	426	16
Function evaluations	41107	923
CPU (s)	2.82	0.5

3.9 Accuracy and the Concept of Differentiation Matrices

In previous sections, we focused attention on stability of the numerical scheme resulting from spatial discretization and time integration (the two fundamental steps in the MOL). In the following, we will assume that stability is achieved by a proper choice of a FD scheme and time integrator, and turn to another consideration: accuracy.

Accuracy can be influenced at two levels: spatial discretization and time integration. Spatial discretization itself has two tuning features: the number of spatial grid points which determines the fineness or resolution of the spatial grid and the *stencil of the FD scheme*, i.e., the pattern of points on which the FD scheme is built. Increasing the number of spatial grid points, a procedure called *h refinement* (as h denotes the spacing between two adjacent grid points in the mathematical literature) will usually enhance the accuracy of the solution. Increasing the order of the finite difference approximation, a procedure called *p refinement*, can also be beneficial to accuracy.

We will first consider a uniform spatial grid and return to the construction of finite difference schemes using linear combination of Taylor series expansions. For instance, a centered approximation of the first derivative of $x(z)$ at z_i can be obtained by subtracting the Taylor series expansions at z_{i+1} and z_{i-1}

$$x_{i+1} = x_i + \frac{\Delta z}{1!}\left.\frac{\mathrm{d}x}{\mathrm{d}z}\right|_{z_i} + \frac{\Delta z^2}{2!}\left.\frac{\mathrm{d}^2x}{\mathrm{d}z^2}\right|_{z_i} + \frac{\Delta z^3}{3!}\left.\frac{\mathrm{d}^3x}{\mathrm{d}z^3}\right|_{z_i} + \frac{\Delta z^4}{4!}\left.\frac{\mathrm{d}^4x}{\mathrm{d}z^4}\right|_{z_i} + \cdots$$

$$x_{i-1} = x_i - \frac{\Delta z}{1!}\left.\frac{\mathrm{d}x}{\mathrm{d}z}\right|_{z_i} + \frac{\Delta z^2}{2!}\left.\frac{\mathrm{d}^2x}{\mathrm{d}z^2}\right|_{z_i} - \frac{\Delta z^3}{3!}\left.\frac{\mathrm{d}^3x}{\mathrm{d}z^3}\right|_{z_i} + \frac{\Delta z^4}{4!}\left.\frac{\mathrm{d}^4x}{\mathrm{d}z^4}\right|_{z_i} - \cdots$$

so as to get

$$\left.\frac{dx}{dz}\right|_z = \frac{x_{i+1} - x_{i-1}}{2\Delta z} + O(\Delta z^2) \tag{3.127}$$

which is a second-order accurate formula.

Combining Taylor series at additional locations $z_{i+1}, z_{i+2}, \ldots, z_{i-1}, z_{i-2}, \ldots$, allows more elaborate formulas to be obtained. For instance,

$$\left.\frac{dx}{dz}\right|_z = \frac{-x_{i-3} + 6x_{i-2} - 18x_{i-1} + 10x_i + 3x_{i+1}}{12\Delta z} + O(\Delta z^4) \tag{3.128}$$

is a fourth-order (five-point) biased-upwind finite difference approximation for the first derivative. This latter formula is particularly useful to approximate strongly convective terms, with a propagation from left to right (i.e., positive flow velocity $v > 0$).

Formulas such as (3.127)–(3.128) can be used at all the grid points, with the exception of a few points near the boundaries (the number of these boundary points depends on the approximation stencil of the formula, i.e., the number and configuration of the grid points involved in the formula). At those points, alternative formulas have to be developed in order to avoid the introduction of fictitious grid points outside of the spatial domain. In the case of (3.127), these latter formulas are derived by considering the Taylor series expansions at the boundary point z_0 and z_N.

Consider first the following expansions at z_0

$$x_1 = x_0 + \frac{\Delta z}{1!}\left.\frac{dx}{dz}\right|_{z_0} + \frac{\Delta z^2}{2!}\left.\frac{d^2x}{dz^2}\right|_{z_0} + \frac{\Delta z^3}{3!}\left.\frac{d^3x}{dz^3}\right|_{z_0} + \frac{\Delta z^4}{4!}\left.\frac{d^4x}{dz^4}\right|_{z_0} + \cdots$$

$$x_2 = x_0 + \frac{2\Delta z}{1!}\left.\frac{dx}{dz}\right|_{z_0} + \frac{(2\Delta z)^2}{2!}\left.\frac{d^2x}{dz^2}\right|_{z_0} + \frac{(2\Delta z)^3}{3!}\left.\frac{d^3x}{dz^3}\right|_{z_0} + \frac{(2\Delta z)^4}{4!}\left.\frac{d^4x}{dz^4}\right|_{z_0} + \cdots$$

These two expressions are combined using coefficients a and b so as to obtain

$$ax_1 + bx_2 = (a+b)x_0 + (a+2b)\left.\frac{dx}{dz}\right|_{z_0} + (a+4b)\left.\frac{d^2x}{dz^2}\right|_{z_0}\frac{\Delta z^2}{2!} + O(\Delta z^2)$$

In order to retain the first-order derivative, the term multiplying it, i.e. $a + 2b$, must be different from zero, for instance

$$a + 2b = 1$$

and to cancel exactly the second-order term, coefficients a and b must be of the form

$$a + 4b = 0$$

The previous equations form a system of linear equations whose solution is $a = 2$, $b = -1/2$. The resulting formula at z_0 is therefore

$$\left.\frac{\mathrm{d}x}{\mathrm{d}z}\right|_{z_0} = \frac{-3x_0 + 4x_1 - x_3}{2\Delta z} + O(\Delta z^2) \tag{3.129}$$

A similar procedure at z_N yields

$$\left.\frac{\mathrm{d}x}{\mathrm{d}z}\right|_{z_N} = \frac{3x_N - 4x_{N-1} + x_{N-2}}{2\Delta z} + O(\Delta z^2) \tag{3.130}$$

These formulas are no longer centered formulas, but are still second order accurate.

For the MATLAB implementation of formulas such as (3.127)–(3.130), the concept of a differentiation matrix is useful, i.e.,

$$\mathbf{x}_z = \mathbf{D}_1\mathbf{x} = \frac{1}{2\Delta z}\begin{bmatrix} -3 & 4 & -1 & 0 & \cdots & \cdots & 0 \\ -1 & 0 & 1 & 0 & \cdots & \cdots & 0 \\ & & & \cdots & & & \\ 0 & \cdots & -1 & 0 & 1 & \cdots & 0 \\ & & & \cdots & & & \\ 0 & \cdots & \cdots & 0 & -1 & 0 & 1 \\ 0 & \cdots & \cdots & 0 & 1 & -4 & 3 \end{bmatrix}\begin{bmatrix} x_0 \\ x_1 \\ \cdots \\ x_i \\ \cdots \\ x_{N-1} \\ x_N \end{bmatrix} \tag{3.131}$$

so that a spatial derivative can be computed by a simple matrix operation (which is the key operation in MATLAB).

Differentiation matrices allow higher-order spatial derivatives to be easily computed in either of the following two ways:

- *Stagewise differentiation:* $\mathbf{x}_z = \mathbf{D}_1\mathbf{x}$ and $\mathbf{x}_{zz} = \mathbf{D}_1\mathbf{x}_z$
- *Direct differentiation:* $\mathbf{x}_{zz} = \mathbf{D}_2\mathbf{x}$

where $\mathbf{D}_1$ and $\mathbf{D}_2$ are matrices for computing first- and second-order derivatives, respectively. Other combinations are possible as well, e.g., $x_{zzzz} = \mathbf{D}_1(\mathbf{D}_1(\mathbf{D}_1(\mathbf{D}_1\mathbf{x})))$ or $x_{zzzz} = \mathbf{D}_2(\mathbf{D}_2\mathbf{x})$ or $x_{zzzz} = \mathbf{D}_4\mathbf{x}$. Stagewise differentiation is a simple and very flexible procedure, which should be applied with care, as an order of accuracy is lost at each stage.

Even though FD schemes and the corresponding differentiation matrices can be computed analytically in a number of cases, the developments become tedious for high-order schemes and it is interesting to look for an automated, numerical procedure. Such a procedure exists and has been proposed by Fornberg [6, 7] for iteratively computing the weighting coefficients of finite difference formulas of arbitrary order of accuracy on arbitrarily spaced spatial grids (nonuniform grids). The starting point of this procedure is the consideration of an interpolation polynomial of a function $x(z)$ which is known only at a set of $N + 1$ arbitrarily distributed points z_i:

z	$x(z)$
z_0	x_0
z_1	x_1
$\vdots$	$\vdots$
z_i	x_i
$\vdots$	$\vdots$
z_N	x_N

Interpolation can be achieved using a Lagrange polynomial given by

$$P_{0\ldots N}(z) = \sum_{k=0}^{N} \phi_{Nk}(z) x_k \tag{3.132}$$

where ϕ_{Nk} is a Lagrange coefficient of the form

$$\phi_{Nk} = \frac{(z - z_0)\cdots(z - z_{k-1})(z - z_{k+1})\cdots(z - z_N)}{(z_k - z_0)\cdots(z_k - z_{k-1})(z_k - z_{k+1})\cdots(z_k - z_N)} \tag{3.133}$$

The approximation of the nth-order derivative of $x(z)$ in z_i, i.e., $\dfrac{\mathrm{d}^n x(z_i)}{dz^n}$ can be obtained by differentiating this interpolation polynomial

$$\left.\frac{\mathrm{d}^n x}{dz^n}\right|_z = \left(\frac{\mathrm{d}^n P_{0\ldots N}(z)}{dz^n}\right)_z = \sum_{k=0}^{n} \left(\frac{\mathrm{d}^n \phi_{Nk}(z)}{dz^n}\right)_z x_k = \sum_{k=0}^{n} c_{Nk}^n x_k \tag{3.134}$$

The idea behind Fornberg's procedure is to derive an iterative algorithm for computing the coefficients c_{Nk}^m for $1 \le m \le n$, which allow the evaluation of the successive derivatives

$$\left.\frac{\mathrm{d}x}{\mathrm{d}z}\right|_z = \sum_{k=0}^{n} c_{Nk}^1 x_k$$

$$\left.\frac{\mathrm{d}^2 x}{\mathrm{d}z^2}\right|_z = \sum_{k=0}^{n} c_{Nk}^2 x_k$$

$$\ldots$$

$$\left.\frac{\mathrm{d}^n x}{dz^n}\right|_z = \sum_{k=0}^{n} c_{Nk}^n x_k$$

From (3.133) we deduce that

$$\phi_{Nk}(z) = \phi_{N-1k}(z) \frac{z - z_N}{z_k - z_N} \tag{3.135}$$

which is valid for all k, except $k = N$. For the latter, another expression is sought starting from

$$\phi_{NN}(z) = \frac{(z - z_0)\cdots(z - z_{N-1})}{(z_N - z_0)\cdots(z_N - z_{N-1})} \tag{3.136}$$

$$\phi_{N-1N-1}(z) = \frac{(z - z_0)\cdots(z - z_{N-2})}{(z_{N-1} - z_0)\cdots(z_{N-1} - z_{N-2})} \tag{3.137}$$

which leads to

$$\phi_{NN}(z) = \phi_{N-1N-1}(z)\frac{(z_{N-1} - z_0)\cdots(z_{N-1} - z_{N-2})}{(z_N - z_0)\cdots(z_N - z_{N-1})}(z - z_{N-1}) \tag{3.138}$$

If we define

$$\omega_N(z) = (z - z_0)\cdots(z - z_N) \tag{3.139}$$

Equation (3.138) can be written in a more compact way as

$$\phi_{NN}(z) = \phi_{N-1N-1}(z)\frac{\omega_{N-2}(z_{N-1})}{\omega_{N-1}(z_N)}(z - z_{N-1}) \tag{3.140}$$

Let us now return to the expression of the polynomial $\phi_{Nk}(z)$ which has degree N and as such can be developed as follows

$$\phi_{Nk}(z) = \alpha_0 + \alpha_1(z - z_i) + \cdots + \alpha_N(z - z_i)^N \tag{3.141}$$

where

$$\alpha_m = \frac{1}{m!}\left(\frac{\mathrm{d}^m\phi_{Nk}}{\mathrm{d}z^m}\right)_{z_i} \tag{3.142}$$

so that, using the notation of Eq. (3.134),

$$\phi_{Nk}(z) = \sum_{m=0}^{N}\frac{1}{m!}\left(\frac{\mathrm{d}^m\phi_{Nk}}{\mathrm{d}z^m}\right)_{z_i}(z - z_i)^m = \sum_{m=0}^{N}\frac{c_{Nk}^m}{m!}(z - z_i)^m \tag{3.143}$$

This result can be used in Eq. (3.135) to give

$$\begin{aligned}\sum_{m=0}^{N}\frac{c_{Nk}^m}{m!}(z - z_i)^m &= \sum_{m=0}^{N-1}\frac{c_{N-1k}^m}{m!}(z - z_i)^m\frac{z - z_N}{z_k - z_N}\\ &= \sum_{m=0}^{N-1}\frac{c_{N-1k}^m}{m!}(z - z_i)^m\frac{z - z_i + z_i - z_N}{z_k - z_N}\end{aligned} \tag{3.144}$$

If we equate the coefficients of the terms in $(z - z_i)^m$, we obtain an iterative formula

$$\frac{c_{Nk}^m}{m!} = \frac{c_{N-1k}^{m-1}}{(m-1)!}\frac{1}{z_k - z_N} + \frac{c_{N-1k}^m}{m!}\frac{z_i - z_N}{z_k - z_N} \tag{3.145}$$

or

$$c_{Nk}^m = \frac{1}{z_k - z_N}\left[m c_{N-1k}^{m-1} + (z_i - z_N)\, c_{N-1k}^m\right] \tag{3.146}$$

The particular case $k = N$ has to be deduced from (3.140)

$$\sum_{m=0}^{N} \frac{c_{NN}^m}{m!}(z - z_i)^m = \sum_{m=0}^{N-1} \frac{c_{N-1N-1}^m}{m!}(z - z_i)^m \frac{\omega_{N-2}(z_{N-1})}{\omega_{N-1}(z_N)} (z - z_i + z_i - z_{N-1}) \tag{3.147}$$

Again, the equality of the coefficients of the terms in $(z - z_i)^m$ in both expressions leads to an iterative formula

$$\frac{c_{NN}^m}{m!} = \frac{\omega_{N-2}(z_{N-1})}{\omega_{N-1}(z_N)}\left[\frac{c_{N-1N-1}^m}{m!}(z_i - z_{N-1}) + \frac{c_{N-1N-1}^{m-1}}{(m-1)!}\right] \tag{3.148}$$

or

$$c_{NN}^m = \frac{\omega_{N-2}(x_{N-1})}{\omega_{N-1}(x_N)}\left[c_{N-1N-1}^m (z_i - z_{N-1}) + m c_{N-1N-1}^{m-1}\right] \tag{3.149}$$

Equations (3.146) and (3.149) are the iterative formulas proposed by Fornberg [6, 7]. We can observe the iterative computation with respect to the order of the derivative (superscript m) and with respect to the number of grid points (subscript N). These formulas are coded in function `weights`.

```
function [w] = weights(zd,zs,ns,m)
% This function computes the weights of a finite difference
% scheme on a nonuniform grid
%
% Input Parameters
%  zd:       location where the derivative is to be computed
%  ns:       number of points in the stencil
%  zs(ns):  stencil of the finite difference scheme
%  m:        highest derivative for which weights are sought
%
% Output Parameter
%  w(1:ns,1:m+1): weights at grid locations z(1:ns) for
%                 derivatives of order 0:m, found in
%                 w(1:ns,1:m+1)

c1 = 1.0;
c4 = zs(1)-zd;
for k = 0 : m
```

```
    for j = 0 : ns-1
        w(j+1,k+1) = 0.0;
    end
end
w(1,1) = 1.0;
for i = 1 : ns-1
    mn = min(i,m);
    c2 = 1.0;
    c5 = c4;
    c4 = zs(i+1)-zd;
    for j = 0 : i-1
        c3 = zs(i+1)-zs(j+1);
        c2 = c2*c3;
        if (j = = i-1)
            for k = mn: -1:1
                w(i+1,k+1) = c1*(k*w(i,k)-c5*w(i,k+1))/c2;
            end
            w(i+1,1) = -c1*c5*w(i,1)/c2;
        end
        for k = mn : -1 : 1
            w(j+1,k+1) = (c4*w(j+1,k+1)-k*w(j+1,k))/c3;
        end
        w(j+1,1) = c4*w(j+1,1)/c3;
    end
    c1 = c2;
end
```

Function weights Function `weights` for the iterative computation of finite difference weights on arbitrarily spaced grids.

This function can be called by more specific functions computing particular differentiation matrices. These latter functions automate the computation of the differentiation matrices, and have names that explicit their purposes. For instance, function `two_point_upwind_D1`, computes a differentiation matrix for a first-order derivative using a 2-point upwind formula. In the calling list, `z` is the vector containing the grid points, which can be uniformly or non uniformly distributed, and `v` is a parameter indicating the direction of the flow (positive from left to right, and negative in the opposite direction).

```
function [D] = two_point_upwind_D1(z,v)
%
% function two_point_upwind_D1 returns the differentiation
% matrix for computing the first derivative, xz, of a
% variable x over a nonuniform grid z from upwind two-point,
% first-order finite difference approximations
%
%  the following parameters are used in the code:
%
%       z                 spatial grid
%
%       v                 fluid velocity (positive from left
%                         to right - only the sign is used)
%
%       n                 number of grid points
%
%       zs(ns)            stencil of the finite difference scheme
%
```

```
%       ns               number of points in the stencil
%
%       zd               location where the derivative is to be
%                        computed
%
%       m                highest derivative for which weights are
%                        sought

m  = 1;
ns = 2;

%  sparse discretization matrix
n = length(z);
D = sparse(n,n);

% (1)  finite difference approximation for positive v
if v > 0
    % boundary point
    zs        = z(1:ns);
    zd        = z(1);
    [w]       = weights(zd,zs,ns,m);
    D(1,1:2)  = w(1:ns,m+1)';
    % interior points
    for i = 2:n,
        zs           = z(i-1:i);
        zd           = z(i);
        [w]          = weights(zd,zs,ns,m);
        D(i,i-1:i)   = w(1:ns,m+1)';
    end;
end;
% (2)  finite difference approximation for negative v
if v < 0
    % interior points
    for i = 1:n-1,
        zs           = z(i:i+1);
        zd           = z(i);
        [w]          = weights(zd,zs,ns,m);
        D(i,i:i+1)   = w(1:ns,m+1)';
    end;
    % boundary point
    zs           = z(n-1: n);
    zd           = z(n);
    [w]          = weights(zd,zs,ns,m);
    D(n,n-1: n)  = w(1:ns,m+1)';
end;
```

Function two_point_upwind_D1 Function to compute a first-order differentiation matrix using a 2-point upwind finite difference scheme.

To conclude this section, and illustrate the use of various differentiation matrices, with different orders of accuracy, we return to Burgers equation already presented in Chap. 1 (Sect. 1.3).

$$\frac{\partial x}{\partial t} = -x\frac{\partial x}{\partial z} + \mu\frac{\partial^2 x}{\partial z^2} \tag{3.150}$$

We consider another exact solution given by

$$x_{\text{exact}}(z, t) = \frac{1}{1 + e^{(z-0.5t)/2\mu}} \tag{3.151}$$

This solution can be used to assess the accuracy of the numerical solution. The code is given in the following listings (`burgers_main`, `burgers_pde` and `burgers_exact`).

```
close all
clear all

% set global variables
global mu;
global z0 zL n D1 D2;

% spatial grid
z0 = 0.0;
zL = 1.0;
n  = 201;
dz = (zL-z0)/(n-1);
z  = [z0:dz:zL]';

% model parameter
mu = 0.001;

% initial conditions
x = burgers_exact(z,0);

% select finite difference (FD) approximation of the spatial
% derivative
method  = 'centered'        % three point centered approximation
% method = 'upwind 2'       % two point upwind approximation
% method  = 'upwind 3'      % three point upwind approximation
% method = 'upwind 4'       % four point upwind approximation
%
switch method
    %
    % three point centered approximation
    case('centered')
        D1 = three_point_centered_D1(z);
        % two point upwind approximation
    case('upwind 2')
        v  = 1;
        D1 = two_point_upwind_D1(z,v);
        % three point upwind approximation
    case('upwind 3')
        v  = 1;
        D1 = three_point_upwind_D1(z,v);
        % four point upwind approximation
    case('upwind 4')
        v  = 1;
        D1 = four_point_upwind_D1(z,v);
end

% differentiation matrix (diffusive term)
%     D2=three_point_centered_D2(z);
D2 = five_point_centered_D2(z);

% call to ODE solver
t = [0:0.1:1.5];
```

```
options = odeset('RelTol',1e-3,'AbsTol',1e-3);
%
[tout, yout] = ode15s('burgers_pde',t,x,options);

% Plot the numerical results
plot(z,yout,'k');
xlabel('z');
ylabel('x');
title('Burgers equation')

% compute exact solution
x_exact = [];
for k = 1:length(tout)
    x = burgers_exact(z,tout(k));
    x_exact = [x_exact ; x'];
end

% compare with numerical solution
hold on
plot(z,x_exact,'r');
```

Script burgers_main Main program for the solution of Burgers equation using various spatial differentiation matrices.

```
function xt = burgers_pde(t,x)

% set global variables
global mu;
global z0 zL n D1 D2;

% boundary conditions at z = 0
x(1) = burgers_exact(z0,t);

% boundary conditions at z = zL
x(n) = burgers_exact(zL,t);

% second - order spatial derivative
xzz = D2*x;

% first - order spatial derivative
xz = D1*x;

% temporal derivatives
xt    = -x.*xz + mu*xzz;
xt(1) = 0;
xt(n) = 0;
```

Function burgers_pde PDE function for the solution of Burgers equation using various spatial differentiation matrices.

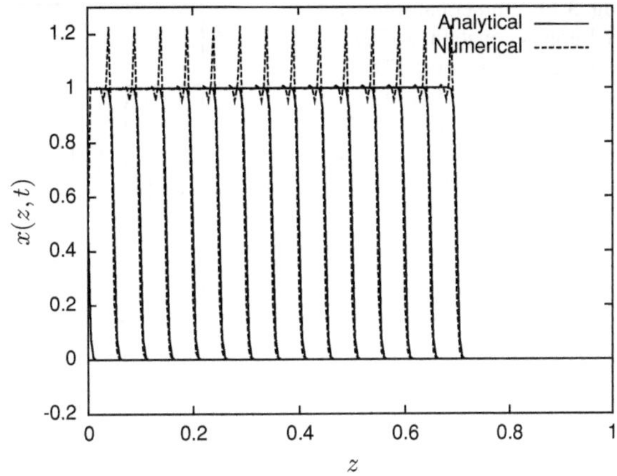

Fig. 3.13 Numerical solution of Burgers equation (3.150) using 3-point centered FDs for the first-order derivative, 5-point centered FDs for the second-order derivative and a spatial grid with 201 points

```
function [x] = burgers_exact(z,t)

% This function computes an exact solution to Burgers' equation

global mu

ex = exp((z-0.5*t)/(2*mu));
x  = 1./(1+ex);
```

Function burgers_exact Exact solution to Burgers equation

Note the possibility of testing different approximations for the first-order derivative. As discussed earlier in the present chapter, the use of a centered scheme leads to spurious oscillations (see Fig. 3.13, which has been produced with a spatial grid of 201 points).

The application of a 2-point upwind scheme avoids oscillations, but does not produce a satisfactory solution as the speed of the wave is not well estimated (the numerical solution lags behind the exact solution—see Fig. 3.14).

A 3-point upwind scheme improves the results (Fig. 3.15), but a 4-point upwind scheme produces much better results as shown in Fig. 3.16.

To get similar results with a 3-point upwind scheme, it is necessary to increase the number of grid points up to 701, thus increasing significantly the computational load.

3.10 Various Ways of Translating the Boundary Conditions

At this stage, we know how to approximate the derivative operators in the partial differential equations using a variety of finite difference schemes, on uniform and nonuniform grids, and with different orders of accuracy. An important aspect of the solution of an initial-boundary value problem that we have already touched upon,

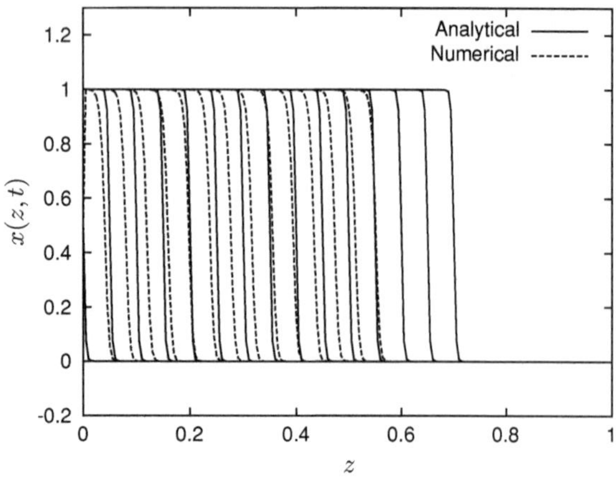

Fig. 3.14 Numerical solution of Burgers equation (3.150) using 2-point upwind FDs for the first-order derivative, 5-point centered FDs for the second-order derivative and a spatial grid with 201 points

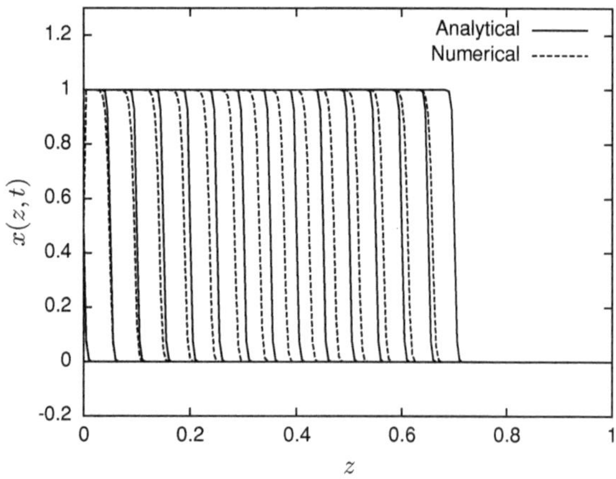

Fig. 3.15 Numerical solution of Burgers equation (3.150) using 3-point upwind FDs for the first-order derivative, 5-point centered FDs for the second-order derivative and a spatial grid with 201 points

but not really studied in details, is the implementation of the boundary conditions. The translation of the boundary conditions into the method of lines is a crucial step, as boundary conditions are a very important part of the problem definition.

There are different approaches to this implementation, some of them are rather crude approximations which can be convenient in some situations, whereas others are more rigorous but require a closer analysis of the problem or a slightly more sophisticated coding.

We first recall the main types of boundary conditions:

- Dirichlet BCs specify the values of the dependent variable at the boundary of the spatial domain:

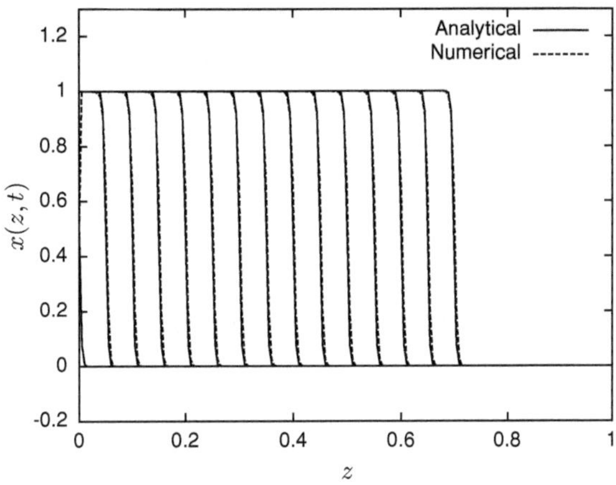

Fig. 3.16 Numerical solution of Burgers equation (3.150) using 4-point upwind FDs for the first-order derivative, 5-point centered FDs for the second-order derivative and a spatial grid with 201 points

$$\begin{aligned} x(z_0, t) &= g_0(t) \\ x(z_L, t) &= g_L(t) \end{aligned} \tag{3.152}$$

- Neumann BCs specify the values of the gradient in the normal direction to the boundary surface, i.e. in one spatial dimension the value of the derivative at the boundary

$$\begin{aligned} \frac{\partial x}{\partial z}(z_0, t) &= g_0(t) \\ \frac{\partial x}{\partial z}(z_L, t) &= g_L(t) \end{aligned} \tag{3.153}$$

- Robin BCs are a combination of the first two:

$$\begin{aligned} \frac{\partial x}{\partial z}(z_0, t) + \alpha_0 x(z_0, t) &= g_0(t) \\ \frac{\partial x}{\partial z}(z_L, t) + \alpha_L x(z_L, t) &= g_L(t) \end{aligned} \tag{3.154}$$

The numerical implementation of the BCs will be discussed through a very simple equation, e.g. a diffusion equation supplemented by Neumann and Robin BCs

$$\frac{\partial x}{\partial t} = \frac{\partial^2 x}{\partial z^2}; \qquad 0 < z < 1; \qquad 0 < t < 2 \tag{3.155}$$

with ICs

$$x(z, 0) = 1 \tag{3.156}$$

and BCs

$$\frac{\partial x}{\partial z}(0,t) = 0 \tag{3.157}$$

$$\frac{\partial x}{\partial z}(1,t) + \alpha x(1,t) = 0 \tag{3.158}$$

3.10.1 Elimination of Unknown Variables

This approach is straightforward in the case of Dirichlet BCs, as they give the values of the independent variables at the boundaries of the spatial domain. These variables therefore do not appear any longer in the vector of variables to be computed through the solution of a system of differential equations. We have already used this approach in the treatment of the advection equation earlier in this chapter. Let us investigate this situation again in the case of our current example, i.e. the diffusion equation, and imagine that the BCs (3.157)–(3.158) would be replaced by

$$x(0,t) = g_0(t) \tag{3.159}$$

$$x(1,t) = g_1(t) \tag{3.160}$$

If the second-order spatial derivative in PDE (3.155) is approximated by a 3-point centered scheme, the MOL yields the following system of ODEs

$$\begin{bmatrix} x_{1,t} \\ x_{2,t} \\ \vdots \\ x_{N-2,t} \\ x_{N-1,t} \end{bmatrix} = \frac{1}{\Delta z^2}\begin{bmatrix} -2 & 1 & & & \\ 1 & -2 & 1 & & \\ & 1 & -2 & 1 & \\ & & 1 & -2 & 1 \\ & & & 1 & -2 \end{bmatrix}\begin{bmatrix} x_1 \\ x_2 \\ \vdots \\ x_{N-2} \\ x_{N-1} \end{bmatrix} + \frac{1}{\Delta z^2}\begin{bmatrix} g_0(t) \\ 0 \\ \vdots \\ 0 \\ g_1(t) \end{bmatrix}$$

Even though this approach seems limited to Dirichlet BCs, it can be adapted to the other type of BCs. Consider now the original BCs (3.157)–(3.158) in discretized form, using simple 2-point schemes

$$\frac{x_1 - x_0}{\Delta z} = 0; \qquad \frac{x_N - x_{N-1}}{\Delta z} + \alpha x_N = 0$$

or

$$x_0 = x_1; \qquad x_N = \frac{x_{N-1}}{1+\alpha\Delta z} \tag{3.161}$$

These expressions can be used in the computation of the finite difference approximation of the second-order derivative operator in node 1 and $N-1$

$$x_{1,t} = \frac{x_0 - 2x_1 + x_2}{\Delta z^2} = \frac{-x_1 + x_2}{\Delta z^2}$$

$$x_{N-1,t} = \frac{x_{N-2} - 2x_{N-1} + x_N}{\Delta z^2} = \frac{1}{\Delta z^2}\left(x_{N-2} + x_{N-1}\left(\frac{1}{1+\alpha\Delta z} - 2\right)\right)$$

$$= \frac{1}{\Delta z^2}\left(x_{N-2} + \gamma x_{N-1}\right)$$

with $\gamma = \left(\frac{1}{1+\alpha\Delta z} - 2\right)$. The MOL therefore yields the following system of ODEs

$$\begin{bmatrix} x_{1,t} \\ x_{2,t} \\ \vdots \\ x_{N-2,t} \\ x_{N-1,t} \end{bmatrix} = \frac{1}{\Delta z^2}\begin{bmatrix} -1 & 1 & & & \\ 1 & -2 & 1 & & \\ & 1 & -2 & 1 & \\ & & 1 & -2 & 1 \\ & & & 1 & \gamma \end{bmatrix}\begin{bmatrix} x_1 \\ x_2 \\ \vdots \\ x_{N-2} \\ x_{N-1} \end{bmatrix} \tag{3.162}$$

This solution is coded in the main program `diffusion_main` and function `diffusion_pde`. This code implements the spatial differentiation matrix at the interior points (1 to $N-1$), solves the system of ODEs (3.162), and finally computes the values at the boundary points using Eq. (3.161).

```
clear all
close all

% set global variables
global alpha
global n D2 dz

% model parameter
alpha = 2;

% spatial domain
z0 = 0.0;
zL = 1.0;

% uniform spatial grid
n  = 201;
dz = (zL - z0)/(n-1);
z  = [z0:dz:zL]';

% initial conditions (interior points)
x = ones(n-2,1);

% differentiation matrix (interior points)
D2              = diag(+1*ones(n-3,1),-1) +...
                  diag(-2*ones(n-2,1),0) +...
                  diag(+1*ones(n-3,1),1);
D2(1,1:2)       = [-1 +1];
gamma           = 1/(1+alpha*dz)-2;
```

```
D2(n-2,n-3:n-2) = [+1 gamma];
D2              = D2/(dz*dz);

% call to ODE solver
options = odeset('RelTol',1e-3,'AbsTol',1e-3);
[tout, xout] = ode15s(@diffusion_pde,[0:0.2:2],x,options);

% boundary conditions at z = z0
x0 = xout(:,1);

% boundary conditions at z = zL
xn = xout(:,n-2)/(1+alpha*dz);

% plot results
xsol = [x0 xout xn];
plot(z,xsol);
xlabel('z');
ylabel('x(z,t)');
title('Diffusion equation')
```

Script diffusion_main Main program for the evaluation of the MOL discretization of IBVP (3.155)–(3.158) using the approach based on the elimination of variables for the treatment of the BCs

```
function xt = diffusion_pde(t,x)

% set global variables
global alpha
global n D2 dz

% second - order spatial derivative
xzz = D2*x;

% temporal derivatives
xt = xzz;
```

Function diffusion_pde Function `diffusion_pde` for the evaluation of the MOL discretization of IBVP (3.155)–(3.158) using the approach based on the elimination of variables for the treatment of the BCs.

Numerical simulation results for $\alpha = 2$ are shown in Fig. 3.17.

More sophisticated schemes than in (3.161) can be used for approximating the BCs, for instance 3-points non-centered schemes

$$\frac{-3x_0 + 4x_1 - x_2}{2\Delta z} = 0; \qquad \frac{3x_N - 4x_{N-1} + x_{N-2}}{2\Delta z} + \alpha x_N = 0 \tag{3.163}$$

leading to

$$x_0 = \frac{1}{3}(4x_1 - x_2); \qquad x_N = \frac{4x_{N-1} - x_{N-2}}{3 + 2\alpha\Delta z} \tag{3.164}$$

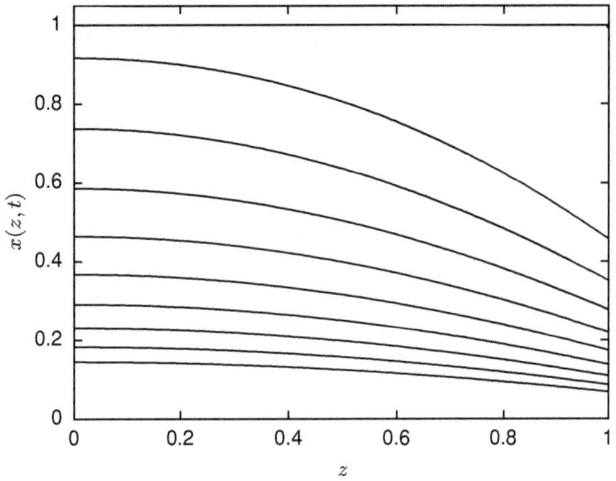

Fig. 3.17 Numerical solution of the diffusion problem (3.155)–(3.158) using 3-point centered FDs for the second-order derivative and elimination of the boundary values, with a spatial grid of 201 points, and a time span of 2 (plot interval of 0.2)

and

$$x_{1t} = \frac{x_0 - 2x_1 + x_2}{\Delta z^2} = \frac{2}{3}\frac{-x_1 + x_2}{\Delta z^2} \tag{3.165}$$

$$\begin{aligned} x_{N-1t} &= \frac{x_{N-2} - 2x_{N-1} + x_N}{\Delta z^2} \\ &= \frac{1}{\Delta z^2}\left(x_{N-2}\left(1 - \frac{1}{3 + 2\alpha\Delta z}\right) - 2x_{N-1}\left(1 - \frac{2}{3 + 2\alpha\Delta z}\right)\right) \\ &= \frac{1}{\Delta z^2}\left(\beta x_{N-2} - \lambda x_{N-1}\right) \end{aligned} \tag{3.166}$$

The MOL then yields the following system of ODEs

$$\begin{bmatrix} x_{1t} \\ x_{2t} \\ \vdots \\ x_{N-2t} \\ x_{N-1t} \end{bmatrix} = \frac{1}{\Delta z^2}\begin{bmatrix} -2/3 & 2/3 & & & \\ 1 & -2 & 1 & & \\ & 1 & -2 & 1 & \\ & & 1 & -2 & 1 \\ & & & \beta & -\lambda \end{bmatrix}\begin{bmatrix} x_1 \\ x_2 \\ \vdots \\ x_{N-2} \\ x_{N-1} \end{bmatrix} \tag{3.167}$$

The only changes in the main program `diffusion_main` are in the computation of the differentiation matrix

```
% differentiation matrix (interior points)
D2          = diag(+1*ones(n-3,1),-1)
                 + diag(-2*ones(n-2,1),0) +...
              diag(+1*ones(n-3,1),1);
D2(1,1:2)   = [-2/3 +2/3];
beta        = 1-1/(3+(2*alpha*dz));
lambda      = 2-4/(3+(2*alpha*dz));
```

```
D2(n-2,n-3:n-2)  = [beta -lambda];
D2               = D2/(dz*dz);
```

and in the explicit computation of the dependent variables at the boundary nodes

```
% boundary conditions at z = z0
x0 = 4*xout(:,1)/3 - xout(:,2)/3;

% boundary conditions at z = zL
xn = (4*xout(:,n-2) - xout(:,n-3))/(3+(2*alpha*dz));
```

More accurate schemes at the boundaries can influence the overall accuracy of the solution, particularly in problems where there are dynamic changes in the value and slope of the solution at the boundaries. However, in our simple IVP problem, the numerical simulation results are basically the same as in Fig. 3.17.

3.10.2 Fictitious Nodes

The idea behind the second approach is to develop an approximation of the PDE at the boundary nodes, using "fictitious nodes" located outside the spatial domain

$$\begin{aligned} x_{0t} &= \frac{1}{\Delta z^2}(x_{-1} - 2x_0 + x_1) \\ x_{Nt} &= \frac{1}{\Delta z^2}(x_{N-1} - 2x_N + x_{N+1}) \end{aligned} \tag{3.168}$$

The fictitious nodes with index -1 and $N+1$, respectively, are associated with unknown values of the independent variables x_{-1} and x_{N+1} which can be eliminated using approximations of the BCs (3.157–3.158) constructed using these fictitious nodes as well

$$\frac{x_1 - x_{-1}}{2\Delta z} = 0; \qquad \frac{x_{N+1} - x_{N-1}}{2\Delta z} + \alpha x_N = 0 \tag{3.169}$$

or

$$x_{-1} = x_1; \qquad x_{N+1} = x_{N-1} - 2\alpha\Delta z x_N \tag{3.170}$$

so that (3.168) can be rewritten as

$$\begin{aligned} x_{0t} &= \frac{2}{\Delta z^2}(-x_0 + x_1) \\ x_{Nt} &= \frac{2}{\Delta z^2}(x_{N-1} - (1 + \alpha\Delta z)x_N) = \frac{2}{\Delta z^2}(x_{N-1} - \delta x_N) \end{aligned} \tag{3.171}$$

The MOL therefore yields the following system of ODEs

$$\begin{bmatrix} x_{0t} \\ x_{1t} \\ \vdots \\ x_{N-1t} \\ x_{Nt} \end{bmatrix} = \frac{1}{\Delta z^2} \begin{bmatrix} -2 & 2 & & & \\ 1 & -2 & 1 & & \\ & 1 & -2 & 1 & \\ & & 1 & -2 & 1 \\ & & & 2 & -2\delta \end{bmatrix} \begin{bmatrix} x_0 \\ x_1 \\ \vdots \\ x_{N-1} \\ x_N \end{bmatrix} \tag{3.172}$$

This method, implemented in `diffusion_main_2`, has an obvious advantage over the former one: the system of ODEs is expressed in all the $N + 1$ nodes of the spatial grid, and it is no longer required to explicitly compute the value of the dependent variables at the boundary nodes.

```
clear all
close all

% set global variables
global alpha
global n D2 dz

% model parameter
alpha = 2;

% spatial domain
z0 = 0.0;
zL = 1.0;

% uniform spatial grid
n  = 201;
dz = (zL - z0)/(n-1);
z  = [z0:dz:zL]';

% initial conditions
x = ones(n,1);

% differentiation matrix
D2             = diag(+1*ones(n-1,1),-1) + diag(-2*ones(n,1),0)+...
                 diag(+1*ones(n-1,1),1);
D2(1,1:2)      = [-2 +2];
delta          = 1 + alpha*dz;
D2(n,n-1: n) = [+2 -2* delta];
D2             = D2/(dz*dz);

% call to ODE solver
options      = odeset('RelTol',1e-3,'AbsTol',1e-3);
[tout, xout] = ode15s(@diffusion_pde,[0:0.2:2],x,options);

% plot results
plot(z,xout);
xlabel('z');
ylabel('x(z,t)');
title('Diffusion equation')
```

Script diffusion_main_2 Main program for the evaluation of the MOL discretization of IVP (3.155)–(3.158) using the approach based on the introduction of fictitious nodes for the treatment of the BCs.

It is also possible to avoid defining explicitly matrix D2 and to use one of the library functions provided in the companion software.

```
% differentiation matrix
D2       = three_point_centered_uni_D2(z0,zL,n);
D2(1,:)  = [-2 +2 zeros(1,n-2)]/(dz*dz);
delta    = 1 + alpha*dz;
D2(n,:)  = [zeros(1,n-2) +2 -2*delta]/(dz*dz);
```

3.10.3 Solving Algebraic Equations

Boundary conditions can be viewed as algebraic equations. Indeed a Dirichlet condition

$$x(z_0, t) = g_0(t) \tag{3.173}$$

can be written as

$$0 = g_0(t) - x(z_0, t) = g_0(t) - x_0(t) \tag{3.174}$$

and a Neumann boundary condition of the form

$$\frac{\partial x}{\partial z}(z_0, t) = g_0(t) \tag{3.175}$$

can be approximated as

$$0 = \frac{\partial x}{\partial z}(z_0, t) - g_0(t) \approx \frac{x_1(t) - x_0(t)}{\Delta z} - g_0(t) \tag{3.176}$$

In our application example, the BCs can therefore be expressed as

$$0 = \frac{x_1(t) - x_0(t)}{\Delta z}; \qquad 0 = -\frac{x_N - x_{N-1}}{\Delta z} - \alpha x_N \tag{3.177}$$

or

$$0 = 0\frac{\mathrm{d}x_0}{\mathrm{d}t}(t) = \frac{-x_0(t) + x_1(t)}{\Delta z}; \qquad 0 = 0\frac{\mathrm{d}x_N}{\mathrm{d}t}(t) = \frac{x_{N-1} - (1 + \alpha\Delta z)x_N}{\Delta z} \tag{3.178}$$

MOL therefore yields the following differential-algebraic system (when using three-point centered FDs)

$$\begin{pmatrix} 0 & & & & \\ & 1 & & & \\ & & \ddots & & \\ & & & 1 & \\ & & & & 0 \end{pmatrix} \begin{pmatrix} x_{0t} \\ x_{1t} \\ \vdots \\ x_{N-1t} \\ x_{Nt} \end{pmatrix} = \frac{1}{\Delta z^2} \begin{pmatrix} -\Delta z & \Delta z & & & \\ 1 & -2 & 1 & & \\ & \ddots & \ddots & \ddots & \\ & & 1 & -2 & 1 \\ & & & \Delta z & -\eta \Delta z \end{pmatrix} \begin{pmatrix} x_0 \\ x_1 \\ \vdots \\ x_{N-1} \\ x_N \end{pmatrix} \tag{3.179}$$

with $\eta = 1 + \alpha \Delta z$.

The system of Eq. (3.179) is in the form

$$\mathbf{M}\dot{\mathbf{x}} = \mathbf{f}(\mathbf{x}) \tag{3.180}$$

which can be solved using a DAE solver such as `ode15s`. Here the RHS $\mathbf{f}(\mathbf{x})$ is a linear expression of the form $\mathbf{Ax}$ since the spatial operator representing diffusion is linear, but more general nonlinear terms could appear (as for instance a chemical reaction term in a tubular reactor; an example is provided later on in Sect. 3.10.5). The coding of the main program requires the definition of a mass matrix $\mathbf{M}$, see the main script `diffusion_main_5` and functions `diffusion_pde_5` and `mass_diffusion`.

```
clear all

% set global variables
global alpha eta
global n D2 dz

% model parameter
alpha = 2;

% spatial domain
z0 = 0.0;
zL = 1.0;

% uniform spatial grid
n   = 201;
dz  = (zL - z0)/(n-1);
z   = [z0:dz:zL]';
eta = 1 + alpha*dz;

% initial conditions
x = ones(n,1);

% differentiation matrix
D2 = three_point_centered_uni_D2(z0,zL,n);
```

```
% call to ODE solver
options      = odeset('RelTol',1e-3,'AbsTol',1e-3,...
                      'Mass',mass_diffusion(n));
[tout,xout] = ode15s(@diffusion_pde_5,0:0.2:2,x,options);

% plot results
plot(z,xout);
xlabel('z');
ylabel('x(z,t)');
title('Diffusion equation')
```

Script diffusion_main_5 Main program for the evaluation of the MOL discretization of IVP (3.155)–(3.158) using the approach based on the formulation of the BCs as AEs.

```
function xt = diffusion_pde_5(t,x)

% set global variables
global alpha eta
global n D2 dz

% second - order spatial derivative
xzz = D2*x;

% temporal derivatives
xt = xzz;

% boundary condition at z = z0
xt(1) = (x(2)-x(1))/dz;

% boundary conditions at z = zL
xt(n) = (x(n-1)-eta*x(n))/dz;
```

Function diffusion_pde_5 Function `diffusion_pde` for the evaluation of the MOL discretization of IVP (3.155)–(3.158) using the approach based on the formulation of the BCs as AEs.

```
function M = mass_diffusion(n)

% Mass matrix
M      = eye(n);
M(1,1) = 0;
M(n,n) = 0;
M      = sparse(M);
```

Function mass_diffusion Function defining the mass matrix.

The equations representing the BCs in function `diffusion_pde_5` require some knowledge about the numerical approximation schemes. A simpler way to represent them is based on their definition (3.157)–(3.158) and the use of the library of differentiation matrices provided in the companion software to compute the spatial operators in the boundary conditions. In our simple application example, this can be accomplished by computing the first-order derivative using, for instance, the function `three_point_centered_uni_D1`, as shown in function `diffusion_pde_5_bis`. To make the code simple, the second-order derivative can be computed using stagewise differentiation (but this is of course not mandatory).

```
function xt = diffusion_pde_5_bis(t,x)

% set global variables
global alpha
global n D1 dz

% first and second - order spatial derivatives
xz  = D1*x;
xzz = D1*xz;

% temporal derivatives
xt = xzz;

% boundary condition at z = z0
xt(1) = xz(1);

% boundary conditions at z = zL
xt(n) = xz(n) + alpha*x(n);
```

Function diffusion_pde_5_bis Function `diffusion_pde` for the evaluation of the MOL discretization of IVP (3.155)–(3.158) using the approach based on the formulation of the BCs as AEs and differentiation matrices.

3.10.4 Tricks Inspired by the Previous Methods

When there is not much temporal and spatial activity at the boundaries, the previous approaches can be simplified, at the price of some approximation, to handle the BCs in a fast, yet effective way.

The first "trick" is based on the first two approaches, i.e., the substitution of the BCs in the expression of the finite difference approximation of the spatial derivatives in the PDEs. This substitution can be effected in a convenient way using stagewise differentiation. To illustrate this, let us consider again our diffusion equation, where the approximation of the second-order derivative can be computed by the successive application of a first-order differentiation matrix; see function `diffusion_pde_6` where the differentiation matrix `D1` is computed using function `three_point_centered_uni_D1`. The Neumann or mixed-type (Robin) BCs can be incorporated between the two differentiation stages. The resulting code is simple, easy to write and to understand for an external reader. However, remember that this approach is an approximation. Indeed, the values of the derivatives at the boundaries are reset to the values imposed by the BCs at intervals depending on time integration (each time the ODE solver calls the function), but the error introduced by this procedure is not controlled.

```
function xt = diffusion_pde_6(t,x)

% set global variables
global alpha
global n D1 dz

% second - order spatial derivative (stagewise)
xz     = D1*x;
xz(1)  = 0;
xz(n)  = -alpha*x(n);
xzz    = D1*xz;

% temporal derivatives
xt = xzz;
```

Function diffusion_pde_6 Function `diffusion_pde` for the evaluation of the MOL discretization of IVP (3.155)–(3.158) using substitution of the BCs in a stagewise differentiation

The second "trick" is based on the third approach and avoids the use of a DAE solver, and the definition of a mass matrix. It considers expression (3.178) and replaces the zero-factor by a small epsilon factor, i.e.

$$0 = 0\frac{\mathrm{d}x_0}{\mathrm{d}t}(t) = \frac{-x_0(t) + x_1(t)}{\Delta z}; \qquad 0 = 0\frac{\mathrm{d}x_N}{\mathrm{d}t}(t) = \frac{x_{N-1} - (1 + \alpha \Delta z) x_N}{\Delta z}$$

become

$$\varepsilon\frac{\mathrm{d}x_0}{\mathrm{d}t}(t) = \frac{-x_0(t) + x_1(t)}{\Delta z}; \qquad \varepsilon\frac{\mathrm{d}x_N}{\mathrm{d}t}(t) = \frac{x_{N-1} - (1 + \alpha \Delta z) x_N}{\Delta z} \tag{3.181}$$

so that the ODE system resulting from the MOL takes the form

$$\begin{pmatrix} x_{0t} \\ x_{1t} \\ \vdots \\ x_{N-1t} \\ x_{Nt} \end{pmatrix} = \frac{1}{\Delta z^2} \begin{pmatrix} -\Delta z/\varepsilon & \Delta z/\varepsilon & & & \\ 1 & -2 & 1 & & \\ & \ddots & \ddots & \ddots & \\ & & 1 & -2 & 1 \\ & & & \Delta z/\varepsilon & -\eta \Delta z/\varepsilon \end{pmatrix} \begin{pmatrix} x_0 \\ x_1 \\ \vdots \\ x_{N-1} \\ x_N \end{pmatrix} \tag{3.182}$$

However this ODE system is stiff and requires an implicit ODE solver. Function `diffusion_pde_5` changes slightly and is given in function `diffusion_pde_7`, where the epsilon factor appears in the equations defining the BCs. The use of a mass function - see function `mass_diffusion`- is no longer required. Of course the use of the more physical representation of the BCs, as in function `diffusion_pde_5_bis`, is also possible.

```
function xt = diffusion_pde_7(t,x)

% set global variables
global alpha eta epsilon
global n D2 dz

% second - order spatial derivative
xzz = D2*x;

% temporal derivatives
xt = xzz;

% boundary condition at z = z0
xt(1)=(x(2)-x(1))/(epsilon*dz);

% boundary conditions at z = zL
xt(n) = (x(n-1)-eta*x(n))/(epsilon*dz);
```

Function diffusion_pde_7 Function `diffusion_pde` for the evaluation of the MOL discretization of IVP (3.155)–(3.158) using the approach based on the formulation of the BCs as singularly perturbed ODEs

3.10.5 An Illustrative Example (with Several Boundary Conditions)

To illustrate the coding of a more complex model, and in particular the implementation of the BCs, we consider a tubular fixed bed reactor consisting of three sections: an inert entrance section, an active catalyst section, and an inert outlet section [8]. A dynamic model is developed to study benzene hydrogenation (an exothermic reaction) and the poisoning kinetics of thiophene on a nickel catalyst. This model includes material balance equations for benzene and thiophene, an energy balance, and an equation accounting for catalyst deactivation. Since hydrogen is in great excess in all the experiments considered in [8], the hydrogen concentration is essentially constant.

$$\frac{\partial c_B}{\partial t} = -v\frac{\partial c_B}{\partial z} + D_B\frac{\partial^2 c_B}{\partial^2 z} + \frac{\rho_c}{\varepsilon} r_B(\theta_A, c_B, T) \tag{3.183}$$

$$\frac{\partial c_T}{\partial t} = -v\frac{\partial c_T}{\partial z} + D_T\frac{\partial^2 c_T}{\partial^2 z} + \frac{\rho_c}{\varepsilon} r_T(\theta_A, c_T, T) \tag{3.184}$$

$$\begin{aligned}\frac{\partial T}{\partial t} &= -\varepsilon v\frac{\rho_G c_{pG}}{\bar{\rho}\bar{c}_p}\frac{\partial T}{\partial z} + \frac{\lambda_{eff}}{\bar{\rho}\bar{c}_p}\frac{\partial^2 T}{\partial^2 z} + \frac{2\alpha}{R\bar{\rho}\bar{c}_p}(T_w - T)\\ &\quad - \frac{(-\Delta H)}{\bar{\rho}\bar{c}_p}\rho_c r_B(\theta_A, c_B, T)\end{aligned} \tag{3.185}$$

$$\frac{\partial \theta_A}{\partial t} = r_d(\theta_A, c_T, T) \tag{3.186}$$

where c_B and c_T are the concentrations of benzene and thiophene, T is the temperature, and θ_A is the catalyst activity.

In these expressions, v is the gas velocity, D_B and D_T are diffusion coefficients, λ_{eff} is the bed effective thermal conductivity, ρ_c and ρ_G are the catalyst and gas densities, $\bar{\rho}\bar{c}_p$ is the average volumetric heat capacity, ε is the bed void fraction, r_B, r_T and r_d are the rates of hydrogenation, thiophene chemisorption and poisoning, $(-\Delta H)$ is the heat of benzene hydrogenation, α is the wall heat transfer coefficient, T_w is the wall temperature, R and L are the reactor radius and length. The numerical values of all these parameters can be found in [8] and in the code `three_zone_reactor_main`.

As the reactor is subdivided into three sections, Eqs. (3.183)–(3.185) have to be solved with $r_B = 0$ and $r_T = 0$ in the inert sections. The complete set of Eqs. (3.183)–(3.186) has to be solved in the catalyst section. A total of 10 PDEs have therefore to be solved simultaneously.

In $z = 0$, and $z = L$ (reactor inlet and outlet), Dirichlet boundary conditions apply

$$c_B(t, z = 0) = c_{B,\text{in}}(t); \qquad c_T(t, z = 0) = c_{T,\text{in}}(t) \tag{3.187}$$

$$T(t, z = 0) = T_{\text{in}}(t) \tag{3.188}$$

$$c_{Bz}(t, z = L) = 0; \qquad c_{Tz}(t, z = L) = 0 \tag{3.189}$$

$$T_z(t, z = L) = 0 \tag{3.190}$$

At the section interfaces $z = L_1$ and $z = L_1 + L_2$ (where L_1 is the length of the entrance section, and L_2 is the length of the catalyst section -note that $L_1/L = 0.28$ and $(L_1 + L_2)/L = 0.47$), special continuity boundary conditions apply, e.g.,

$$c_B(t, z = L_1^-) = c_B(t, z = L_1^+); \qquad c_T(t, z = L_1^-) = c_T(t, z = L_1^+) \tag{3.191}$$

$$T(t, z = L_1^-) = T(t, z = L_1^+) \tag{3.192}$$

$$c_{Bz}(t, z = L_1^-) = c_{Bz}(t, z = L_1^+); \qquad c_{Tz}(t, z = L_1^-) = c_{Tz}(t, z = L_1^+) \tag{3.193}$$

$$T_z(t, z = L_1^-) = T_z(t, z = L_1^+) \tag{3.194}$$

where L_1^- denotes the rightmost limit of the entrance section, and L_1^+ denotes the leftmost limit of the catalyst section. Equations (3.191)–(3.192) express concentration and temperature continuity between sections, whereas Eqs. (3.193)–(3.194) express the continuity of mass and energy fluxes (here, diffusion and heat conduction are the same in both sections so that the diffusion and thermal conductivity coefficients do not appear). Similar expressions hold at $z = L_1 + L_2$.

If abrupt time variations are expected in the inlet (forcing) variables $c_{B,\text{in}}(t)$, $c_{T,\text{in}}(t)$, $T_{\text{in}}(t)$, then BCs (3.187)–(3.188) are best implemented as ODEs, e.g.

$$\frac{\mathrm{d}T}{\mathrm{d}t}(z = 0, t) = \frac{1}{\tau}(T_{\text{in}}(t) - T(z = 0, t)) \tag{3.195}$$

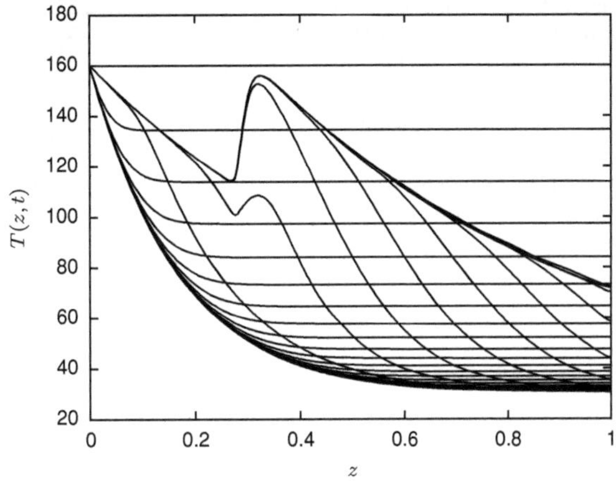

Fig. 3.18 Numerical solution of the IBVP (3.183)–(3.194) using 5-point centered FDs for the second-order derivative and 5-point biased upwind FDs for the first-order derivative—preheating followed by normal operation (benzene hydrogenation)

where τ is a small time constant (equivalent to the epsilon factor introduced earlier) representing the dynamics of a temperature change at the reactor inlet. In fact, abrupt time variations are always the results of mathematical idealization, and τ models, in a very simple way, the actuator dynamics.

The 15 other boundary and continuity conditions can be implemented as algebraic equations, leading, after approximation of the spatial derivatives in the PDEs and BCs, to a large system of DAEs. Here, finite differences (five-point biased-upwind FDs for first derivatives and five-point centered FDs for second derivatives) are used with $n_1 = 71, n_2 = 71$ and $n_3 = 21$ grid points in the first, second and third reactor section, respectively. The resulting 560 DAEs can be solved efficiently using `ode15s`.

Figure 3.18 shows the evolution of the temperature profile inside the reactor operated in two distinct modes: (a) reactor pre-heating with $T_{\text{in}}(t) = 160\ °C$ (for $0 < t \leq 20$ min), and (b) benzene hydrogenation with $T_{\text{in}}(t) = 160\ °C$ and $c_{B,\text{in}}(t)$ corresponding to a mole fraction $x_{B,\text{in}}(t) = 0.066$ (for $20 < t \leq 40$ min). During phase (a), the temperature gradually decreases from the inlet to the outlet of the reactor, due to heat exchange with the reactor jacket. During phase (b), an exothermic reaction takes place, leading to the formation of a "hot-spot" at about one third of the reactor length.

Figures 3.19, 3.20 and 3.21 illustrate the effect of catalyst poisoning. From $t =$ 40 min, a small quantity of thiophene is fed to the reactor. Consequently, catalyst activity decays (Fig. 3.21), and traveling waves of concentration and temperature form (Figs. 3.19 and 3.20).

The MATLAB codes for solving this problem are `three_zone_reactor_main`, `three_zone_reactor_pdes` and `mass_react`.

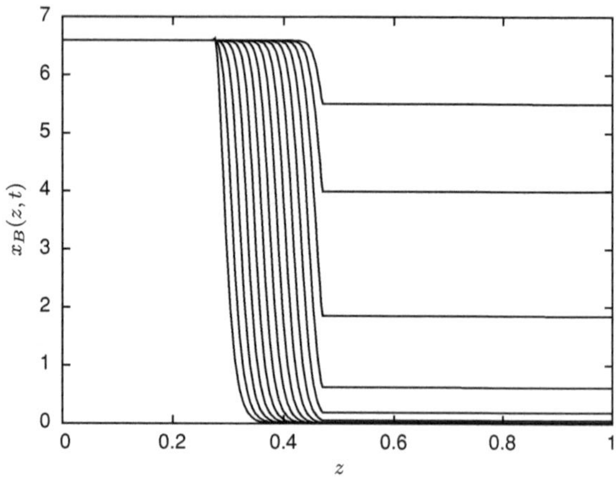

Fig. 3.19 Numerical solution of the IBVP (3.183)–(3.194) using 5-point centered FDs for the second-order derivative and 5-point biased upwind FDs for the first-order derivative—evolution of concentration during catalyst poisoning

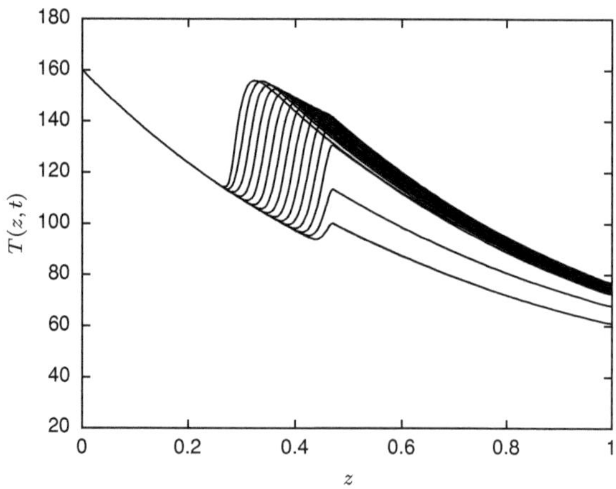

Fig. 3.20 Numerical solution of the IBVP (3.183)–(3.194) using 5-point centered FDs for the second-order derivative and 5-point biased upwind FDs for the first-order derivative—evolution of temperature during catalyst poisoning

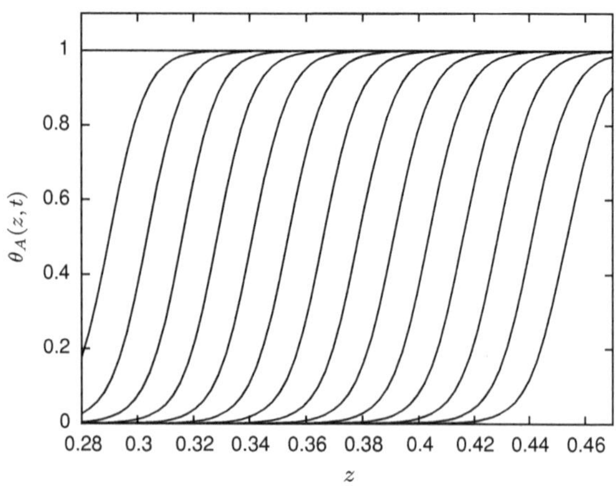

Fig. 3.21 Numerical solution of the IBVP (3.183)–(3.194) using 5-point centered FDs for the second-order derivative and 5-point biased upwind FDs for the first-order derivative—evolution of catalyst activity during catalyst poisoning

```
close all
clear all

% start a stopwatch timer
tic

% set global variables
global v DB DT rhocp cpg rhog Tw Rr DH lH2 Dp leff eps;
global ki0 K0 Q EB R_kcal R_J kd0 Ed MT;
global Ptot cBin cTin Tin cHin xHin;
global z01 zL1 z02 zL2 z03 zL3 n1 n2 n3 z1 z2 z3;
global D1_1 D2_1 D1_2 alpha rhoc;
global D2_2 D1_3 D2_3;

% model parameters
% experiment G3 (catalyst pretreatment first)
Ptot   = 1e5;                    % total pressure (Pa)
Tin    = 160 + 273.16;           % inlet temperature (K)
T0     = 160 + 273.16;           % initial temperature (K)
Tw     = 29 + 273.16;            % wall temperature  (K)
xBin   = 0;                      % mole fraction benzene
R_kcal = 1.987;                  % gas constant (kcal/kmol K)
R_J    = 8.314e3;                % gas constant (J/kmol K)
cBin   = xBin*Ptot/(R_J*Tin);
xHin   = 1;                      % mole fraction hydrogen
cHin   = xHin*Ptot/(R_J*Tin);
xTin   = 0;                      % mole fraction thiophene
cTin   = xTin*Ptot/(R_J*Tin);
flow   = 1551.5e-6/60;           % total flow rate (m^3/s)
theta0 = 1;
%
DB = 4e-5;                  % diffusion coefficients (m^2/s)
DT = 4e-5;
eps = 0.6;                  % void fraction
%
MW = 2.106*xHin + 78.12*xBin; % molar weight of the gas
                              % mixture  (kg/kmol)
rhog = MW*273.16*Ptot*0.0075/(22.41*760*Tin); % gas dens
                                              % (kg/m^3)
cpg = 6.935*xHin + 23.15*xBin;  % heat capacity of the gas
                                % (kcal/kg K)
rhocp = 175;                    % average heat capacity
                                % (kcal/m^3 K)
lH2 = 733*0.019/(100*3600); % thermal conductivity of the
                            % gas  (kcal/s m K)
rhoc = 354*2/3; % catalyst density (kg/m^3) with a
                % dilution factor  of 2
Dp = 1.25e-3;   % particle diameter (m) (average between
                % 0.75 and 1.8 mm)
Rr = 0.822e-2;              % reactor radius (m)
SA = pi*Rr^2;               % cross-section area (m^2)
v = flow/(eps*SA);          % intersticial velocity (m/s)
leff = 7*lH2 + 0.8*rhog*cpg*v*eps*Dp; % effective thermal
                              % conductivity (kcal/s m K)
alpha = 2.6e-3;  % heat transfer coefficient (kcal/m^2 K)
%
% benzene hydrogenation kinetics
EB = 13770;                 % activation energy (kcal/kmol)
```

```
ki0 = 4.22*0.0075;        % rate constant (kmol/kg s Pa)
Q = -16470;               % heat of adsorption (kcal/kmol)
K0  = 4.22e-11*0.0075;    % adsorption constant (1/torr)
%
% thiophene poisoning kinetics
Ed  = 1080;               % activation energy (kcal/kmol)
kd0 = (2.40e-2)*0.0075;   % pre-exponential factor (1/Pa s)
MT  = 1.03e-3; % catalyst adsorption capacity for
               % thiophene (kmol/kg)
DH = 49035;    % heat of reaction (kcal/kmole)

% spatial grid
L   = 0.5;                      % reactor length (m)
z01 = 0.0;
zL1 = 0.14;                     % entrance (inert) section
z02 = zL1;
zL2 = z02 + 0.095;              % catalyst section
z03 = zL2;
zL3 = L;                        % exit (inert) section
n1  = 71;
n2  = 71;
n3  = 71;
dz1 = (zL1 - z01)/(n1-1);
dz2 = (zL2 - z02)/(n2-1);
dz3 = (zL3 - z03)/(n3-1);
z1  = [z01:dz1:zL1]';
z2  = [z02:dz2:zL2]';
z3  = [z03:dz3:zL3]';
z   = [z1;z2;z3];

% differentiation matrix in zone 1 (convective and
% diffusive terms)
D1_1 = five_point_biased_upwind_D1(z1,v);
D2_1 = five_point_centered_D2(z1);

% differentiation matrix in zone 2 (convective and
% diffusive terms)
D1_2 = five_point_biased_upwind_D1(z2,v);
D2_2 = five_point_centered_D2(z2);

% differentiation matrix in zone 3 (convective and
% diffusive terms)
D1_3 = five_point_biased_upwind_D1(z3,v);
D2_3 = five_point_centered_D2(z3);

% initial conditions
cB1   = cBin*ones(size(z1));
cT1   = cTin*ones(size(z1));
T1    = T0*ones(size(z1));
cB2   = cBin*ones(size(z2));
cT2   = cTin*ones(size(z2));
T2    = T0*ones(size(z2));
theta = theta0*ones(size(z2));
cB3   = cBin*ones(size(z3));
cT3   = cTin*ones(size(z3));
T3    = T0*ones(size(z3));

% initial conditions in x
x(1:3:3*n1-2,1)                   = cB1;
```

```
x(2:3:3*n1-1,1)                        = cT1;
x(3:3:3*n1,1)                          = T1;
x(3*n1+1:4:3*n1+4*n2-3,1)              = cB2;
x(3*n1+2:4:3*n1+4*n2-2,1)              = cT2;
x(3*n1+3:4:3*n1+4*n2-1,1)              = T2;
x(3*n1+4:4:3*n1+4*n2,1)                = theta;
x(3*n1+4*n2+1:3:3*n1+4*n2+3*n3-2,1) = cB3;
x(3*n1+4*n2+2:3:3*n1+4*n2+3*n3-1,1) = cT3;
x(3*n1+4*n2+3:3:3*n1+4*n2+3*n3,1)   = T3;

% Time span
t = 60*[0:1:30]; % sec

% call to ODE solver
M       = mass_react;
options = odeset('Mass',M,'MassSingular','yes',...
                 'RelTol', 1e-3, 'AbsTol',1e-3);
[tout,yout] = ode15s(@three_zone_reactor_pdes_preheat,...
                     t,x,options);

% Recover the temperature in the three zones
T1_out  = yout(:,3:3:3*n1);
T2_out  = yout(:,3*n1+3:4:3*n1+4*n2-1);
T3_out  = yout(:,3*n1+4*n2+3:3:3*n1+4*n2+3*n3);

% Assembly of temperature and spatial coordinates
T_out   = [T1_out T2_out T3_out]-273.16;
zz      = [z1; z2; z3]/L;

% Plot the solution
plot(zz,T_out)

% read the stopwatch timer
tcpu=toc;
```

Script three_zone_reactor_main Main program for the solution of IVP (3.183) using the approach based on the formulation of the BCs as AEs and singularly perturbed ODEs.

```
function xt = three_zone_reactor_pdes_preheat(t,x)

% set global variables
global v DB DT rhocp cpg rhog Tw Rr DH lH2 Dp leff eps;
global ki0 K0 Q EB R_kcal R_J kd0 Ed MT;
global Ptot cBin cTin Tin cHin xHin;
global z01 zL1 z02 zL2 z03 zL3 n1 n2 n3 z1 z2 z3;
global alpha rhoc D1_1 D2_1 D1_2;
global D2_2 D1_3 D2_3;

% Transfer dependent variables
cB1 = x(1:3:3*n1-2,1);
cT1 = x(2:3:3*n1-1,1);
T1  = x(3:3:3*n1,1);
%
cB2   = x(3*n1+1:4:3*n1+4*n2-3,1);
cT2   = x(3*n1+2:4:3*n1+4*n2-2,1);
T2    = x(3*n1+3:4:3*n1+4*n2-1,1);
theta = x(3*n1+4:4:3*n1+4*n2,1);
```

```
%
cB3 = x(3*n1+4*n2+1:3:3*n1+4*n2+3*n3-2,1);
cT3 = x(3*n1+4*n2+2:3:3*n1+4*n2+3*n3-1,1);
T3  = x(3*n1+4*n2+3:3:3*n1+4*n2+3*n3,1);

% spatial derivatives - 1st zone
cB1z = D1_1*cB1;
cT1z = D1_1*cT1;
T1z  = D1_1*T1;
%
cB1zz = D2_1*cB1;
cT1zz = D2_1*cT1;
T1zz  = D2_1*T1;

% spatial derivatives - 2nd zone
cB2z = D1_2*cB2;
cT2z = D1_2*cT2;
T2z  = D1_2*T2;
%
cB2zz = D2_2*cB2;
cT2zz = D2_2*cT2;
T2zz  = D2_2*T2;

% spatial derivatives - 3rd zone
cB3z = D1_3*cB3;
cT3z = D1_3*cT3;
T3z  = D1_3*T3;
%
cB3zz = D2_3*cB3;
cT3zz = D2_3*cT3;
T3zz  = D2_3*T3;

% several operating conditions

% 1) catalyst pretreatment with hydrogen at 160 Celsius
%    (for 20 min)
rB = 0;          % reaction rates in 2nd zone
rd = 0;
rT = 0;

% 2) benzene hydrogenation (experiment G3)
if (t > 20*60) & (t < 30*60)
    %
    Tin  = 160 + 273.16;          % inlet temperature (K)
    xBin = 2*0.033;               % mole fraction benzene
    cBin = xBin*Ptot/(R_J*Tin);
    xHin = 1-xBin;                % mole fraction hydrogen
    cHin = xHin*Ptot/(R_J*Tin);
    xTin = 0;                     % mole fraction thiophene
    cTin = xTin*Ptot/(R_J*Tin);
    %
    MW = 2.106*xHin + 78.12*xBin; % gas mixture molar
                                  % weight (kg/kmol)
    rhog = MW*273.16*Ptot*0.0075/(22.41*760*Tin); % density
    cpg = 6.935*xHin + 23.15*xBin; % heat capacity of the
                                   % gas (kcal/kg K)
    leff = 7*lH2 + 0.8*rhog*cpg*v*eps*Dp;
    %
    xB2 = cB2./(cB2+cT2+cHin); % reaction rates 2nd zone
    rB = ki0*K0*Ptot^2*(xHin*xB2.*theta.*exp((-Q - EB)./...
         (R_kcal*T2)))./(1+K0*Ptot*xB2.*...
         exp(-Q./(R_kcal*T2)));
    rd = 0;
    rT = 0;

    % 3) catalyst poisoning (experiment G3)
```

```
elseif t > 60*120
    Tin = 160 + 273.16;              % inlet temperature (K)
    xBin = 2*0.033;                  % mole fraction benzene
    cBin = xBin*Ptot/(R_J*Tin);
    xHin = 1-xBin;                  % mole fraction hydrogen
    cHin = xHin*Ptot/(R_J*Tin);
    xTin = 1.136*xBin/100;         % mole fraction thiophene
    cTin = xTin*Ptot/(R_J*Tin);
    %
    MW = 2.106*xHin + 78.12*xBin; % gas mixture molar
                                   % weight (kg/kmol)
    rhog = MW*273.16*Ptot*0.0075/(22.41*760*Tin); % gas
                                                  % density
    cpg = 6.935*xHin + 23.15*xBin;  % heat capacity of the
                                    % gas (kcal/kg K)
    leff = 7*lH2 + 0.8*rhog*cpg*v*eps*Dp;
    xB2 = cB2./(cB2+cT2+cHin); % reaction rates 2nd zone
    xT2 = cT2./(cB2+cT2+cHin);
    rB = ki0*K0*Ptot^2*(xHin*xB2.*theta.*exp((-Q-...
        EB)./(R_kcal*T2)))./(1+K0*Ptot*xB2.*...
        exp(-Q./(R_kcal*T2)));
    rd = kd0*Ptot*xT2.*theta.*exp(-Ed./(R_kcal*T2));
    rT = MT*rd;
end

% temporal derivatives
cB1t = -v*cB1z + DB*cB1zz;
cT1t = -v*cT1z + DT*cT1zz;
T1t  = -((eps*v*rhog*cpg)/rhocp)*T1z +...
       (leff/rhocp)*T1zz + 2*alpha*(Tw - T1)/(Rr*rhocp);
%
cB2t = -v*cB2z + DB*cB2zz - rhoc*rB/eps;
cT2t = -v*cT2z + DT*cT2zz - rhoc*rT/eps;
T2t = -((eps*v*rhog*cpg)/rhocp)*T2z +...
      (leff/rhocp)*T2zz + 2*alpha*(Tw - T2)/(Rr*rhocp) +...
      (DH/rhocp)*rhoc*rB;
thetat = -rd;
%
cB3t = -v*cB3z + DB*cB3zz;
cT3t = -v*cT3z + DT*cT3zz;
T3t = -((eps*v*rhog*cpg)/rhocp)*T3z +...
      (leff/rhocp)*T3zz + 2*alpha*(Tw - T3)/(Rr*rhocp);

% boundary conditions at z = z01
cB1t(1) = cBin - cB1(1);
cT1t(1) = cTin - cT1(1);
T1t(1) = Tin - T1(1);

% boundary conditions at z = zL1 = z02
cB1t(n1) = cB2z(1) - cB1z(n1);
cT1t(n1) = cT2z(1) - cT1z(n1);
T1t(n1) = T2z(1) - T1z(n1);
%
cB2t(1) = cB1(n1) - cB2(1);
cT2t(1) = cT1(n1) - cT2(1);
T2t(1) = T1(n1) - T2(1);

% boundary conditions at z = zL2 = z03
cB2t(n2) = cB3z(1) - cB2z(n2);
cT2t(n2) = cT3z(1) - cT2z(n2);
T2t(n2) = T3z(1) - T2z(n2);
%
cB3t(1) = cB2(n2) - cB3(1);
cT3t(1) = cT2(n2) - cT3(1);
T3t(1) = T2(n2) - T3(1);
```

```
% boundary conditions at z = zL
cB3t(n3) = - cB3z(n3);
cT3t(n3) = - cT3z(n3);
T3t(n3) = - T3z(n3);

% Transfer temporal derivatives
xt(1:3:3*n1-2,1) = cB1t;
xt(2:3:3*n1-1,1) = cT1t;
xt(3:3:3*n1,1) = T1t;
%
xt(3*n1+1:4:3*n1+4*n2-3,1) = cB2t;
xt(3*n1+2:4:3*n1+4*n2-2,1) = cT2t;
xt(3*n1+3:4:3*n1+4*n2-1,1) = T2t;
xt(3*n1+4:4:3*n1+4*n2,1) = thetat;
%
xt(3*n1+4*n2+1:3:3*n1+4*n2+3*n3-2,1) = cB3t;
xt(3*n1+4*n2+2:3:3*n1+4*n2+3*n3-1,1) = cT3t;
xt(3*n1+4*n2+3:3:3*n1+4*n2+3*n3,1) = T3t;
```

Function three_zone_reactor_pdes Function for the for the solution of IBVP (3.155)–(3.158) using the approach based on the formulation of the BCs as AEs and singularly perturbed ODEs.

```
function M = mass_react

% set global variables
global n1 n2 n3;

% Assemble mass matrix
M = diag([1 1 1 ones(1,3*n1-6) 0 0 0 0 0 0 1 ...
        ones(1,4*n2-8) 0 0 0 1 0 0 0 ones(1,3*n3-6) ...
        0 0 0],0);
```

Function mass_react Mass matrix used in the solution of IBVP (3.155)–(3.158) using the approach based on the formulation of the BCs as AEs and singularly perturbed ODEs.

3.11 Computing the Jacobian Matrix of the ODE System

At this stage, we know how to solve an IBVP using the MOL and finite difference approximations. As we have seen, the semi-discrete system of ODEs resulting from the MOL can be stiff and require the use of an implicit solver such as `ode15s`. This kind of solvers make use of the Jacobian matrix of the ODE system, and it is interesting to take a closer look at the structure of this matrix and at how it can be computed in practice. To this end, we will consider a simple example taken from [9], which represents a tubular bioreactor where bacteria (biomass X) grows on a single substrate S while producing a product P. Biomass mortality is also taken into account. The macroscopic reaction scheme is given by

$$\nu_1 S \xrightarrow{\varphi_1} X + \nu_2 P$$

$$X \xrightarrow{\varphi_2} X_d$$

The biomass grows on a fixed bed while the substrate S is in solution in the flowing medium. Mass balance PDEs can be written as

$$
\begin{aligned}
S_t &= -vS_z - \nu_1 \frac{1-\varepsilon}{\varepsilon}\varphi_1 \\
& \qquad\qquad\qquad\qquad 0 \le z \le L; \quad t \ge 0 \\
X_t &= \varphi_1 - \varphi_2
\end{aligned}
\tag{3.196}
$$

In the substrate mass balance PDE, $v = F/(\varepsilon A)$ is the superficial velocity of the fluid flowing through the bed, ε is the total void fraction (or bed porosity), and φ_1 is the growth rate given by a model of Contois

$$
\varphi_1 = \mu_{\max} \frac{S}{k_c X + S} \tag{3.197}
$$

where $\mu_{\max}$ is the maximum specific growth rate and k_c is the saturation coefficient. This kinetic model represents the effects of substrate limitation and biomass inhibition. In the biomass balance ODE (note that it is an ODE and not a PDE as biomass is fixed on the bed, and there is neither convection nor diffusion), φ_2 is the death rate given by a simple linear law

$$
\varphi_2 = k_d X \tag{3.198}
$$

The model is supplemented by a Dirichlet boundary condition in $z = 0$

$$
S(t, z = 0) = S_{\text{in}}(t) \tag{3.199}
$$

and initial conditions

$$
S(t = 0, z) = S^0(z) \tag{3.200}
$$

$$
X(t = 0, z) = x^0(z) \tag{3.201}
$$

The numerical values of the model parameters are: $L = 1\,\text{m}$, $A = 0.04\,\text{m}^2$, $\varepsilon = 0.5$, $F = 2l \cdot h^{-1}$, $\nu_1 = 0.4$, $\mu_{\max} = 0.35\,\text{h}^{-1}$, $k_c = 0.4$, $k_d = 0.05\,\text{h}^{-1}$.

This problem is solved using a spatial grid with 101 points and 2-point upwind finite differences. The code of this example is available in the companion software. Here, we focus attention on the call to ode15s in the main program:

```
% compute the Jacobian matrix
options = odeset(options,'Jacobian',@jacobian_num);
% options = odeset(options,'Jacobian',@jacobian_complex
                    _diff);

% or compute the sparcity pattern
% S = jpattern_num
% S = jpattern_complex_diff
% options = odeset(options,'JPattern',S)
```

```
%
[tout, yout] = ode15s(@bioreactor_pdes,[t0:Dtplot:tf],
                        x,options);
```

Several options are coded. The first options consists in providing a Jacobian to ode15s, which is evaluated numerically in function `jacobian_num`.

```
function Jac = jacobian_num (t,x)

% numerical Jacobian matrix
xt         = bioreactor_pdes(t,x);
fac        = [];
thresh     = 1e-6;
threshv    = thresh*ones(length(x),1);
vectorized = 0;
[Jac, fac] = numjac(@bioreactor_pdes,t,x,xt,threshv,...
                    fac,vectorized);
```

Function jacobian_num Function to numerically evaluate the Jacobian.

This function calls the PDE function `bioreactor_pdes` to evaluate the RHS of the ODE system at a base point (t, x) and then calls the MATLAB function `numjac` to compute numerically the Jacobian matrix $\partial f(t,x)/\partial x$ using forward finite difference approximations

$$\partial f(t,x)/\partial x = \frac{f(t, x+\Delta) - f(t,x)}{\Delta}$$

with adaptive step Δ. The coding of `numjac` is rather complex and not very well documented as it is intended to be called by implicit solvers.

The second option uses an alternative method for evaluating the Jacobian matrix based on complex step differentiation [10, 11] inspired from a code by Cao [12] (see function `jacobian_complex_diff`).

```
function Jac = jacobian_complex_diff(t,x)

% Jacobian through complex step differentiation

n = length(x);
Jac = zeros(n,n);
h = n*eps;
for k = 1:n
    x1 = x;
    x1(k) = x1(k)+h*i;
    xt = bioreactor_pdes(t,x1);
    Jac(:,k)=imag(xt)/h;
end
```

Function jacobian_complex_diff Function to numerically evaluate the Jacobian.

In finite difference approximations, such as the one used in numjac, a compromise has to be reached between a small step Δ to minimize the truncation error and a larger step Δ to avoid large subtractive errors (as the step Δ is reduced, $f(t, x+\Delta)$ and $f(t, x)$ get very close). The idea of complex step differentiation is to replace the real step Δ by an imaginary number ih (where $i^2 = -1$). If we consider the Taylor series expansion around x.

$$f(t, x+ih) = f(t, x) + ih\frac{\partial f(t, x)}{\partial x} - \frac{h^2}{2}\frac{\partial^2 f(t, x)}{\partial x^2} - \frac{ih^3}{6}\frac{\partial^3 f(t, x)}{\partial x^3} + \cdots$$

Taking the imaginary part of both sides and dividing by h leads to

$$\frac{\mathrm{Im}\left[f(t, x+ih)\right]}{h} = \frac{\partial f(t, x)}{\partial x} - \frac{h^2}{6}\frac{\partial^3 f(t, x)}{\partial x^3} + \cdots$$

So that a second-order approximation of the first-order derivative is given by

$$\frac{\partial f(t, x)}{\partial x} = \frac{\mathrm{Im}\left[f(t, x+ih)\right]}{h} + O(h^2)$$

which is the formula implemented in function jacobian_complex_diff. The main advantage of this formula is that it is not subject to subtractive cancellation so that a very small step h can be taken (for instance related to the spacing of floating point numbers defined in MATLAB as eps).

A third option consists of computing the sparcity pattern of the Jacobian, i.e. compute a map of the Jacobian where nonzero elements are indicated by “1” and zero elements by “0”. This map can be used by ode15s to compute more efficiently the Jacobian (see function jpattern_num).

```
function S = jpattern_num

% Set global variables
global nz

% sparsity pattern of the Jacobian matrix based on a
% numerical evaluation
tstar = 0;
Sstar = linspace(5,3,nz);
Xstar = linspace(0,100,nz);
xstar(1:2:2*nz-1,1) = Sstar';
xstar(2:2:2*nz,1) = Xstar';
xtstar = bioreactor_pdes(tstar,xstar);
fac = [];
thresh = 1e-6;
vectorized = 0;
[Jac, fac] = numjac(@bioreactor_pdes,tstar,xstar,...
                    xtstar,thresh,fac,vectorized);

% replace nonzero elements by "1"
% (so as to create a "0-1" map of the Jacobian matrix)
```

```
S = spones(sparse(Jac));

% plot the map
figure
spy(S,'k');
xlabel('dependent variables');
ylabel('semi-discrete equations');

% compute the percentage of non - zero elements
[njac,mjac]      = size(S);
ntotjac          = njac*mjac;
non_zero         = nnz(S);
non_zero_percent = non_zero/ntotjac*100;
stat             =...
sprintf('Jacobian sparsity pattern-nonzeros %d (%.3f)',...
        non_zero,non_zero_percent);
title(stat);
```

Function jpattern_num Function to numerically evaluate the Jacobian.

This function is called once in the main program, before the call to ode15s, so as to define the structure of the Jacobian matrix. It uses the PDE function `bioreactor_pdes` to evaluate the RHS of the ODE system at a base point (t^*, x^*)—nonzero values are arbitrarily defined for the biomass and substrate spatial profiles using `linspace`—and then calls the MATLAB function `numjac` to compute numerically the Jacobian matrix $\partial f(t,x)/\partial x$ using forward finite difference approximations. The MATLAB function `spones` is then used to create the "0-1" map. A plot of the matrix can be obtained using the function `spy` (see Fig. 3.22).

As we can see, this matrix has a nice banded structure, i.e., the nonzero elements are concentrated near the main diagonal. This is due to the fact that the dependent variables containing the biomass and substrate values are stored according to the grid point order, i.e., $S(1)$, $X(1)$, $S(2)$, $X(2)$, $S(3)$, $X(3)$,..., as indicated in the PDE function `bioreactor_pdes`.

```
function xt = bioreactor_pdes(t,x)

% set global variables
global v nu1 eps mumax Kc kd;
global Sin gamma
global z0 zL z nz D1z;

% transfer dependent variables
S = x(1:2:2*nz-1,1);
X = x(2:2:2*nz,1);

% Temporal derivatives
%
% Entering conditions (z = 0)
S_t(1,1) = gamma*(Sin - S(1));
%
% kinetics
phi1 = mumax*S.*X./(Kc*X+S);
phi2 = kd*X;

% spatial derivatives
```

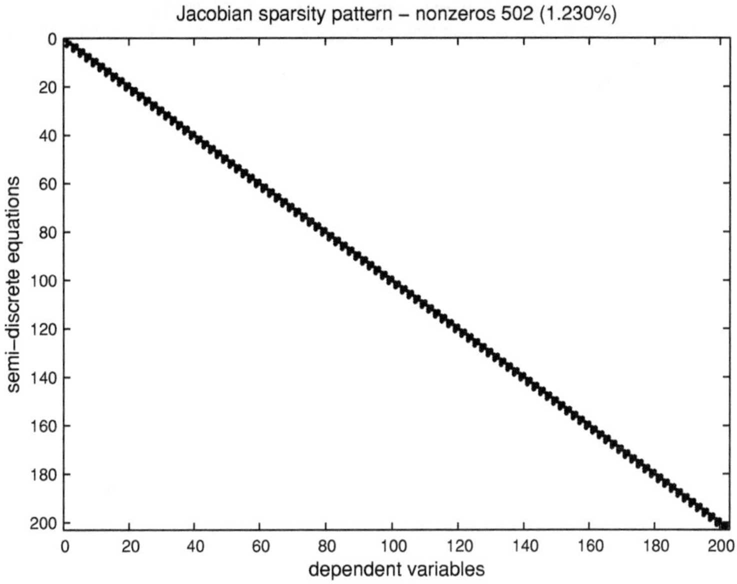

Fig. 3.22 Sparcity pattern of the Jacobian matrix of the bioreactor example when using variable storage according to the spatial grid points

```
%
% Sz  (by two - point upwind approximations)
Sz = D1z*S;

% Rest of bioreactor
S_t(2:nz,1) = -v*Sz(2:nz) - nu1*(1-eps)*phi1(2:nz)/eps;
X_t = phi1 - phi2;

% transfer temporal derivatives
xt(1:2:2*nz-1,1) = S_t;
xt(2:2:2*nz,1)   = X_t;
```

Function bioreactor_pdes Programming of the PDE equations for the bioreactor system

If these variables would have been stored in another way, for instance one vector after the other

```
S = x(1:nz,1);
X = x(nz+1:2*nz,1);
```

then the structure of the Jacobian would have been altered (see Fig. 3.23).

The first coding gives a more elegant structure to the Jacobian matrix which could be more easily exploited in ODE solvers with computational procedures adapted to banded Jacobian matrices.

In all cases, note that the Jacobian matrix is a quite empty matrix, i.e., most of the elements are zero, and it is of course advantageous to take this fact into account when

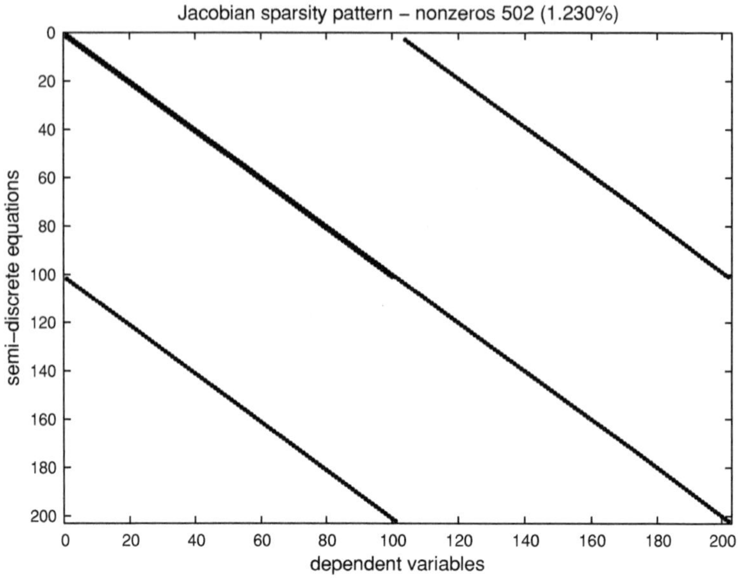

Fig. 3.23 Sparcity pattern of the Jacobian matrix of the bioreactor example when using storage of the variable vectors one after the other

evaluating the Jacobian. It is indeed a waste of computational resources to evaluate repeatedly elements that are known in advance to be zero.

We conclude this section in mentioning that the Jacobian information can also be exploited in our implementation of the Rosenbrock method `ros23p`:

```
% call to ODE solver
%
t0 = 0.0;
tf = 50.0;
hmin = 1e-3;
nstepsmax = 1000;
abstol = 1e-3;
reltol = 1e-3;
Dtplot = 1.0;
%
[tout, yout] = ros23p_solver(@bioreactor_pdes,...
                    @jacobian_complex_diff,@ft,t0,tf,
                     x,hmin,...
                    nstepsmax, abstol, reltol,Dtplot);
```

3.12 Solving PDEs Using SCILAB and OCTAVE

The tubular bioreactor example (see Eqs. (3.196)–(3.201)) is now used to illustrate the solution of PDE models using SCILAB and OCTAVE.

Most of our home-made methods for time integration and spatial discretization can can easily be translated to SCILAB and OCTAVE, as, for instance, the library of FD differentiation matrices or the complex differentiation method introduced in the previous sections. OCTAVE is very close to MATLAB so that the translation usually requires minimal efforts. The syntax of SCILAB departs more from MATLAB and the translation can be facilitated using a tool called `mfile2sci`.

Let us first consider the SCILAB implementation. The main script `bioreactor_main.sci`, which shares many of the features already presented in Sect. 2.6, has the following structure:

- The script begins by setting the global variables to be passed to the SCILAB functions
- The functions are loaded with the SCILAB command `exec`
- The main difference with respect to the example in Sect. 2.6 is that now we are considering a PDE system instead of an ODE system. Therefore, a spatial grid has to be defined, which is used by function `two_point_upwind_uni_D1.sci` to obtain the first order differentiation matrix.
- Next, the model parameters and initial conditions are defined. Since dependent variables are spatially distributed, initial conditions must be defined for each point of the spatial grid. Also, the dependent variables are stored according to the spatial grid point order so as to obtain a banded sparcity pattern of the Jacobian matrix.
- Once the initial conditions are imposed, the time span is defined and the ODE integrator is called. In the previous chapter, we made use of our basic time integrators. Now, we explore the possibilities offered by the ODE solvers (ODEPACK) already implemented in SCILAB. The calling sequence of the function `ode` is somewhat different from the Matlab counterpart.

```
yout = ode(x0,tl(1),tl,list(bioreactor_pdes),...
        jacobian_complex_diff);
```

The first input parameter corresponds to the initial conditions. Then, initial time and the list of times at which the solution has to be returned are passed to the function. The next input argument is the name of the function where the PDEs are defined (in this example `bioreactor_pdes`). The last argument in this call corresponds to the function where the jacobian matrix is computed through complex differentiation.

By default, the `ode` function makes use of the `lsoda` solver (that automatically selects between Adams and BDF methods). Other solvers like Runge-Kutta of order 4 or the Shampine and Watts solver can be used. In this case, the call is

```
yout = ode("method",x0,tl(1),tl,list(bioreactor_pdes),
         ... jacobian_complex_diff);
```

where `method` can be: `adams`, `stiff` (that uses the BDF method), `rk`, `rkf`, among others. Other options include the relative and absolute integration tolerances, maximum and minimum step sizes, etc. It is not the intent of this example to describe all of the possibilities available for calling the `ode` function and the reader is referred to the SCILAB help.

- Finally, the solution is drawn using the `plot` function.

```
clear

// Display mode
mode(-1);

// set global variables
global("v","nu1","peps","mumax","Kc","kd");
global("Sin","pgamma")
global("z0","zL","z","nz","D1z");

// Load the subroutines
exec('bioreactor_pdes.sci');
exec('two_point_upwind_uni_D1.sci');
exec('jacobian_complex_diff.sci');

// model parameters
peps  = 0.5;
A     = 0.04;
F     = 0.002;
v     = F/(peps*A);
nu1   = 0.4;
mumax = 0.35;
Kc    = 0.4;
kd    = 0.05;

// inlet concentration
Sin   = 5;
pgamma = 10;

// grid in axial direction
z0 = 0;
zL = 1;
nz = 101;
dz = (zL - z0)/(nz-1);
z = (z0:dz:zL)';

// differentiation matrix in z (convective term)
D1z = two_point_upwind_uni_D1(z0,zL,nz,A);

// initial conditions
S0 = 0;
X0 = 10;
```

```
// Distribution of initial conditions
S = S0*ones(nz,1);
X = X0*ones(nz,1);

// transfer dependent variables
x0(1:2:2*nz-1) = S;
x0(2:2:2*nz)    = X;

// call to ODE solver
t0       = 0;
tf       = 50;
Dtplot   = 1;
tl       = t0:Dtplot:tf;

// read the stopwatch timer
tic

// Call the integrator
yout = ode("stiff",x0,tl(1),tl,list(bioreactor_pdes),...
           jacobian_complex_diff);

// read the stopwatch timer
tcpu = toc();

// Plot the solution
plot(tl,yout(1:2:2*nz-1,:))
```

Script bioreactor_main.sci Main program for the implementation of the bioreactor problem, Eqs. (3.196)–(3.201), in SCILAB.

The programming in SCILAB of functions `two_point_upwind_uni_D1.sci`, `bioreactor_pdes.sci` and `jacobian_complex_diff.sci` is similar to their MATLAB counterparts.

Regarding the OCTAVE codes for simulation, as explained in the previous chapter, the only difference with respect to MATLAB is the call to the ODE solver. Using MATLAB, the call to `ode45` is of the form

```
options        = odeset('RelTol',1e-3,'AbsTol',1e-3);
tout           = [t0:Dtplot:tf];
[tout, yout] = ode45(@bioreactor_pdes,tout,x,options);
```

while in OCTAVE the call to `lsode` reads as

```
lsode_options('absolute tolerance',1e-3);
lsode_options('relative tolerance',1e-3);
tout                  = [t0:Dtplot:tf];
[xout, istat, msg] = lsode(@bioreactor_pdes_oct,
                               x, tout);
```

Table 3.4 Performance of different IVP solvers in several environments for the distributed bioreactor problem

	N	MATLAB		SCILAB		OCTAVE	
		ode45	ode15s	rkf	Adams	lsode	dassl
CPU time	100	1.50	1.50	1.00	1.75	2.25	2.75
	300	3.25	5.25	3.75	10.50	11.00	12.25
	700	8.00	40.25	12.50	47.25	68.75	48.50

Times have been normalized with respect to the most efficient case, i.e., rkf in SCILAB

Note also that the order of the input variables in the ODE function changes from MATLAB to OCTAVE

```
function xt = bioreactor_pdes(t,x)
function xt = bioreactor_pdes_oct(x,t)
```

To conclude this section, a comparison of the performance of different solvers in MATLAB, SCILAB and OCTAVE, for the tubular bioreactor problem—Eqs. (3.196)–(3.201)—is included in Table 3.4. Different numbers of grid points were used in the comparison. MATLAB appears as the most efficient alternative, especially with the solver ode45. This indicates that the ODE system resulting from the application of the MOL to the tubular bioreactor problem is non stiff. Indeed ode45 is five times faster than ode15s whith $N = 700$. For a small number of discretization points, i.e. 101 grid points, the rkf solver of SCILAB is faster than its MATLAB counterpart. However, as N increases SCILAB becomes gradually less efficient. On the other hand, OCTAVE, which was the most efficient environment for the example of Chap. 2 (see Table 2.13), shows now the worst performance. This issue becomes more evident as N increases.

It is of course impossible to draw general conclusions from a single example, and for other PDE problems, the selection of an appropriate solver and spatial discretization method can lead to different comparison results. A recent comparative evaluation of MATLAB, SCILAB, OCTAVE as well as FreeMat [13], concludes that OCTAVE is an excellent alternative to MATLAB, being able to solve problems of the same size and with equivalent efficiency.

3.13 Summary

In this chapter, finite difference (FD) approximations and the method of lines, which combine FD with available time integrators, are discussed. First, a convection-diffusion-reaction PDE is used to introduce a few basic FD schemes and address the concept of stability of the numerical scheme. As a rule of thumb, centered FD schemes appear as good choices for approximating second-order (diffusion) operators, whereas upwind FD schemes are preferable for first-order (convection) operators. Once stability is ensured, accuracy can be adjusted by selecting appropriately the fineness of the spatial grid and the order of the FD approximation (i.e. the stencil of points on which the FD approximation is built). The calculation of FD approximation can be conveniently organized using differentiation matrices, which are easy to manipulate in MATLAB, SCILAB or OCTAVE. FD can also be efficiently computed on arbitrarily spaced grids using an algorithm due to Fornberg [6, 7]. We continue this chapter with a presentation of different methods to take the boundary conditions into account. Boundary conditions are an essential part of the IBVP definition, and several methods for translating the BCs in the code implementation are presented. Finally, attention is paid to the computation of the Jacobian matrix of the ODE system, which is used by various solvers.

References

1. Schiesser WE (1991) The numerical method of lines. Academic Press, San Diego
2. Owens DH (1981) Multivariable and optimal systems. Academic Press, London
3. Varga RS (1981) Matrix iterative analysis. Prentice-Hall, New York
4. Glansdorff P, Prigogine I (1971) Thermodynamic theory of structure, stability and fluctuations. Wiley-Interscience, New York
5. Hairer E, Wanner G (1996) Solving ordinary differential equations II: stiff and differential algebraic problems. Springer, Berlin
6. Fornberg B (1988) Generation of finite difference formulas on arbitrarily spaced grids. Math Comput 51(184):699–706
7. Fornberg B (1998) Calculation of weights in finite difference formulas. SIAM Rev 40(3):685–691
8. Weng HS, Eigenberger G, Butt JB (1975) Catalyst poisoning and fixed bed reactor dynamics. Chem Eng Sci 30(11):1341–1351
9. Tali-Maamar N, Damak T, Babary JT, Nihtilä MT (1993) Application of a collocation method for simulation of distributed parameter bioreactors. Math Comp Sim 35:303–319
10. Lyness JN, Moler CB (1967) Numerical differentiation of analytic functions. SIAM J Numer Anal 4(2):202–210
11. Squire W, Trapp G (1998) Using complex variables to estimate derivatives of real functions. SIAM Rev 40(1):110–112
12. Cao Y (2008) Complex step jacobian. Technical Report 18176, Matlab Central
13. Brewster MW, Gobber MK (2011) A comparative evaluation of matlab, octave, freemat, and scilab on tara. Technical Report HPCF-2011-10, Department of Mathematics and Statistics, University of Maryland, Baltimore County

Further Reading

14. Vande Wouwer A, Saucez P, Schiesser WE (2001) Adaptive method of lines. Chapman Hall/CRC, Boca Raton

Chapter 4
Finite Elements and Spectral Methods

In Chap. 3, one of the most popular techniques for solving PDE systems, the FD method, has been presented. Finite differences are easy to use and allow to address a large class of IBVPs, but also suffer from a few drawbacks. In particular, FDs are not very convenient when dealing with problems in several spatial dimensions with irregular domains (in this book, we mostly consider 1-D problems so that this drawback is not apparent, but more on 2-D problems will be said in Chap. 6). In this chapter, a new family of techniques, the *methods of weighted residuals* (WRM), will be described paying special attention to the *finite element method* (FEM), *orthogonal collocation* and the *proper orthogonal decomposition*.

The starting point of this family of methods is the *generalized Fourier series* theorem. Essentially, this theorem establishes that, given an orthogonal basis set on a Hilbert space $\mathscr{L}_2$, $\Phi = \{\phi_i(z)\}_{i=0}^{\infty}$, any function $f(z, t) \in \mathscr{L}_2(\mathscr{V})$ can be expanded in convergent series of the form:

$$f(z, t) = \sum_{i=0}^{\infty} m_i(t)\phi_i(z). \tag{4.1}$$

There exists a large range of options for the selection of the basis set, for instance, Legendre, Hermite, or Chebyshev polynomials; the eigenfunctions obtained from the laplacian operator or the proper orthogonal decomposition, to name a few. Depending on the problem under consideration and on the desired accuracy some of them will be more appropriate than others. Let us first illustrate this concept with a simple example: *the heat equation in 1D*.

For the sake of clarity, we will focus on the simplest form of the heat equation where the spatial domain is isolated (leading to homogeneous Neumann boundary conditions) and no source term is considered. Under these conditions, the mathematical equation describing the system is:

$$\frac{\partial T}{\partial t} = \alpha \frac{\partial^2 T}{\partial z^2} \tag{4.2}$$

A. Vande Wouwer et al., *Simulation of ODE/PDE Models with MATLAB®, OCTAVE and SCILAB*, DOI: 10.1007/978-3-319-06790-2_4,

with boundary and initial conditions of the form:

$$\left.\frac{\partial T}{\partial z}\right|_{z=0} = \left.\frac{\partial T}{\partial z}\right|_{z=L} = 0 \tag{4.3}$$

$$T(z,0) = g(z), \tag{4.4}$$

where L is the length of the spatial domain, α is the diffusivity parameter, and $T(z,t)$ is the temperature field. The analytical solution to this problem can be computed using the method of *separation of variables* [1]. In this method, we will assume that the solution, which depends on time and spatial coordinates, can be expressed as the product of functions depending only on time and functions depending only on the spatial coordinates so that:

$$T(z,t) = R(z)S(t) \tag{4.5}$$

It follows that:

$$\frac{\partial T(z,t)}{\partial t} = R(z)\frac{\mathrm{d}S(t)}{\mathrm{d}t} \quad \text{and} \quad \frac{\partial^2 T(z,t)}{\partial z^2} = \frac{\mathrm{d}^2 R(z)}{\mathrm{d}z^2}S(t) \tag{4.6}$$

Introducing these expressions into Eq. (4.2), we have

$$R(z)\frac{\mathrm{d}S(t)}{\mathrm{d}t} = \alpha\frac{\mathrm{d}^2 R(z)}{\mathrm{d}z^2}S(t)$$

Rearranging terms and using the compact notation for the derivatives

$$\frac{S_t}{\alpha S} = \frac{R_{zz}}{R}$$

Since the LHS does not depend on z and the RHS does not depend on t, both sides must be independent of z and t and equal to a constant:

$$\frac{S_t}{\alpha S} = \frac{R_{zz}}{R} = -\lambda \tag{4.7}$$

where the minus sign is introduced for convenience, anticipating the fact that the constant $(-\lambda)$ will be negative as we explain in the following. Considering the part of Eq. (4.7) involving R, which is an eigenvalue problem, we have

$$R_{zz} + \lambda R = 0 \tag{4.8}$$

and we distinguish three cases:

- $\lambda < 0$.
  Equation (4.8) has a characteristic equation $r^2 + \lambda = 0$ whose solutions are real $r = \pm\sqrt{-\lambda}$. The general solution is $R(z) = A\exp\left(\sqrt{-\lambda}z\right) + B\exp\left(-\sqrt{-\lambda}z\right)$. Computing the first derivatives in $z = 0$ and $z = L$ and using the boundary conditions leads to

$$R_z(0) = A\sqrt{-\lambda} - B\sqrt{-\lambda} = 0$$

$$R_z(L) = A\sqrt{-\lambda}\exp\left(\sqrt{-\lambda}L\right) - B\sqrt{-\lambda}\exp-\left(\sqrt{-\lambda}L\right) = 0$$

  The first expression shows that $A = B$, whereas the second then imposes $A = B = 0$. The solution of (4.8) is thus trivial $R \equiv 0$.
- $\lambda = 0$.
  Equation (4.8) simplifies to $R_{zz} = 0$ which has the general solution $R(z) = Az + B$. Using the boundary conditions, we find that $A = 0$, and the solution reduces to $R \equiv B$.
- $\lambda > 0$.
  The characteristic equation has imaginary roots $r = \pm j\sqrt{\lambda}$ (with $j^2 = -1$) and the general solution of (4.8) is $R(z) = A\cos\left(\sqrt{\lambda}z\right) + B\sin\left(\sqrt{\lambda}z\right)$. Again, computing the first derivatives in $z = 0$ and $z = L$ and using the boundary conditions, we obtain:

$$R_z(0) = B\sqrt{\lambda} = 0$$

$$R_z(L) = -A\sqrt{\lambda}\sin\left(\sqrt{\lambda}L\right) + B\sqrt{\lambda}\cos\left(\sqrt{\lambda}L\right) = 0$$

  The first condition implies that $B = 0$ and the second that either $A = 0$ (trivial solution) or $\lambda_i = i\pi/L$, with $i = 1, 2, 3, \ldots$ These latter values define the eigenvalues corresponding to the eigenfunctions $R_i(z) = A_i\cos(i\pi/Lz)$.

The part of Eq. (4.7) involving variable S, i.e., $S_t = -\lambda\alpha S$, has solutions of the form

$$S_i(t) = \exp(-\lambda_i\alpha t) = \exp\left(-\left(\frac{i\pi}{L}\right)^2\alpha t\right)$$

Accordingly, the temperature field is given by:

$$T(z,t) = \sum_{i=0}^{\infty} T_i(z,t) = \sum_{i=0}^{\infty}\exp\left(-\frac{i^2\pi^2}{L^2}\kappa t\right)A_i\cos\left(\frac{i\pi}{L}z\right) \tag{4.9}$$

Note that this expression is equivalent to (4.1) with $\phi_i(z) = A_i\cos\left(\frac{i\pi}{L}z\right)$ and $m_i(t) = \exp\left(-\frac{i^2\pi^2}{L^2}\kappa t\right)$. From the initial condition $T(z,0) = g(z)$ and

Table 4.1 Maximum and mean relative errors for the analytical solution of the heat equation using different number of terms N in the series (4.9)

N	Maximum error (%)	Mean error (%)
1	3.30	2.8
2	0.59	3.0×10^{-2}
3	0.20	6.2×10^{-3}
4	9.8×10^{-2}	2.0×10^{-3}

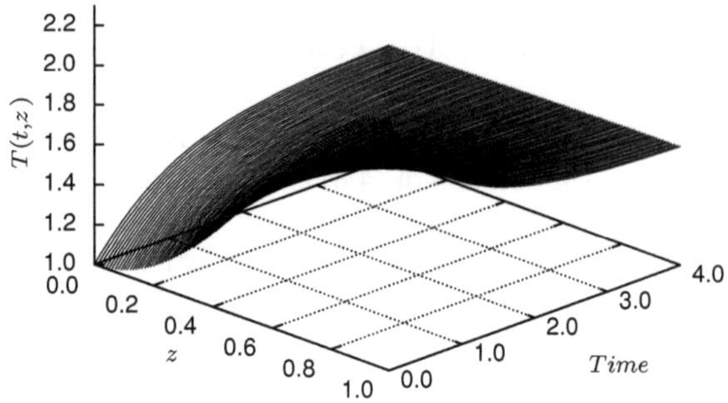

Fig. 4.1 Analytical solution of the heat equation computed using the method of separation of variables

$$T(z,0) = A_0 + \sum_{i=1}^{\infty} A_i \cos\left(\frac{i\pi}{L} z\right) \tag{4.10}$$

we deduce that

$$A_0 = \frac{1}{L}\int_0^L g(z)\mathrm{d}z; \quad A_i = \frac{2}{L}\int_0^L g(z)\cos\left(\frac{i\pi z}{L}\right)\mathrm{d}z \tag{4.11}$$

As shown in Table 4.1, for $L = 1$, the series converge very rapidly to the exact solution and using only four terms in Eq. (4.9) is enough to obtain relative errors below 0.1 %.

Figure 4.1 represents the analytical solution of the heat equation computed by means of the method of separation of variables. As shown in this figure, the initial temperature spatial distribution becomes homogeneous with time as a result of the effects of the diffusion operator.

In the previous example, the basis functions in expansion (4.9) were obtained analytically (they correspond to the eigenfunctions). However, in general, it will not be possible to compute an analytical solution. In those cases, the functions for the series expansion must be selected a priori. To illustrate this point, let us use, for instance, the Legendre polynomials whose recursive formula in the spatial domain $\Omega(z) \in [0, 1]$ is given by the *Rodrigues representation*:

$$\phi_i(z) = \frac{1}{i!}\frac{\mathrm{d}}{\mathrm{d}z}\left(z^2 - z\right)^i \tag{4.12}$$

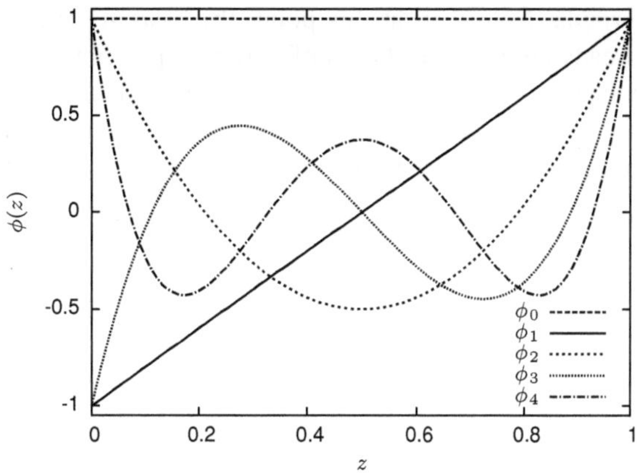

Fig. 4.2 Shape of the first five Legendre polynomials

So the first five polynomials, represented in Fig. 4.2, are of the form:

$$\phi_0 = 1; \quad \phi_1 = 2z - 1; \quad \phi_2 = 6z^2 - 6z + 1;$$

$$\phi_3 = 20z^3 - 30z^2 + 12z - 1; \quad \phi_4 = 70z^4 - 140z^3 + 90z^2 - 20z + 1$$

The expansion (4.1) is exact for smooth functions, however it requires an infinite number of terms. In practice, the series are truncated so that function $f(z,t)$ is approximated as:

$$f(z,t) \approx \sum_{i=0}^{N} m_i(t)\phi_i(z). \tag{4.13}$$

In order to compute the coefficients m_i of the series expansion, (4.13), we project the nonlinear function $f(z,t)$ onto the polynomials $\phi_i(z)$.

Let us first define the inner product and the norm of two functions $g(z)$ and $h(z)$ as:

$$\langle g(z), h(z)\rangle = \int_V g(z)h(z)\mathrm{d}z; \quad \|g\|_2 = \sqrt{\langle g(z), g(z)\rangle} \tag{4.14}$$

The projection is, then, expressed by the inner product as:

$$m_i(t) = \int_V \phi_i(z)f(z,t)\mathrm{d}z = \langle \phi_i(z), f(z,t)\rangle \tag{4.15}$$

Depending on the form of $f(z,t)$ it may be quite difficult or even impossible, to find an analytical solution for this integral. Therefore, we will have to make use of numerical techniques such as the Simpson's rule or the Gauss-Legendre quadrature formula. Note that, at this point, the basis set ($\{\phi_i(z)\}_{i=0}^{\infty}$) is known (as we have already defined it as the set of Legendre polynomials) and the coefficients $m_i(t)$ are computed through (4.15) so function $f(z,t)$ can be obtained using (4.13).

We now apply the series expansion (4.13) to the heat equation using the previously presented Legendre polynomials as basis functions. Substitution of Eq. (4.13) into (4.2) leads to:

$$\frac{\partial \sum_{i=0}^{N} m_i(t)\phi_i(z)}{\partial t} = \alpha \frac{\partial^2 \sum_{i=0}^{N} m_i(t)\phi_i(z)}{\partial z^2} \tag{4.16}$$

or, rearranging terms

$$\sum_{i=0}^{N} \phi_i(z)\frac{\mathrm{d}m_i(t)}{\mathrm{d}t} = \alpha \sum_{i=0}^{N} m_i(t)\frac{\mathrm{d}^2\phi_i(z)}{\mathrm{d}z^2} \tag{4.17}$$

The only information required to obtain the solution are the time dependent coefficients $m_i(t)$. These are computed by projecting (4.17) onto the first N Legendre polynomials, i.e.,

$$\begin{aligned} \int_{\Omega(z)} \phi_j \sum_{i=0}^{N} \phi_i(z)\frac{\mathrm{d}m_i(t)}{\mathrm{d}t}\mathrm{d}z &= \alpha \int_{\Omega(z)} \phi_j \sum_{i=0}^{N} m_i(t)\frac{\mathrm{d}^2\phi_i(z)}{\mathrm{d}z^2}\mathrm{d}z \\ \Rightarrow \sum_{i=0}^{N} \frac{\mathrm{d}m_i}{\mathrm{d}t} \int_{\Omega(z)} \phi_j\phi_i \mathrm{d}z &= \alpha \sum_{i=0}^{N} m_i \int_{\Omega(z)} \phi_j \frac{\mathrm{d}^2\phi_i}{\mathrm{d}z^2}\mathrm{d}z \end{aligned} \tag{4.18}$$

To include the boundary conditions in this computation we make use of *Green's first identity* which expresses that, for a twice continuously differentiable function $f(z)$ and a once continuously differentiable function $g(z)$:

$$\int_{\Omega(z)} g(z)\frac{\mathrm{d}^2 f(z)}{\mathrm{d}z^2}\mathrm{d}z = \int_{\Gamma(z)} g(z)\frac{\mathrm{d}f(z)}{\mathrm{d}n}\mathrm{d}z - \int_{\Omega(z)} \frac{\mathrm{d}g(z)}{\mathrm{d}z}\frac{\mathrm{d}f(z)}{\mathrm{d}z}\mathrm{d}z \tag{4.19}$$

where $\Gamma(z)$ is the boundary of the spatial domain $\Omega(z)$ and n represents a unitary vector pointing outwards the spatial domain.

The use of *Green's first identity* in the integral term in the RHS of Eq. (4.18) leads to:

$$\int_{\Omega(z)} \phi_j \frac{\mathrm{d}^2\phi_i}{\mathrm{d}z^2}\mathrm{d}z = \int_{\Gamma(z)} \phi_j \frac{\mathrm{d}\phi_i}{\mathrm{d}n}\mathrm{d}z - \int_{\Omega(z)} \frac{\mathrm{d}\phi_j}{\mathrm{d}z}\frac{\mathrm{d}\phi_i}{\mathrm{d}z}\mathrm{d}z$$

and since homogeneous Neumann BCs (4.3) are considered in this problem:

Table 4.2 Maximum and mean relative errors for the numerical solution of the heat equation using Legendre polynomials

N	Maximum error (%)	Mean error (%)
3	10.7	1.97
5	0.18	5.3×10^{-2}
7	1.7×10^{-2}	1.4×10^{-3}

$$\int_{\Omega(z)} \phi_j \frac{\mathrm{d}^2\phi_i}{\mathrm{d}z^2}\mathrm{d}z = -\int_{\Omega(z)} \frac{\mathrm{d}\phi_j}{\mathrm{d}z}\frac{\mathrm{d}\phi_i}{\mathrm{d}z}\mathrm{d}z \tag{4.20}$$

Substituting Eq. (4.20) into (4.18) and taking into account that Legendre polynomials are orthonormal over the spatial domain $\Omega(z) = [0, 1]$, we have that:

$$\frac{\mathrm{d}\mathbf{m}}{\mathrm{d}t} = -\alpha\mathscr{A}\mathbf{m} \tag{4.21}$$

where $\mathbf{m} = [m_1, m_2, \ldots, m_N]^T$ and each element of matrix $\mathscr{A}$ is of the form $\mathscr{A}_{i,j} = \int_{\Omega(z)} \frac{\mathrm{d}\phi_j}{\mathrm{d}z}\frac{\mathrm{d}\phi_i}{\mathrm{d}z}\mathrm{d}z$. The coefficients of the series expansion are then computed by solving the ODE system (4.21).

The maximum and mean relative errors between the analytical solution and the series approximation using Legendre polynomials are presented in Table 4.2. As shown in the table, the number of elements required to obtain a given accuracy is larger than with the problem eigenfunctions (see Table 4.1). The advantage of using a previously defined basis set is that no analytical solution is required.

We now turn our attention to a more challenging example, e.g., Burgers' equation already introduced in Chap. 1. Depending on the initial conditions, different analytical solutions are available, as already discussed in Chaps. 1 and 3. For convenience, we rewrite the one considered in Chap. 1 (using the notation $x(z,t)$ for the solution instead of $u(z,t)$):

$$x_\mathrm{a}(z,t) = \frac{0.1e^a + 0.5e^b + e^c}{e^a + e^b + e^c} \tag{4.22}$$

where

$$a = -(0.05/\mu)(z - 0.5 + 4.95t);$$

$$b = -(0.25/\mu)(z - 0.5 + 0.75t);$$

$$c = -(0.5/\mu)(z - 0.375)$$

Depending on the value of μ different solution behaviors are obtained. In this connection, the larger the value of μ, the smoother the resulting spatial profiles whereas low values for μ will result into steep profiles which are a challenge for any numerical technique.

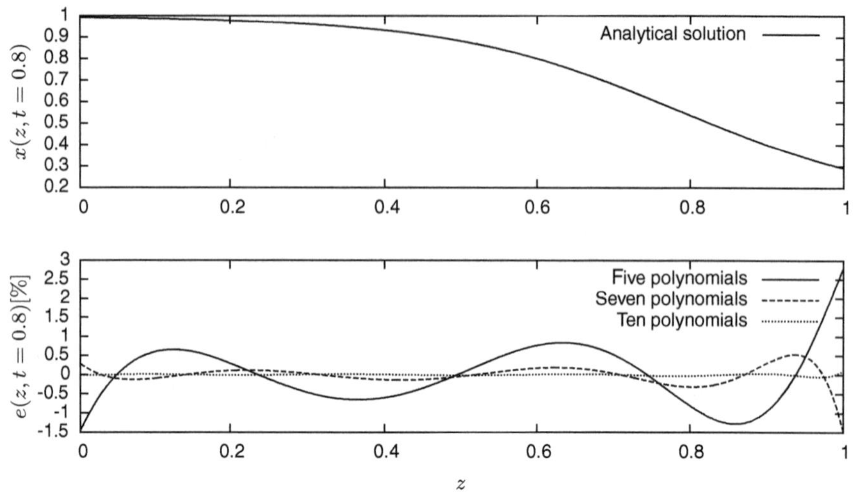

Fig. 4.3 Analytical solution of Burgers' equation at $t = 0.8$ with $\mu = 0.05$ (*top figure*). Relative errors between the analytical solution and the Fourier series approximation with different numbers of terms (*bottom figure*)

Let us compare the analytical solution with the truncated series expansion. We start with $\mu = 0.05$ which will result into smooth profiles and we simplify the problem by considering the solution of Burgers' equation at time $t = 0.8$ (in this way, we eliminate the time dependency). As in (4.15) the coefficients are computed by:

$$m_i = \int_0^1 \phi_i x_a(z, 0.8)\mathrm{d}z$$

For this particular example, the spatial integral will be computed using the trapezoidal rule.

Figure 4.3 presents the analytical solution (top figure) and the relative errors between such solution and the series expansion approximation (bottom figure). As we can see, with only five terms in the series expansion, the maximum relative error is lower than 3 % and, as we increase the number of elements, the error decreases very rapidly. In fact, with 10 elements the error is around 0.1 %.

Let us now consider the solution of Burgers' equation using $\mu = 0.001$ (this solution develops front sharpening and steep moving fronts) at time $t = 0.8$. A comparison between the analytical solution and the series expansion is represented in Fig. 4.4. Note that a larger number of terms are required (as compared with the smooth profile case) to reproduce with satisfactory accuracy the solution. Also, even if we increase the number of terms, the oscillations near the steep front remain. These oscillations, which appear in steep fronts or in discontinuities, are known as the *Gibbs phenomenon*. They will not disappear even if we use an infinite number of elements in the series.

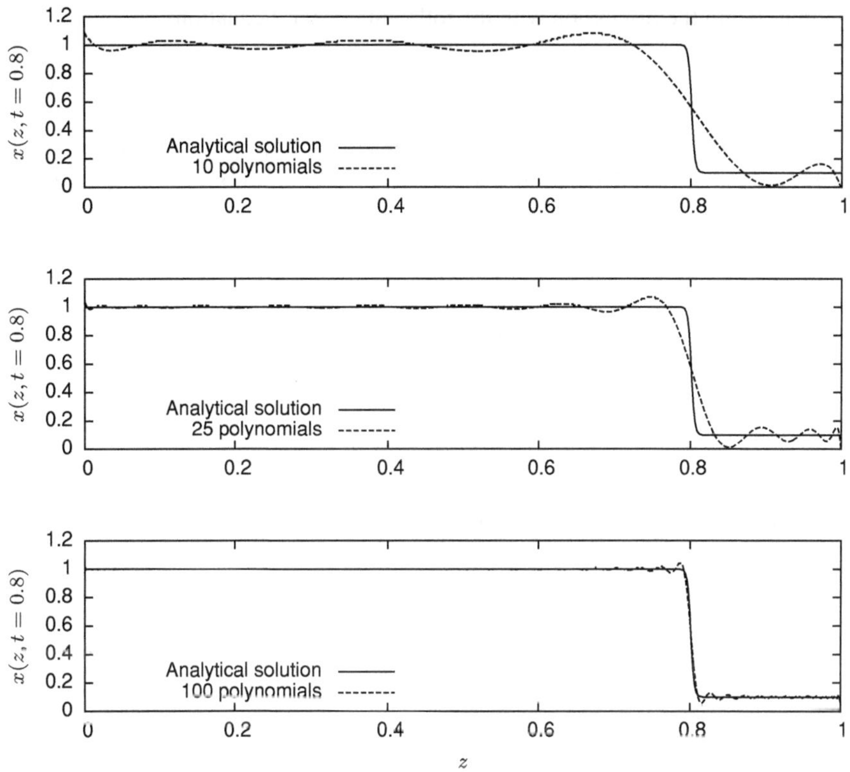

Fig. 4.4 Comparison between the analytical solution of Burgers' equation with $\mu = 0.001$ at $t = 0.8$ and the series approximation with different number of elements

One possible solution to this problem is to discretize the spatial domain into a number of smaller spatial domains, also called *elements*, and to use functions locally defined on these elements (this will become clearer later on when describing the finite element method).

Let us now, before starting with the description of the finite element method, introduce the basics of the *weighted residuals methods* (WRM) for solving PDE systems.

4.1 The Methods of Weighted Residuals

The methods of weighted residuals constitute a whole family of techniques for developing approximate solutions of operator equations. The starting point for all these methods is the series expansion approximation (4.13). The basis functions are chosen based on various criteria to give the "best" approximation for the selected family. These methods are applicable for arbitrary operators and boundary conditions.

Let us begin the discussion with the following general equation:

$$x_t = \alpha_0 x + \alpha_1 x_z + \alpha_2 x_{zz} + f(x) \tag{4.23}$$

where the first three terms on the RHS constitute the linear part of the equation and, as we will show, their treatment by the finite element method leads to the computation of matrix operators analogous to the differentiation matrices used in the finite difference method. The last term, $f(x)$, known as *source term* will require further numerical treatment. The system must be completed with appropriate boundary and initial conditions. Denoting by Γ the boundary of the spatial domain Ω, the following general expression can be used to define the boundary conditions:

$$\beta_1 \frac{\partial x}{\partial z} + \beta_0 x + g(x) = 0\Big|_{\Gamma} \tag{4.24}$$

where $g(x)$ is a given nonlinear function.

The first step is to approximate the original solution $x(z,t)$ by a truncated series expansion of the form

$$x(z,t) \approx \tilde{x}(z,t) = \sum_{i=1}^{N} m_i(t)\phi_i(z)$$

Since $\tilde{x}(z,t)$ is only an approximation, its substitution into Eq. (4.23) will result into the following *residual*:

$$\tilde{x}_t - (\alpha_0 \tilde{x} + \alpha_1 \tilde{x}_z + \alpha_2 \tilde{x}_{zz} + f(\tilde{x})) = R.$$

The best approximation can be found by making the residual R close to zero in some average sense. To this end, N weighting functions ($w(z) : \mathbb{R} \to \mathbb{R}$) are introduced, and weighted residuals are forced to zero. This procedure leads to the following system of ODEs:

$$\int_{\Omega} R(z,t) w_i(z) dz = 0; \qquad i = 1, \ldots, N. \tag{4.25}$$

The name of the family of methods based on *weighted residuals* (WR) now becomes clear, i.e., the residual resulting from the substitution of the truncated Fourier series into the original PDE is multiplied by a number of weighting functions.

We are concerned with the problem of choosing a linear space in which the WRM is formulated. The WRM should apply to any linear space $\mathfrak{X}$ to which the solution $x(z,t)$ belongs and on which the weighted integrals make sense.

Depending on the selection of the weighting functions different methods arise. Among them, the most popular ones are briefly described below.

4.1.1 Interior Method

The basis functions $\phi_i(z)$ are chosen such that the approximation $\tilde{x}$ automatically satisfies the boundary conditions. However, it does not immediately satisfy the differential equation. Substitution of the approximation $\tilde{x}$ into the differential equation produces an error function $R_E(\tilde{x})$, which is called the residual equation:

$$\tilde{x}_t - (\alpha_0\tilde{x} + \alpha_1\tilde{x}_z + \alpha_2\tilde{x}_{zz} + f(\tilde{x})) = R_E(\tilde{x}) \tag{4.26}$$

and substitution into the initial conditions yields an initial residual:

$$x_0(z) - \sum_{i=1}^{N} m_i(0)\phi_i(z) = R_I(\tilde{x}) \tag{4.27}$$

These residuals are measures of how well the approximation $\tilde{x}(z,t)$ satisfies the equation and the initial conditions, respectively. They are equal to zero for the exact solution but not for the approximation. The functions $m_i(t)$ are determined in such a way that the residuals are zero in some average sense: we set the N weighted integrals of the equation and initial residuals to zero, i.e.

$$\langle R_E(\tilde{x}), w_j\rangle = 0; \quad j = 1, 2, \ldots, N \tag{4.28}$$

$$\langle R_I(\tilde{x}), w_j\rangle = 0$$

Equation (4.28) represents a system of N differential equations while (4.27) represents the initial conditions of coefficients m_i, i.e., $m_i(0)$.

4.1.2 Boundary Method

If the approximate solution, also called trial solution, is selected so as to satisfy the differential equation but not the boundary conditions, the procedure is called the boundary method. Substitution of $\tilde{x}$ into the boundary conditions yields a boundary residual $R_B(\tilde{x})$. As in the interior method, the residual is forced to be zero in an average sense:

$$\langle R_B(\tilde{x}), w_j\rangle = 0; \quad j = 1, 2, \ldots, N \tag{4.29}$$

4.1.3 Mixed Method

The trial solution does not satisfy neither the equation nor the boundary conditions. Substitution of $\tilde{x}$ into the initial boundary value problem produces now an equation, a boundary and an initial residual. We can attempt to balance the equation residual and the boundary residual:

$$\langle R_E(\tilde{x}), w_j \rangle = \langle R_B(\tilde{x}), w_j \rangle; \quad j = 1, 2, \ldots, N \tag{4.30}$$

4.1.4 Galerkin Method

In the *Galerkin* scheme, the weighting functions w_i are the same as the basis functions ϕ_i which, recall, form a complete set for the N dimensional subspace where the approximated solution is sought. Imposing that the weighted residuals vanish thus makes the residual error orthogonal to the basis function space. The larger the number N, the better the approximation, so that in the limit when $N \to \infty$ it follows that $x = \tilde{x}$. The basis functions in this method can be locally or globally defined. Among the Galerkin schemes with locally defined functions, the most popular is the *finite element method* (FEM). Flexibility when solving problems with irregular spatial domains or with non homogeneous boundary conditions makes the FEM a versatile and an efficient method. This technique will be described in more details in Sect. 4.2.

4.1.5 Collocation Method

Whereas all the methods introduced so far require the computation of spatial integrals, the collocation method drastically simplifies this procedure.

Indeed, the weighting functions are selected as the Dirac delta, i.e., $w_j = \delta(z - z_j)$, thus:

$$\langle R_E, w_j \rangle = \int_{\Omega} R_E \delta\left(z - z_j\right) d\Omega = R_E\left(\tilde{x}\left(z_j, t\right)\right) = 0 \quad j = 1, \ldots, N$$

Therefore, the collocation method appears as the simplest WRM to apply. The main issue in this technique is the selection of the collocation points. Even though an obvious choice is to take N points evenly distributed in the spatial domain, this choice is not the best possible regarding the accuracy of the approximation and the possible occurrence of oscillations (a phenomenon highlighted in 1901 by Carl Runge [2] that we will discuss more in the section on orthogonal collocation).

4.1.6 Orthogonal Collocation Method

It is possible to speed up the convergence (and to make the collocation method competitive with the Galerkin method) by carefully selecting the collocation points at the roots of orthonormal polynomials (i.e., polynomials forming an orthonormal basis, such as the Legendre or Chebyshev polynomials). The procedure is then called orthogonal collocation.

The method of orthogonal collocation dates back at least to Lanczos [3]. Chebyshev polynomials and collocation at Legendre points were used to approximate solutions to initial value problems. Chemical engineers have used the method extensively to solve initial and boundary value problems arising in reactor dynamics and other systems (see [4]). Existence, uniqueness, and error estimates for one-dimensional orthogonal collocation using splines rather than full polynomials was first presented in the paper of de Boor and Swartz [5]. We will come back to this method later on in Sect. 4.9.

4.2 The Basics of the Finite Element Method

As the finite difference method, the FEM is based on the discretization of the spatial domain (see Fig. 4.5) to transform the original PDE into a set of ODEs. Each of the segments e_i obtained after discretization is known as *finite elements* and the points where two elements coincide are the *nodes*.

As mentioned in the introduction of this chapter, the FEM belongs to the family of *methods of weighted residuals* (MWR) and as such, the implementation makes use of different concepts, namely, truncated Fourier series, weighted residuals and the weighting/basis functions. The underlying idea of FEM is that it will be easier to represent parts of the solution of the PDE, corresponding to each of the finite elements, using truncated Fourier series. For instance, if polynomials are used, lower-order polynomials will probably be able to do the job of representing parts of the solution, whereas high-order polynomials would be required to represent the full solution, with the associated risk of oscillation (Gibbs phenomenon). In this context, the basis functions will therefore have a compact support (i.e., they are defined on each element).

- *Truncated Fourier series* The first step is to approximate the solution of the PDE (4.23) by a truncated Fourier series of the form:

$$x(z,t) \approx \tilde{x}(z,t) = \sum_{i=1}^{N} X_i(t)\phi_i(z) \tag{4.31}$$

where the time-dependent coefficients $X_i(t)$ correspond with the values of variable $\tilde{x}(z,t)$ at nodes z_i. The basis functions $\phi_i(z)$ are defined a priori and, in the case of the FEM are selected as low-order polynomials on a compact support. The substitution of approximation $\tilde{x}(z,t)$ into Eq. (4.23) yields an equation similar to (4.26):

$$\tilde{x}_t - (\alpha_0\tilde{x} + \alpha_1\tilde{x}_z + \alpha_2\tilde{x}_{zz} + f(\tilde{x})) = R_E(z,t) \tag{4.32}$$

that we have to minimize.

The above expression is an equation residual, which implicitly assumes that the boundary conditions will be satisfied in some way. Alternatively, a mixed method

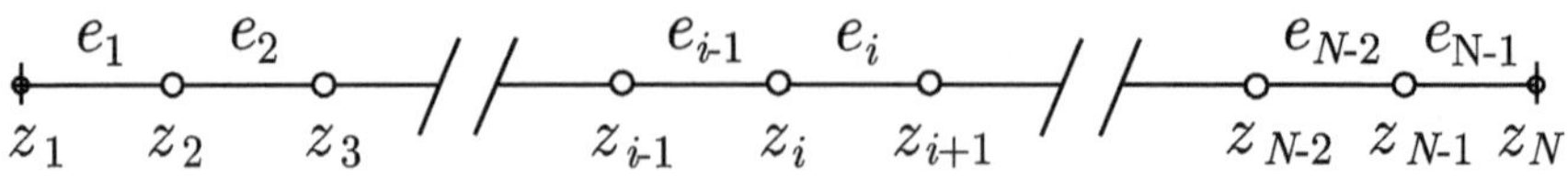

Fig. 4.5 Spatial discretization of a one-dimensional spatial domain

could be considered according to (4.30), which involves the sum of equation and boundary residuals. The treatment of boundary conditions deserves special attention and will be discussed later on.

- *Weighted residuals* The residual R_E is multiplied by N weighting functions $\{w_i\}_{i=1}^N$ and the result is integrated over the spatial domain $\Omega(z)$:

$$\int_{\Omega(z)} w_i \tilde{x}_t \mathrm{d}z = \alpha_0 \int_{\Omega(z)} w_i \tilde{x} \mathrm{d}z + \alpha_1 \int_{\Omega(z)} w_i \tilde{x}_z \mathrm{d}z + \alpha_2 \int_{\Omega(z)} w_i \tilde{x}_{zz} \mathrm{d}z + \int_{\Omega(z)} w_i f(\tilde{x}) \mathrm{d}z; \quad i = 1, \ldots, N \tag{4.33}$$

Note that $\tilde{x} = \sum_{i=1}^N X_i(t)\phi_i(z)$, so that the problem translates into finding the time-dependent coefficients $X_i(t)$ which fulfill Eq. (4.33).

- *Weighting function selection* In the FEM, as in other methods known as Galerkin methods, the weighting functions are the same as the basis functions, i.e., $W_i = \phi_i$, $\forall i = 1, \ldots, N$. In the framework of the method of lines, we will consider the following cases:

  - Galerkin method over linear Lagrangian elements
  - Galerkin method over linear Hermitian elements of class C^1 on two nodes.

4.3 Galerkin Method Over Linear Lagrangian Elements

The basis functions in the FEM are locally defined, in other words, a given basis function $\phi_i(z)$ takes a given value (typically 1) at node z_i (in contrast with the notation used for introducing finite difference schemes the numbering of the nodes is from 1 to N) and zero in the rest of the nodes (see Fig. 4.6).

Mathematically, this is expressed by:

$$\phi_i(z_j) = \delta_{i,j} \tag{4.34}$$

where $\delta_{i,j}$ is the Kronecker delta.

With linear Lagrangian basis, only two basis functions, namely ϕ_i and ϕ_{i+1}, are defined (with values different from zero) at element e_i. Therefore, for convenience a *local notation* will be used. Let us denote by $\phi_1^{e_i}$ and $\phi_2^{e_{i-1}}$ the parts of function ϕ_i

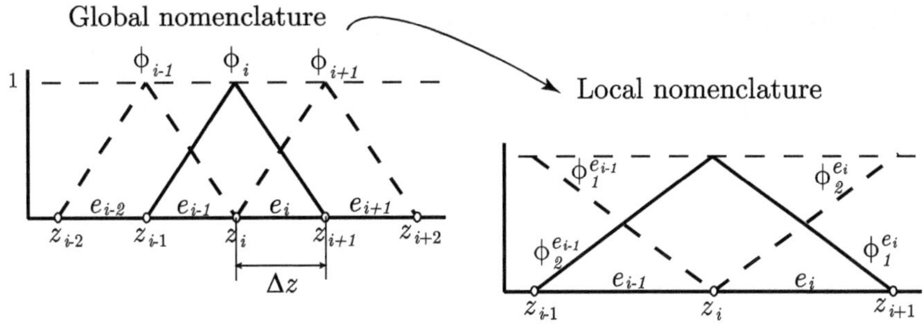

Fig. 4.6 Linear basis functions at the finite element e_i

which are defined over elements e_i and e_{i-1}, respectively. Over the element e_i the following functions are defined as:

$$\phi_1^{e_i}(z) = \frac{z_{i+1} - z}{z_{i+1} - z_i} = \frac{z_{i+1} - z}{\Delta z}; \quad \phi_2^{e_i}(z) = \frac{z - z_i}{z_{i+1} - z_i} = \frac{z - z_i}{\Delta z} \tag{4.35}$$

with Δz being the length of element e_i.

In order to simplify the formulation, particularly in view of the weighted residual (integral) computation, the following coordinate change is proposed:

$$\xi = \frac{2}{\Delta z}(z - z_c) \Longrightarrow \frac{d\xi}{dz} = \frac{2}{\Delta z} \tag{4.36}$$

where z_c is the abscissa of the midpoint of e_i. In this new coordinate system, the basis functions are of the form:

$$\phi_1^{e_i}(\xi) = \frac{1-\xi}{2}; \quad \phi_2^{e_i}(\xi) = \frac{1+\xi}{2} \tag{4.37}$$

It is clear that $\phi_1^{e_1} = \phi_1^{e_2} = \cdots = \phi_1^{e_{i-1}} = \phi_1^{e_i} = \cdots$ and $\phi_2^{e_1} = \phi_2^{e_2} = \cdots = \phi_2^{e_{i-1}} = \phi_2^{e_i} = \cdots$. To further simplify the notation, we define $\phi_1 = \phi_1^{e_i}$ and $\phi_2 = \phi_2^{e_i}$ for all $i = 1, \ldots, N$ so that the solution can be expressed, within element e_i, as:

$$\tilde{x}^{e_i}(\xi, t) = \phi_1(\xi)X_i(t) + \phi_2(\xi)X_{i+1}(t) \tag{4.38}$$

With some abuse of notation let us write $X_1^{e_i}(t) = X_i(t)$ and $X_2^{e_i}(t) = X_{i+1}(t)$, so that Eq. (4.38) becomes:

$$\tilde{x}^{e_i}(\xi, t) = \phi_1(\xi)X_1^{e_i}(t) + \phi_2(\xi)X_2^{e_i}(t) \tag{4.39}$$

We will now analyze separately the different terms of Eq. (4.33) starting with the LHS of the equation.

4.3.1 LHS of the Weighted Residual Solution

We take advantage of the fact that, inside each finite element, only two basis functions have values different from zero. As a consequence, all the integrals $i = 1, \ldots, N$ in the LHS term of Eq. (4.33) will be zero except two of them (those corresponding to the projection onto the basis functions $W_i = \phi_1$ and $W_{i+1} = \phi_2$). Therefore, one element (for instance e_i) will contribute to two rows (rows i and $i+1$) of the equation system resulting from the method of lines. Let us denote by $\Omega^{e_i}(z)$ the portion of the spatial domain $\Omega(z)$ in which element e_i is defined. The projection of the LHS term of Eq. (4.33) onto ϕ_1, leads to:

$$\int\limits_{\Omega^{e_i}(z)} \phi_1(z)\tilde{x}_t^{e_i} \mathrm{d}z = \int\limits_{-1}^{1} \phi_1(\xi) \left(\phi_1(\xi)\frac{\mathrm{d}X_1^{e_i}}{\mathrm{d}t} + \phi_2(\xi)\frac{\mathrm{d}X_2^{e_i}}{\mathrm{d}t} \right) \frac{\Delta z}{2} \mathrm{d}\xi \tag{4.40}$$

The limits of the integral come from the fact that in the new coordinate system $\xi_1^{e_i} = \frac{2}{\Delta z}(z_i - z_c) = -1$ and $\xi_2^{e_i} = \frac{2}{\Delta z}(z_{i+1} - z_c) = 1$. The projection onto ϕ_2 yields

$$\int\limits_{\Omega^{e_i}(z)} \phi_2(z)\tilde{x}_t^{e_i} \mathrm{d}z = \int\limits_{-1}^{1} \phi_2(\xi) \left(\phi_1(\xi)\frac{\mathrm{d}X_1^{e_i}}{\mathrm{d}t} + \phi_2(\xi)\frac{\mathrm{d}X_2^{e_i}}{\mathrm{d}t} \right) \frac{\Delta z}{2} \mathrm{d}\xi \tag{4.41}$$

We will use, for convenience, an abbreviated notation for the spatial integrals:

$$\langle \phi_1, \phi_2 \rangle = \int\limits_{-1}^{1} \phi_1(\xi)\phi_2(\xi) d\xi \tag{4.42}$$

With this notation, the previous two equations can be rewritten as:

$$\frac{\Delta z}{2} \begin{pmatrix} \langle \phi_1, \phi_1 \rangle & \langle \phi_1, \phi_2 \rangle \\ \langle \phi_2, \phi_1 \rangle & \langle \phi_2, \phi_2 \rangle \end{pmatrix} \begin{pmatrix} \frac{\mathrm{d}X_1^{e_i}}{\mathrm{d}t} \\ \frac{\mathrm{d}X_2^{e_i}}{\mathrm{d}t} \end{pmatrix} = \frac{\Delta z}{6} \begin{pmatrix} 2 & 1 \\ 1 & 2 \end{pmatrix} \begin{pmatrix} \frac{\mathrm{d}X_1^{e_i}}{\mathrm{d}t} \\ \frac{\mathrm{d}X_2^{e_i}}{\mathrm{d}t} \end{pmatrix} \tag{4.43}$$

As mentioned previously, Eq. (4.43) is the contribution of one element to the whole system of equations. Now we have to take account of all elements of the spatial domain. This procedure is known as the *assembly* and it is depicted in Fig. 4.7. The black and gray squares represent, respectively, the contribution of odd and even elements to the whole matrix. Such contribution corresponds to the matrix appearing in Eq. (4.43). As shown in the figure, black and gray squares overlap as odd and even elements share one spatial node. For example, the nodes defining element e_1 are z_1

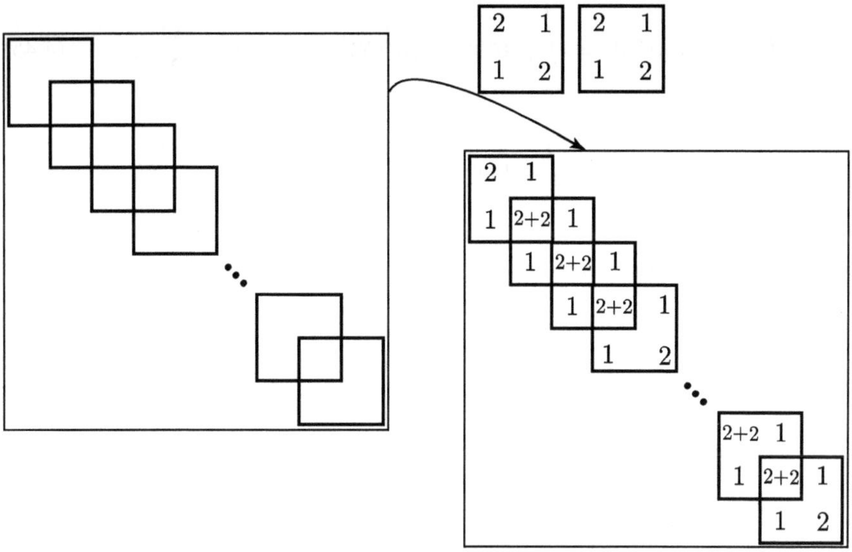

Fig. 4.7 Assembly of the contribution of all elements to the mass matrix

and z_2 while those defining element e_2 are z_2 and z_3, so these two elements share z_2. In the common points, the contribution of both elements is added.

Noting that $X_1^{e_1} = X_1$, $X_2^{e_1} = X_1^{e_2} = X_2$, $X_2^{e_2} = X_1^{e_3} = X_3$, etc. and after the assembly, the discrete FEM version of the LHS of Eq. (4.33), using linear Lagrangian elements, is:

$$\mathbf{M}\frac{\mathrm{d}\tilde{\mathbf{x}}}{\mathrm{d}t} - \frac{\Delta z}{6}\begin{pmatrix} 2 & 1 & & & \\ 1 & 4 & 1 & & \\ & \ddots & \ddots & \ddots & \\ & & 1 & 4 & 1 \\ & & & 1 & 2 \end{pmatrix}\begin{pmatrix} X_1 \\ X_2 \\ \vdots \\ X_{N-1} \\ X_N \end{pmatrix}_t \tag{4.44}$$

where matrix $\mathbf{M}$ is known as the *mass matrix*.

4.3.2 First Term in the RHS of the Weighted Residual Solution

In a similar way, one element will contribute to two rows (rows i and $i+1$) of the equation system resulting from the method of lines. In this case, the projection of this term onto ϕ_1 and ϕ_2 yields

$$\int_{\Omega^{e_i}(z)} \phi_1(z)\alpha_0\tilde{x}^{e_i}\mathrm{d}z = \int_{-1}^{1} \phi_1(\xi)\alpha_0\left(\phi_1(\xi)X_1^{e_i} + \phi_2(\xi)X_2^{e_i}\right)\frac{\Delta z}{2}\mathrm{d}\xi \tag{4.45}$$

$$\int_{\Omega^{e_i}(z)} \phi_2(z)\alpha_0\tilde{x}^{e_i}\mathrm{d}z = \int_{-1}^{1} \phi_2(\xi)\alpha_0\left(\phi_1(\xi)X_1^{e_i} + \phi_2(\xi)X_2^{e_i}\right)\frac{\Delta z}{2}\mathrm{d}\xi \tag{4.46}$$

or, in the compact notation defined in (4.42):

$$\frac{\Delta z}{2}\begin{pmatrix} \langle\phi_1,\phi_1\rangle & \langle\phi_1,\phi_2\rangle \\ \langle\phi_2,\phi_1\rangle & \langle\phi_2,\phi_2\rangle \end{pmatrix}\begin{pmatrix} X_1^{e_i} \\ X_2^{e_i} \end{pmatrix} = \frac{\Delta z}{6}\begin{pmatrix} 2 & 1 \\ 1 & 2 \end{pmatrix}\begin{pmatrix} X_1^{e_i} \\ X_2^{e_i} \end{pmatrix} \tag{4.47}$$

The extension to all the finite elements (*assembly*) leads to the zeroth-order differentiation matrix $\mathbf{D}_0$:

$$\mathbf{D}_0\tilde{\mathbf{x}} = \frac{\Delta z}{6}\begin{pmatrix} 2 & 1 & & & \\ 1 & 4 & 1 & & \\ & \ddots & \ddots & \ddots & \\ & & 1 & 4 & 1 \\ & & & 1 & 2 \end{pmatrix}\begin{pmatrix} X_1 \\ X_2 \\ \vdots \\ X_{N-1} \\ X_N \end{pmatrix} \tag{4.48}$$

which coincides with the mass matrix computed in the previous section, i.e., $\mathbf{D}_0 = \mathbf{M}$.

4.3.3 Second Term in the RHS of the Weighted Residual Solution

We now consider the first spatial derivative $\tilde{x}_z$, whose series expansion is given by:

$$\tilde{x}_z = \frac{\partial}{\partial z}\left(\sum_{i=1}^{N} X_i(t)\phi_i(z)\right) = \sum_{i=1}^{N} X_i(t)\phi_{i,z}(z) = X_1(t)\phi_{1,z}(z) + X_2(t)\phi_{2,z}(z)$$

Using the transformed coordinates, we obtain:

$$\frac{\mathrm{d}\phi_i}{\mathrm{d}z} = \frac{\mathrm{d}\phi_i}{\mathrm{d}\xi}\frac{\mathrm{d}\xi}{\mathrm{d}z} = \frac{\mathrm{d}\phi_i}{\mathrm{d}\xi}\frac{2}{\Delta z} \Longrightarrow \phi_{i,z} = \phi_{i,\xi}\frac{2}{\Delta z}$$

Noting that $dz = \frac{\Delta z}{2}d\xi$, the projection of the second term on the RHS of Eq. (4.33) onto the basis functions $\phi_1(\xi)$ and $\phi_2(\xi)$ can be expressed as:

$$\int_{\Omega^{e_i}(z)} \phi_1(z)\alpha_1 \tilde{x}_z^{e_i} dz = \int_{-1}^{1} \phi_1(\xi)\alpha_1 \left(\phi_{1,\xi}(\xi)X_1^{e_i} + \phi_{2,\xi}(\xi)X_2^{e_i}\right) d\xi \tag{4.49}$$

$$\int_{\Omega^{e_i}(z)} \phi_2(z)\alpha_1 \tilde{x}_z^{e_i} dz = \int_{-1}^{1} \phi_2(\xi)\alpha_1 \left(\phi_{1,\xi}(\xi)X_1^{e_i} + \phi_{2,\xi}(\xi)X_2^{e_i}\right) d\xi \tag{4.50}$$

which, using the compact notation, becomes:

$$\frac{1}{2}\begin{pmatrix} \langle\phi_1, \phi_{1,\xi}\rangle & \langle\phi_1, \phi_{2,\xi}\rangle \\ \langle\phi_2, \phi_{1,\xi}\rangle & \langle\phi_2, \phi_{2,\xi}\rangle \end{pmatrix}\begin{pmatrix} X_1^{e_i} \\ X_2^{e_i} \end{pmatrix} = \frac{1}{2}\begin{pmatrix} -1 & 1 \\ -1 & 1 \end{pmatrix}\begin{pmatrix} X_1^{e_i} \\ X_2^{e_i} \end{pmatrix} \tag{4.51}$$

The assembly leads to the *first-order differentiation matrix* $\mathbf{D}_1$:

$$\mathbf{D}_1\tilde{\mathbf{x}} = \frac{1}{2}\begin{pmatrix} -1 & 1 & & & \\ -1 & 0 & 1 & & \\ & \ddots & \ddots & \ddots & \\ & & -1 & 0 & 1 \\ & & & -1 & 1 \end{pmatrix}\begin{pmatrix} X_1 \\ X_2 \\ \vdots \\ X_{N-1} \\ X_N \end{pmatrix} \tag{4.52}$$

The form of matrix $\mathbf{D}_1$ is exactly the same as a three-point centered finite difference method for the first derivative.

4.3.4 Third Term in the RHS of the Weighted Residual Solution

Since $\tilde{x}$ is a piece-wise linear function, it would in principle not be possible to use it to represent the solution of PDEs with second-order derivative terms (since these terms would vanish as a result of the approximation). To alleviate this apparent impossibility, the PDE can be expressed in a so-called *weak form* where the second-order derivative no longer appears. To this end, Green's first identity, which has already been introduced in (4.19), is applied to the third term of Eq. (4.33):

$$\begin{aligned} \alpha_2 \int_{\Omega^{e_i}(z)} \phi_i(z)\tilde{x}_{zz}^{e_i} \mathrm{d}z &= \alpha_2 \left(\int_{\Gamma(z)} \phi_i \frac{\partial \tilde{x}^{e_i}}{\partial n} \mathrm{d}z - \int_{\Omega(z)} \frac{\mathrm{d}\phi_i}{\mathrm{d}z}\frac{\partial \tilde{x}^{e_i}}{\partial z}\mathrm{d}z \right) \\ &= \alpha_2 \left(\left[\phi_i \frac{\partial \tilde{x}^{e_i}}{\partial n} \right]_{z_L}^{z_R} - \int_{z_L}^{z_R} \frac{\mathrm{d}\phi_i}{\mathrm{d}z}\frac{\partial \tilde{x}^{e_i}}{\partial z}\mathrm{d}z \right) \end{aligned} \tag{4.53}$$

The first term in the RHS of Eq. (4.53) refers to the boundary points. Let us, for the moment, focus our attention on the second term in the RHS of Eq. (4.53). In order to compute the contribution of one element e_i to this term, it is convenient to use the transformed variable ξ so that

$$\int_{\Omega(z)} \frac{d\phi_i}{dz}\frac{\partial \tilde{x}}{\partial z}dz = \int_{-1}^{1} \frac{d\phi_i}{d\xi}\frac{\partial \xi}{\partial z}\;\frac{\partial \tilde{x}}{\partial \xi}\frac{d\xi}{dz}\;\frac{\Delta z}{2}d\xi = \int_{-1}^{1} \frac{2}{\Delta z}\frac{d\phi_i}{d\xi}\frac{\partial \tilde{x}}{\partial \xi}d\xi \quad \text{with} \quad i = 1, 2$$

Therefore, the following expression can be obtained

$$-\alpha_2 \int_{\Omega(z)} \frac{d\phi_i}{dz}\frac{\partial \tilde{x}}{\partial z}dz = -\alpha_2 \int_{-1}^{1} \frac{2}{\Delta z}\frac{d\phi_i}{d\xi}\left(X_1\frac{d\phi_1}{d\xi} + X_2\frac{d\phi_2}{d\xi}\right)d\xi \tag{4.54}$$

or, using the compact notation

$$-\frac{2\alpha_2}{\Delta z}\begin{pmatrix} \langle \phi_{1,\xi}, \phi_{1,\xi}\rangle & \langle \phi_{1,\xi}, \phi_{2,\xi}\rangle \\ \langle \phi_{2,\xi}, \phi_{1,\xi}\rangle & \langle \phi_{2,\xi}, \phi_{2,\xi}\rangle \end{pmatrix}\begin{pmatrix} X_1^{e_i} \\ X_2^{e_i} \end{pmatrix} = -\frac{\alpha_2}{\Delta z}\begin{pmatrix} 1 & -1 \\ -1 & 1 \end{pmatrix}\begin{pmatrix} X_1^{e_i} \\ X_2^{e_i} \end{pmatrix}$$

The assembly of this matrix yields:

$$\mathbf{D}_2^{\text{int}} = \frac{1}{\Delta z}\begin{pmatrix} -1 & 1 & & & \\ 1 & -2 & 1 & & \\ & \ddots & \ddots & \ddots & \\ & & 1 & -2 & 1 \\ & & & 1 & -1 \end{pmatrix} \tag{4.55}$$

where the super-index, int, calls for interior points.

A popular approach to take account of the boundary conditions is to substitute them, whenever possible, in the first term of the RHS of Eq. (4.53), i.e.,

$$\left[\phi_i \frac{\partial \tilde{x}^{e_i}}{\partial n}\right]_{z_L}^{z_R} \tag{4.56}$$

Boundary conditions that specify the value of the gradient (Neumann and Robin) enter "naturally" into the formulation and are often referred to as *natural boundary conditions*. For instance, the following boundary conditions which are common in heat and mass transfer problems:

$$\left.\frac{\partial x}{\partial n} = q(x - x^{\text{inf}})\right|_{z=z_L} \quad ; \quad \left.\frac{\partial x}{\partial n} = q(x - x^{\text{inf}})\right|_{z=z_R}$$

where x^{inf} is the value of the dependent variable in the surrounding media. This expression can be substituted into Eq. (4.56) to yield:

$$\phi_N(z_R)q(x_R - x^{\mathrm{inf}}) - \phi_1(z_L)q(x_L - x^{\mathrm{inf}})$$

This latter expression can easily be introduced into the FEM formulation (for a detailed explanation about this way of implementing the boundary condition the reader is referred to [6]). However, handling Dirichlet boundary conditions (so-called *essential boundary conditions*) can be a bit more tricky, and they will be explained through different examples in Sect. 4.11.3 and Chap. 6.

We will prefer, for the moment, an approach where the expression of the boundary conditions is not introduced at this stage, but rather a generic expression of the second-order spatial differentiation operator is developed. The boundary conditions will be enforced at a later stage using a mixed method according to Eq. (4.30).

Let us consider the first element e_1, i.e., the one located at the left end of the spatial domain. The term corresponding to the boundary point is—see Eq. (4.56)

$$\left[\phi_i \frac{\partial \tilde{x}}{\partial n}\right]_{z_L} = -\left[\phi_i \frac{\partial \tilde{x}}{\partial z}\right]_{z_L} \quad \text{with} \quad i = 1, 2$$

The minus sign comes from the fact that n is a unitary vector pointing outwards the spatial domain. On a 1D spatial domain, positively oriented from left to right, it points to the left at the left end and to the right at the right end. Hence, we have the following equivalences:

$$\left.\frac{\partial}{\partial n}\right|_{z_L} = -\left.\frac{\partial}{\partial z}\right|_{z_L} \tag{4.57}$$

$$\left.\frac{\partial}{\partial n}\right|_{z_R} = \left.\frac{\partial}{\partial z}\right|_{z_R} \tag{4.58}$$

The contribution of element e_1 is given by

$$-\left[\phi_i \frac{\partial \tilde{x}}{\partial z}\right]_{z_L} = -\frac{2}{\Delta z}\left[\phi_i \frac{\partial \tilde{x}}{\partial \xi}\right]_{-1} = -\frac{2}{\Delta z}\left(X_1^{e_1}\frac{\mathrm{d}\phi_1}{\mathrm{d}\xi} + X_2^{e_1}\frac{\mathrm{d}\phi_2}{\mathrm{d}\xi}\right)_{\xi=-1}$$

$$\Rightarrow -\left[\phi_i \frac{\partial \tilde{x}}{\partial z}\right]_{z_L} = -\frac{1}{\Delta z}\begin{pmatrix}-1 & 1\end{pmatrix}\begin{pmatrix}X_1^{e_1} \\ X_2^{e_1}\end{pmatrix} \tag{4.59}$$

The same steps can be followed to find the contribution of the last element e_{N-1} to the boundary conditions:

$$\left[\phi_i \frac{\partial \tilde{x}}{\partial z}\right]_{z_R} = -\frac{1}{\Delta z}\begin{pmatrix}1 & -1\end{pmatrix}\begin{pmatrix}X_1^{e_{N-1}} \\ X_2^{e_{N-1}}\end{pmatrix} \tag{4.60}$$

Now Eqs. (4.59) and (4.60) must be added, respectively, to the first and last rows of matrix $\mathbf{D}_2^{\text{int}}$ defined in (4.55). As a result the *second-order differentiation matrix* $\mathbf{D}_2$ is obtained:

$$\mathbf{D}_2\tilde{\mathbf{x}} = \frac{1}{\Delta z}\begin{pmatrix} 0 & 0 & & & \\ 1 & -2 & 1 & & \\ & \ddots & \ddots & \ddots & \\ & & 1 & -2 & 1 \\ & & & 0 & 0 \end{pmatrix}\begin{pmatrix} X_1 \\ X_2 \\ \vdots \\ X_{N-1} \\ X_N \end{pmatrix} \tag{4.61}$$

Again, we stress the fact that the specificities of the boundary conditions have not been addressed yet. This step will be described in Sect. 4.4 using a weighted residual formulation.

4.3.5 Fourth Term in the RHS of the Weighted Residual Solution

The function $f(x)$ appearing in the last term of the RHS can be any given nonlinear function. In principle, the integrals involved in this term can be numerically computed using any quadrature formula such as Simpson's rule or the Gauss formula with n points, to name a few. However, a more natural scheme can be derived based on the finite element discretization and functional approximation itself, see Fig. 4.8.

In the same way as x is approximated by $\tilde{x} = \sum_{i=1}^{n} \phi_i X_i$, the nonlinear function $f(x)$ can be approximated by $f(\tilde{x}) = f(\sum_{i=1}^{n} \phi_i X_i)$. On the element e_i, $f(\tilde{x}) = f(\phi_i X_i + \phi_{i+1} X_{i+1})$ which, as shown in the figure, is equivalent to $f(\tilde{x}) = \phi_i f(X_i) + \phi_{i+1} f(X_{i+1})$. The projection of $f(x)$ onto the basis functions $\phi_i(x)$ and $\phi_{i+1}(x)$, therefore yields, using the transformed coordinates ξ and the local notation:

$$\begin{pmatrix} \displaystyle\int_{-1}^{1} \phi_1(\xi) f(\tilde{x}) \frac{\Delta z}{2} d\xi \\ \displaystyle\int_{-1}^{1} \phi_2(\xi) f(\tilde{x}) \frac{\Delta z}{2} d\xi \end{pmatrix} = \frac{\Delta z}{2}\begin{pmatrix} \langle \phi_1(\xi), f(\tilde{x}) \rangle \\ \langle \phi_2(\xi), f(\tilde{x}) \rangle \end{pmatrix} = \frac{\Delta z}{2}\begin{pmatrix} \langle \phi_1, \phi_1 f_1 + \phi_2 f_2 \rangle \\ \langle \phi_2, \phi_1 f_1 + \phi_2 f_2 \rangle \end{pmatrix}$$
$$= \frac{\Delta z}{2}\begin{pmatrix} \langle \phi_1, \phi_1 \rangle & \langle \phi_1, \phi_2 \rangle \\ \langle \phi_2, \phi_1 \rangle & \langle \phi_2, \phi_2 \rangle \end{pmatrix}\begin{pmatrix} f_1 \\ f_2 \end{pmatrix} = \frac{\Delta z}{6}\begin{pmatrix} 2 & 1 \\ 1 & 2 \end{pmatrix}\begin{pmatrix} f_1 \\ f_2 \end{pmatrix} \tag{4.62}$$

where $f_1 = f\left(X_1^{e_i}\right)$ and $f_2 = f\left(X_2^{e_i}\right)$. The assembly over the whole domain gives:

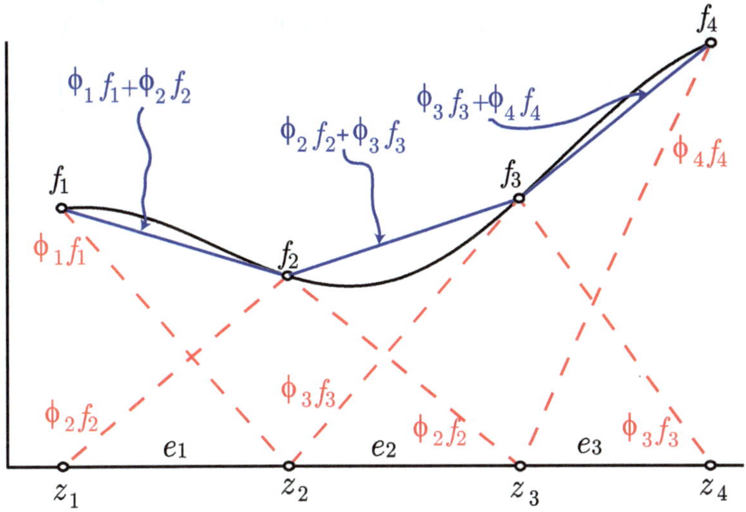

Fig. 4.8 FEM approximation of a given nonlinear function $f(x)$

$$\frac{\Delta z}{6}\begin{pmatrix} 2 & 1 & & & \\ 1 & 4 & 1 & & \\ & \ddots & \ddots & \ddots & \\ & & 1 & 4 & 1 \\ & & & 1 & 2 \end{pmatrix}\begin{pmatrix} f_1 \\ f_2 \\ \vdots \\ f_{N-1} \\ f_N \end{pmatrix} = \mathbf{M}\tilde{f}_{NL} \tag{4.63}$$

With this new result, Eq. (4.33) can be rewritten as:

$$\mathbf{M}\frac{\partial \tilde{\mathbf{x}}}{\partial t} = \alpha_0 \mathbf{D}_0 \tilde{\mathbf{x}} + \alpha_1 \mathbf{D}_1 \tilde{\mathbf{x}} + \alpha_2 \mathbf{D}_2 \tilde{\mathbf{x}} + \mathbf{M}\tilde{f}_{NL} \tag{4.64}$$

It is important to mention that in some occasions, for instance with highly non-linear functions, a more accurate numerical integration method may be required. If we return to the first part of Eq. (4.62)

$$\begin{pmatrix} \int_{-1}^{1} \phi_1(\xi) f(\tilde{x}) \frac{\Delta z}{2} d\xi \\ \\ \int_{-1}^{1} \phi_2(\xi) f(\tilde{x}) \frac{\Delta z}{2} d\xi \end{pmatrix} = \frac{\Delta z}{2}\begin{pmatrix} \langle \phi_1(\xi), f(\tilde{x}) \rangle \\ \\ \langle \phi_2(\xi), f(\tilde{x}) \rangle \end{pmatrix} = \frac{\Delta z}{2}\begin{pmatrix} \langle \phi_1, f(\phi_1 \tilde{x}_1 + \phi_2 \tilde{x}_2) \rangle \\ \\ \langle \phi_2, f(\phi_1 \tilde{x}_1 + \phi_2 \tilde{x}_2) \rangle \end{pmatrix}$$

the assembly over the whole domain results in

$$\frac{\Delta z}{2}\begin{pmatrix} \langle\phi_1, f(\phi_1\tilde{x}_1+\phi_2\tilde{x}_2)\rangle \\ \langle\phi_2, f(\phi_1\tilde{x}_1+\phi_2\tilde{x}_2)\rangle+\langle\phi_1, f(\phi_1\tilde{x}_2+\phi_2\tilde{x}_3)\rangle \\ \langle\phi_2, f(\phi_1\tilde{x}_2+\phi_2\tilde{x}_3)\rangle+\langle\phi_1, f(\phi_1\tilde{x}_3+\phi_2\tilde{x}_4)\rangle \\ \vdots \\ \langle\phi_2, f(\phi_1\tilde{x}_{N-2}+\phi_2\tilde{x}_{N-1})\rangle+\langle\phi_1, f(\phi_1\tilde{x}_{N-1}+\phi_2\tilde{x}_{N-2})\rangle \\ \langle\phi_2, f(\phi_1\tilde{x}_{N-1}+\phi_2\tilde{x}_N)\rangle \end{pmatrix} \tag{4.65}$$

We can distinguish two terms inside this vector: those including the projection onto ϕ_1 and those with the projection performed onto ϕ_2. A quadrature formula such as the Gauss-Legendre (see function Gauss-Legendre) quadrature can be used to compute the integrals.

```
function I = gauss(f,n)
% This code computes the integral of a given function f over the
% interval x = [-1,1] using the Gauss-Legendre quadrature
% Input parameters:
% f: matlab function with the expression of the function to be
%    integrated
% n: Number of points to be employed in the formula
% Output parameter:
% I: Numerical value of the integral

% Computation of the abcissa
beta        = 0.5./sqrt(1-(2*(1:n)).^(-2));
T           = diag(beta,1) + diag(beta,-1);
[V,D]       = eig(T);
xquad       = diag(D);
[xquad,i]   = sort(xquad);

% Computation of the weights
w           = 2*V(1,i).^2;

% Integral value
I           = w*feval(f,xquad);
```

Function Gauss-Legendre Function `Gauss-Legendre` to numerically compute integrals

Table 4.3 shows a comparison between the Gauss-Legendre quadrature and the integral computed using the finite element structure. Although the integration based on the FEM structure is more natural, it also requires a much larger number of points to achieve the same accuracy. For the integration of smooth nonlinear functions, the FEM approach will be quite accurate with a small number of points. Therefore, since its implementation is easier, it will be preferable to the Gauss-Legendre quadrature. When the results with the FEM approach are not satisfactory, the Gauss-Legendre procedure should be used.

Table 4.3 Comparison between the Gauss-Legendre quadrature and the finite element method

Method	Number of points in the scheme	$\int_{-1}^{1} \exp(\mathbf{z}) \sin(\mathbf{3z})\mathbf{dz} = \mathbf{0.7416161}$
FEM	6	0.6532869
	25	0.7378264
	100	0.7413935
	400	0.7416024
Gauss-Legendre	2	0.6926873
	4	0.7416127
	5	0.7416154
	6	0.7416161

In order to close the system and obtain the final version of the equations, boundary conditions must be introduced into Eq. (4.64).

4.4 Galerkin Method Over Linear Lagrangian Elements: Contribution of the Boundary Conditions

We now introduce the boundary conditions into Eq. (4.33) using the mixed weighted residuals approach—see Eq. (4.30):

$$\int_{\Omega(z)} w_i \tilde{x}_t \mathrm{d}z = \alpha_0 \int_{\Omega(z)} w_i \tilde{x} \mathrm{d}z + \alpha_1 \int_{\Omega(z)} w_i \tilde{x}_z \mathrm{d}z + \alpha_2 \int_{\Omega(z)} w_i \tilde{x}_{zz} \mathrm{d}z + \int_{\Omega(z)} w_i f(\tilde{x}) \mathrm{d}z - \int_{\Gamma(z)} w_i \left(\beta_1 \frac{\partial \tilde{x}}{\partial n} + \beta_0 \tilde{x} + g(\tilde{x}) \right) \mathrm{d}z; \quad i = 1, \ldots, N \tag{4.66}$$

Since the spatial domain (Ω) is in 1D and using the Lagrange polynomials (ϕ_i) as the weighting functions (w_i), Eq. (4.66) can be rewritten as:

$$\int_{\Omega(z)} \phi_i \tilde{x}_t dz = \alpha_0 \int_{\Omega(z)} \phi_i \tilde{x} dz + \alpha_1 \int_{\Omega(z)} \phi_i \tilde{x}_z dz + \alpha_2 \int_{\Omega(z)} \phi_i \tilde{x}_{zz} dz + \int_{\Omega(z)} \phi_i f(\tilde{x}) dz - \phi_i \left(\beta_1 \frac{\partial \tilde{x}}{\partial n} + \beta_0 \tilde{x} + g(\tilde{x}) \right) \bigg|_{z=z_L} - \phi_i \left(\beta_1 \frac{\partial \tilde{x}}{\partial n} + \beta_0 \tilde{x} + g(\tilde{x}) \right) \bigg|_{z=z_R}; \quad i = 1, \ldots, N \tag{4.67}$$

Note that the only nonzero basis function at z_L is ϕ_1, and $\phi_1(z = z_L) = 1$. Therefore, the term $\phi_i \left(\beta_1 \frac{\partial \tilde{x}}{\partial n} + \beta_0 \tilde{x} + g(\tilde{x}) \right) \Big|_{z=z_L} = \beta_1 \frac{\partial \tilde{x}_L}{\partial n} + \beta_0 \tilde{x}_L + g(\tilde{x}_L)$ affects only the first row of system (4.64). The only nonzero basis function at z_R is $\phi_N(z_R) = 1$ so that $\phi_i \left(\beta_1 \frac{\partial \tilde{x}}{\partial n} + \beta_0 \tilde{x} + g(\tilde{x}) \right) \Big|_{z=z_R} = \beta_1 \frac{\partial \tilde{x}_R}{\partial n} + \beta_0 \tilde{x}_R + g(\tilde{x}_R)$ affects only the last row of system (4.64).

4.4.1 Dirichlet Boundary Conditions

As mentioned before, this kind of boundary conditions specifies the value of the dependent variable at the boundary (in expression (4.24) they are defined by setting $\beta_1 = 0$ and $\beta_0 = 1$), i.e.:

$$x(z_L, t) = g_L(t); \quad x(z_R, t) = g_R(t) \tag{4.68}$$

The term $\tilde{x}_L + g(\tilde{x}_L)$ in Eq. (4.67) can be substituted with $-(x_1 - g_L)$ and added to the first row of system (4.64). Similarly, the last equation is modified by including the term $-(x_N - g_R)$ which comes from $\tilde{x}_R + g(\tilde{x}_R)$ in Eq. (4.67). Defining the vector:

$$\mathbf{g} = \begin{pmatrix} -(x_1 - g_L) \\ 0 \\ \vdots \\ 0 \\ -(x_N - g_R) \end{pmatrix} \tag{4.69}$$

the final FEM system of equations can be written as

$$\mathbf{M} \frac{\partial \tilde{\mathbf{x}}}{\partial t} = (\alpha_0 \mathbf{D}_0 + \alpha_1 \mathbf{D}_1 + \alpha_2 \mathbf{D}_2) \tilde{\mathbf{x}} + \mathbf{M} \tilde{f}_{NL} + \mathbf{g} \tag{4.70}$$

4.4.2 Neumann Boundary Conditions

In this case, the boundary conditions specify the value of the gradient, i.e.:

$$\left. \frac{\partial \tilde{x}}{\partial n} \right|_{z_L} = g_L(t); \quad \left. \frac{\partial \tilde{x}}{\partial n} \right|_{z_R} = g_R(t) \tag{4.71}$$

where n is a normal vector pointing outwards the boundary. At the left side boundary n and the gradient are vectors with opposite directions. On the other hand, at right side boundary both vectors are in the same direction. As a consequence, $\left. \frac{\partial}{\partial z} \right|_{z_L} = - \left. \frac{\partial}{\partial n} \right|_{z_L}$ and $\left. \frac{\partial}{\partial z} \right|_{z_R} = \left. \frac{\partial}{\partial n} \right|_{z_R}$ and the previous expressions can be rewritten as:

$$\left. \frac{\partial \tilde{x}}{\partial z} \right|_{z_L} = -g_L(t); \quad \left. \frac{\partial \tilde{x}}{\partial z} \right|_{z_R} = g_R(t) \tag{4.72}$$

The approach to implement this kind of boundary conditions is very similar to the Dirichlet case. The main difference is that an expression for the derivative is required.

Using the expression for the first derivative obtained using the Lagrangian elements:

$$-\left.\frac{\partial \tilde{x}}{\partial z}\right|_{z_1} - g_L(t) = -\frac{\tilde{x}_2 - \tilde{x}_1}{\Delta z} - g_L(t) \tag{4.73}$$

$$\left.\frac{\partial \tilde{x}}{\partial z}\right|_{z_N} - g_R(t) = \frac{\tilde{x}_N - \tilde{x}_{N-1}}{\Delta z} - g_R(t) \tag{4.74}$$

The vector **g** in Eq. (4.70) becomes:

$$\mathbf{g} = \begin{pmatrix} -\dfrac{\tilde{x}_2 - \tilde{x}_1}{\Delta z} - g_L \\ 0 \\ \vdots \\ 0 \\ \dfrac{\tilde{x}_N - \tilde{x}_{N-1}}{\Delta z} - g_R \end{pmatrix} \tag{4.75}$$

and the complete system of equations can be rewritten as:

$$\mathbf{M}\frac{\partial \tilde{\mathbf{x}}}{\partial t} = (\alpha_0 \mathbf{D}_0 + \alpha_1 \mathbf{D}_1 + \alpha_2 \mathbf{D}_2)\,\tilde{\mathbf{x}} + \mathbf{M}\tilde{f}_{NL} + \mathbf{g}. \tag{4.76}$$

4.5 The Finite Element Method in Action

In this section, we will use Burgers' equation to illustrate the MATLAB implementation of the finite element method. Let us rewrite for convenience the set of equations:

$$\frac{\partial x}{\partial t} = \mu \frac{\partial^2 x}{\partial z^2} - x\frac{\partial x}{\partial z} = \mu x_{zz} - x x_z \tag{4.77}$$

defined over the spatial domain $z \in [0, 1]$ with $t \geqslant 0$. Initial and boundary conditions are of the form:

$$x(z, 0) = x_a(z, 0) \tag{4.78}$$

$$x(0, t) = x_a(0, t) \tag{4.79}$$

$$x(1, t) = x_a(1, t) \tag{4.80}$$

where x_a is the analytical solution—see Eq. (4.22).

The application of the finite element method yields a system of ordinary differential equations of the form (4.70):

$$\mathbf{M}\frac{\partial \tilde{\mathbf{x}}}{\partial t} = \mu \mathbf{D}_2 \tilde{\mathbf{x}} - \tilde{f}_{NL} + \mathbf{g} \tag{4.81}$$

where $\tilde{f}_{NL}$ represents the nonlinear term xx_z. Let us now describe the different elements required for the implementation of this set of equations. We will start, as in Sect. 1.3, with the function implementing the ODE set which is listed in burgers_pdes.

```
function xt = burgers_pdes(t,x)
global mu
global wquad xquad
global n z0 zL D2 xx

% Second spatial derivative xzz computed through the
% second differentiation matrix
xzz = D2*x;

% Rename state variable x
xx = x;

% Spatial integration of the integrands in the nonlinear
% term of Burgers equation. Nonlinear term: fx = x*xz
y1 = feval('integrand1',xquad')*wquad';
y2 = feval('integrand2',xquad')*wquad';
fx = y1+y2;

% Boundary conditions
gL      = burgers_exact(z0,t);
gR      = burgers_exact(zL,t);
gv(1,1) = x(1)-gL;
gv(n,1) = x(n)-gR;

% System of ordinary differential equations
xt = mu*xzz - fx - gv;
```

Function burgers_pdes Function `burgers_pdes` to define the ODE set for Burgers equation

We can note the following details in the implementation of Eq. (4.81):

- The function starts with the definition of the global variables.
- First, the second derivative of the dependent variable `xzz` is computed using the second-order differentiation matrix $\mathbf{D}_2$—see (4.61).
- Then, the current value of the dependent variable x is stored into a new vector xx which is passed to other functions as a global variable.
- The most involved part of the implementation is the computation of the finite element version of the nonlinear term (remember that it requires the computation of spatial integrals). In this case, the Gauss-Legendre quadrature is used to compute such integrals and it is divided into two parts:
  `y1 = feval('integrand1',xquad')*wquad'`
  where the projection onto the basis functions ϕ_1 is computed and
  `y2 = feval('integrand2',xquad')*wquad';`
  where the projection onto the basis functions ϕ_2 is computed. Functions integrand1 and integrand2 compute the terms inside the integral (4.65).

```
function out = integrand1(z)
% Computation of the first integrand on the nonlinear term
% for Burgers equation. This term corresponds with
%       y1 = phi_1*(xfem*xfem_z)

global xx n

% Dependent variable values with the FEM local
% nomenclature
x1 = xx(1:n-1);
x2 = xx(2:n);

% Preallocation of memmory for the output
out = zeros(n,length(z));

% Computation of the integrand
for i = 1:length(z)
    % Basis function computation
    [phi_1, phi_2] = trialfunctionslagrlin(z(i));
    % Finite element approximation of the dependent
    % variable
    xfem   = phi_1*x1 + phi_2*x2;
    % Finite element approximation of the first spatial
    % derivative of the dependent variable
    xfem_z = 1/2*(-x1 + x2);
    % Final output
    out(1:n-1,i) = phi_1*((xfem).*(xfem_z));
end
```

Function integrand1 Function `integrand1` to evaluate the integrand in (4.65) dealing with the projection onto the basis function ϕ_1

```
function out = integrand2(z)
% Computation of the sencond integrand on the nonlinear
% term for Burgers' equation. This term correspond with
%       y2 = phi_2*(xfem*xfem_z)

global xx n

% Dependent variable values with the FEM local
% nomenclature
x1 = xx(1:n-1);
x2 = xx(2:n);

% Preallocation of memmory for the output
out = zeros(n,length(z));

% Computation of the integrand
for i = 1:length(z)
    % Basis function computation
    [phi_1, phi_2] = trialfunctionslagrlin(z(i));
    % Finite element approximation of the dependent
    % variable
    xfem   = phi_1*x1 + phi_2*x2;
    % Finite element approximation of the first spatial
    % derivative of the dependent variable
    xfem_z = 1/2*(-x1 + x2);
```

```
    % Final output
    out(2:n,i) = phi_2*((xfem).*(xfem_z));
end
```

Function integrand2 Function integrand2 to evaluate the integrand in (4.65) dealing with the projection onto the basis function ϕ_2

- Next, the boundary conditions are implemented as in (4.69)
- Finally, the set of ODEs is developed. As in Sect. 1.3, we can notice a close resemblance of the PDE $x_t = \mu x_{zz} - x x_z$ and the programming of the PDE `xt = mu*xzz-fx-gv`

The main program that calls the MATLAB function burgers_pdes is listed in the script main_burgers_FEM.

```
% Main program. FEM solution of Burgers' equation

close all
clear all

% Global variables (shared by other files)
global mu
global wquad xquad
global n dz z0 zL D2

% Finite element spatial grid
z0  = 0;
zL  = 1;
n   = 201;
nel = n-1;
dz  = (zL - z0)/(n-1);
z   = (z0:dz:zL)';

% Second order differentiation matrix
D2 = lagrlinD2(dz,n);

% Problem parameters
mu = 0.01;

% Computation of the weigths and the abcissa used in the
% numerical integration via the Gauss-Legendre quadrature
nquad      = 2;
beta       = 0.5./sqrt(1-(2*(1:nquad)).^(-2));
T          = diag(beta,1) +diag(beta,-1);
[V,D]      = eig(T);
xquad      = diag(D);
[xquad,i]  = sort(xquad);
wquad      = 2*V(1,i).^2;

% Initial conditions
x = zeros(1,n);
for ii = 1:n
    x(ii) = burgers_exact(z(ii),0);
end
```

```
% Time instants at which the IVP solver will save the
% solution
dt   = 0.1;
time = (0:dt:1);
nt   = length(time);

% Time integration with the IVP solver
ne         = 1;
M          = lagrlinmass(dz,n,ne); % FEM mass matrix
options    = odeset('Mass',M,'RelTol',1e-6,'AbsTol',1e-6);
[tout,yout]= ode15s(@burgers_pdes, time, x, options);


% Plot the results
figure
hold
plot(z,yout,'.-k')
yexact = zeros(n,length(tout));
for k = 1:length(tout)
    for i = 1:n
        yexact(i,k) = burgers_exact(z(i),tout(k));
    end
end
plot(z,yexact,'r')
```

Script main_burgers_FEM Main script to call function `burgers_pdes`

We can note the following points about this main program:

- The program begins with the definition of the global variables which will be passed to other functions.
- As in Sect. 1.3, the spatial grid is defined. In this case, the number of discretization points is set to 201, i.e., the number of finite elements is 200. All the discretization points are equidistant.
- The next step is the construction of the second-order differentiation matrix (4.61) using function lagrlinD2.

```
function out = lagrlinD2(h,n)
% Computation of the second-order differentiation matrix
% of the finite element method with linear lagrangian
% elements

% Main diagonal of the second-order differentiation matrix
d0 = [0 -2*ones(1,n-2) 0];
% Upper first diagonal of the second-order differentiation
% matrix
dp1 = [0 ones(1,n-2)];
% Lower first diagonal of the second-order differentiation
% matrix
dm1 = [ones(1,n-2) 0];
% Second order differentiation matrix
out = sparse((1/h)*(diag(d0,0)+diag(dp1,1)+diag(dm1,-1)));
```

Function lagrlinD2 Function `lagrlinD2` to compute the finite element second-order differentiation matrix using linear Lagrangian elements

- The model parameter μ is defined
- The abscissae xquad and the weights wquad of the Gauss-Legendre quadrature (to be used in the projection of the nonlinear term) are computed
- The initial conditions $x(z, 0)$ are defined for all the discretization points
- The time-span for the output solution is defined
- Then, the finite element mass matrix (4.44) is constructed using the MATLAB function lagrlinmass and is passed to the ODE solver through the MATLAB function odeset.

```
function M = lagrlinmass(h,n,ne)
% Computation of the mass matrix of the finite element
% method with linear Lagrangian elements

% Main diagonal of the mass matrix
d0 = diag(repmat(([2 repmat(4,1,n-2) 2]),1,ne),0);

% First upper diagonal of the mass matrix
d1 = diag([repmat([ones(1,n-1) 0],1,ne-1) ones(1,n-1)],1);

% First lower diagonal of the mass matrix
dm1 = diag([repmat([ones(1,n-1) 0],1,ne-1),...
           ones(1,n-1)],-1);

% Mass matrix
M = sparse((h/6)*(d0 + d1 + dm1));
```

Function lagrlinmass Function lagrlinmass to compute the finite element mass matrix using linear Lagrangian elements

The third input argument of this latter function, ne, is the number of PDEs to be solved. The output will be the mass matrix of the whole system of equations. We will come back to this feature later on when a system with more than one variable is considered.

- The integration is performed with the IVP solver ode15s which requires the definition of the output time instants, the initial conditions and a set of options which in this case include the mass matrix.

The analytical and numerical (using the codes just presented) solutions to Burgers' equation are plotted in Fig. 4.9 showing that this numerical scheme produces very satisfactory results. Depending on the value of parameter μ in Burgers' equation, it might be required to change the number of discretization points and the type of basis functions in order to obtain a good approximation.

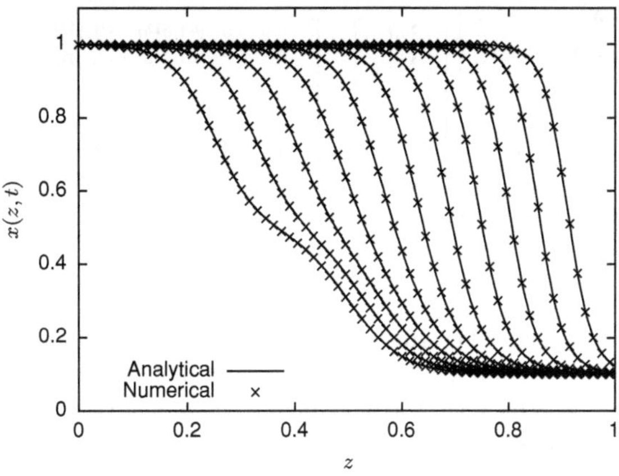

Fig. 4.9 Comparison between the analytical solution (*continuous lines*) and numerical solution (*marks*) computed using the finite element method with Lagrangian elements for Burgers' equation

4.6 The Finite Element Method Applied to Systems of PDEs

Consider now the following system of PDEs:

$$\begin{aligned} x_{1,t} &= \alpha_{1,0}x_1 + \alpha_{1,1}x_{1,z} + \alpha_{1,2}x_{1,zz} + f_1(x_1, x_2, \ldots, x_{n_e}) \\ x_{2,t} &= \alpha_{2,0}x_2 + \alpha_{2,1}x_{2,z} + \alpha_{2,2}x_{2,zz} + f_2(x_1, x_2, \ldots, x_{n_e}) \\ &\vdots \\ x_{n_e,t} &= \alpha_{n_e,0}x_{n_e} + \alpha_{n_e,1}x_{n_e,z} + \alpha_{n_e,2}x_{n_e,zz} + f_{n_e}(x_1, x_2, \ldots, x_{n_e}) \end{aligned} \tag{4.82}$$

with given boundary and initial conditions. Applying the previous procedure to each of these PDEs, the following system of ODEs is obtained:

$$\begin{aligned} \mathbf{M}\frac{\mathrm{d}\tilde{\mathbf{x}}_1}{\mathrm{d}t} &= \left(\alpha_{1,0}\mathbf{D}_0 + \alpha_{1,1}\mathbf{D}_1 + \alpha_{1,2}\mathbf{D}_2\right)\tilde{\mathbf{x}}_1 + \mathbf{M}\tilde{f}_{1,NL} + \mathbf{g}_1 \\ \mathbf{M}\frac{\mathrm{d}\tilde{\mathbf{x}}_2}{\mathrm{d}t} &= \left(\alpha_{2,0}\mathbf{D}_0 + \alpha_{2,1}\mathbf{D}_1 + \alpha_{2,2}\mathbf{D}_2\right)\tilde{\mathbf{x}}_2 + \mathbf{M}\tilde{f}_{2,NL} + \mathbf{g}_2 \\ &\vdots \\ \mathbf{M}\frac{\mathrm{d}\tilde{\mathbf{x}}_{n_e}}{\mathrm{d}t} &= \left(\alpha_{n_e,0}\mathbf{D}_0 + \alpha_{n_e,1}\mathbf{D}_1 + \alpha_{n_e,2}\mathbf{D}_2\right)\tilde{\mathbf{x}}_{n_e} + \mathbf{M}\tilde{f}_{n_e,NL} + \mathbf{g}_{n_e} \end{aligned} \tag{4.83}$$

or, in a more compact form

$$\underbrace{\begin{bmatrix} \mathbf{M} & 0 & 0 & \cdots & 0 \\ 0 & \mathbf{M} & 0 & \cdots & 0 \\ 0 & 0 & \ddots & \ddots & 0 \\ 0 & 0 & \ddots & \mathbf{M} & 0 \\ 0 & 0 & 0 & \cdots & \mathbf{M} \end{bmatrix}}_{\mathbf{M}^s} \begin{bmatrix} \tilde{\mathbf{x}}_{1,t} \\ \tilde{\mathbf{x}}_{2,t} \\ \vdots \\ \tilde{\mathbf{x}}_{ne-1,t} \\ \tilde{\mathbf{x}}_{ne,t} \end{bmatrix} = \begin{bmatrix} (\alpha_{1,0}\mathbf{D}_0 + \alpha_{1,1}\mathbf{D}_1 + \alpha_{1,2}\mathbf{D}_2)\,\tilde{\mathbf{x}}_1 + \mathbf{M}\tilde{f}_{1,NL} + \mathbf{g}_1 \\ (\alpha_{2,0}\mathbf{D}_0 + \alpha_{2,1}\mathbf{D}_1 + \alpha_{2,2}\mathbf{D}_2)\,\tilde{\mathbf{x}}_2 + \mathbf{M}\tilde{f}_{2,NL} + \mathbf{g}_2 \\ \vdots \\ (\alpha_{ne-1,0}\mathbf{D}_0 + \alpha_{ne-1,1}\mathbf{D}_1 + \alpha_{ne-1,2}\mathbf{D}_2)\,\tilde{\mathbf{x}}_{ne-1} + \mathbf{M}\tilde{f}_{ne-1,NL} + \mathbf{g}_{ne-1} \\ (\alpha_{ne,0}\mathbf{D}_0 + \alpha_{ne,1}\mathbf{D}_1 + \alpha_{ne,2}\mathbf{D}_2)\,\tilde{\mathbf{x}}_{ne} + \mathbf{M}\tilde{f}_{ne,NL} + \mathbf{g}_{ne} \end{bmatrix} \tag{4.84}$$

where matrix $\mathbf{M}^s$ is the mass matrix of the whole system which can be directly constructed using function lagrlinmass, included in the companion software. This matrix has to be passed to the ODE solver as already highlighted in the previous application example. For instance, if two PDEs ($n_e = 2$) are considered:

```
ne = 2;
M  = lagrlinmass(dz,n,ne);
options = odeset('Mass',M);
```

An alternative way is to use function lagrlinmass with $n_e = 1$ to find the mass matrix $\mathbf{M}$ corresponding to a single equation

```
ne = 1;
M  = lagrlinmass(dz,n,ne);
```

The inverse ($\mathbf{M}^{-1}$) can then be computed. Multiplying both sides of each of Eq. (4.83) by $\mathbf{M}^{-1}$, the following expression is obtained:

$$\begin{aligned} \frac{d\tilde{\mathbf{x}}_1}{dt} &= \mathbf{M}^{-1}\left(\alpha_{1,0}\mathbf{D}_0 + \alpha_{1,1}\mathbf{D}_1 + \alpha_{1,2}\mathbf{D}_2\right)\tilde{\mathbf{x}}_1 + \tilde{f}_{1,NL} + \mathbf{M}^{-1}\mathbf{g}_1 \\ \frac{d\tilde{\mathbf{x}}_2}{dt} &= \mathbf{M}^{-1}\left(\alpha_{2,0}\mathbf{D}_0 + \alpha_{2,1}\mathbf{D}_1 + \alpha_{2,2}\mathbf{D}_2\right)\tilde{\mathbf{x}}_2 + \tilde{f}_{2,NL} + \mathbf{M}^{-1}\mathbf{g}_2 \\ &\vdots \\ \frac{d\tilde{\mathbf{x}}_{n_e}}{dt} &= \mathbf{M}^{-1}\left(\alpha_{n_e,0}\mathbf{D}_0 + \alpha_{n_e,1}\mathbf{D}_1 + \alpha_{n_e,2}\mathbf{D}_2\right)\tilde{\mathbf{x}}_{n_e} + \tilde{f}_{n_e,NL} + \mathbf{M}^{-1}\mathbf{g}_{n_e} \end{aligned} \tag{4.85}$$

The mass matrix does not need to be passed to the ODE solver any longer, and more basic time integrators could therefore be used, which would not be capable of solving linearly implicit system of ODEs. The dependent variables can also be stored in various ways so as to confer the Jacobian matrix a desirable structure. In particular, the discretized variables can be stored in the solution vector according to the order of the FEM nodes. In this case, the first elements of the solution vector correspond to the values of the dependent variables at the first node. The next elements correspond to the variable values at the second node and so on, i.e., $[\tilde{x}_{1,1}, \tilde{x}_{2,1}, \ldots, \tilde{x}_{n_e,1}, \tilde{x}_{1,2}, \tilde{x}_{2,2}, \ldots, \tilde{x}_{n_e,2}, \tilde{x}_{1,N}, \tilde{x}_{2,N}, \ldots, \tilde{x}_{n_e,N}]^T$. This way of sorting the solution vector may appear a bit intricate, but as shown in Sect. 3.11, the Jacobian matrix will have a banded structure which could be efficiently exploited by specific time integrators.

4.7 Galerkin Method Over Hermitian Elements

Linear Lagrangian elements provide the simplest form of finite element method. The basis functions linearly interpolate the function values in the several discretization points. Is it possible to use more sophisticated (and hopefully more accurate) interpolation functions? The answer is of course positive and a possible option is given by Hermite polynomials. Four basis functions are used to approximate the solution on one element so that, in terms of the transformed variable ξ, we have:

$$\tilde{x}(\xi, t) = X_1(t)\phi_1(\xi) + X_{1,\xi}(t)\phi_2(\xi) + X_2(t)\phi_3(\xi) + X_{2,\xi}(t)\phi_4(\xi) \tag{4.86}$$

where the terms $X_{1,\xi}$ and $X_{2,\xi}$ are of the form:

$$X_{1,\xi} = \left.\frac{\partial X}{\partial \xi}\right|_{\xi=-1} \quad ; \quad X_{2,\xi} = \left.\frac{\partial X}{\partial \xi}\right|_{\xi=1}$$

The Hermitian basis functions ϕ_i with $i = 1, 2, 3, 4$ are given by:

$$\phi_1(\xi) = \frac{1}{4}(1-\xi)^2(2+\xi) \tag{4.87}$$

$$\phi_2(\xi) = \frac{1}{4}(1-\xi^2)(1-\xi) \tag{4.88}$$

$$\phi_3(\xi) = \frac{1}{4}(1+\xi)^2(2-\xi) \tag{4.89}$$

$$\phi_4(\xi) = \frac{1}{4}(-1+\xi^2)(1+\xi) \tag{4.90}$$

A remarkable feature of the Hermite basis functions is that they make use not only of the function values in the several nodes but also of the solution slopes. We will follow here the same procedure as in the case of linear Lagrangian elements to develop the different terms of Eq. (4.33).

4.7.1 *LHS Term of the Weighted Residual Solution*

Recall that, when considering Lagrangian elements, only two basis functions were different from zero inside each finite element e^i. Now, with Hermitian elements, four basis functions have values different from zero inside each element. Therefore, one element (for instance e_i) will contribute to four rows of the equation system resulting from the method of lines.

This contribution is of the form $\int_{\Omega(z)} w_i \tilde{x}_t dz$. Substituting the weighting function w_i with each of the nonzero basis functions (ϕ_1, ϕ_2, ϕ_3 and ϕ_4) and using the series expansion (4.86) we have, for element e_i:

$$\begin{bmatrix} \int_{-1}^{1} \phi_1 \left(\frac{dX_1^{e_i}}{dt}\phi_1 + \frac{dX_{1,\xi}^{e_i}}{dt}\phi_2 + \frac{dX_2^{e_i}}{dt}\phi_3 + \frac{dX_{2,\xi}^{e_i}}{dt}\phi_4 \right) \frac{\Delta z}{2} d\xi \\ \int_{-1}^{1} \phi_2 \left(\frac{dX_1^{e_i}}{dt}\phi_1 + \frac{dX_{1,\xi}^{e_i}}{dt}\phi_2 + \frac{dX_2^{e_i}}{dt}\phi_3 + \frac{dX_{2,\xi}^{e_i}}{dt}\phi_4 \right) \frac{\Delta z}{2} d\xi \\ \int_{-1}^{1} \phi_3 \left(\frac{dX_1^{e_i}}{dt}\phi_1 + \frac{dX_{1,\xi}^{e_i}}{dt}\phi_2 + \frac{dX_2^{e_i}}{dt}\phi_3 + \frac{dX_{2,\xi}^{e_i}}{dt}\phi_4 \right) \frac{\Delta z}{2} d\xi \\ \int_{-1}^{1} \phi_4 \left(\frac{dX_1^{e_i}}{dt}\phi_1 + \frac{dX_{1,\xi}^{e_i}}{dt}\phi_2 + \frac{dX_2^{e_i}}{dt}\phi_3 + \frac{dX_{2,\xi}^{e_i}}{dt}\phi_4 \right) \frac{\Delta z}{2} d\xi \end{bmatrix}$$

or, using the compact notation and rearranging terms:

$$\frac{\Delta z}{2} \begin{pmatrix} \langle\phi_1,\phi_1\rangle & \langle\phi_1,\phi_2\rangle & \langle\phi_1,\phi_3\rangle & \langle\phi_1,\phi_4\rangle \\ \langle\phi_2,\phi_1\rangle & \langle\phi_2,\phi_2\rangle & \langle\phi_2,\phi_3\rangle & \langle\phi_2,\phi_4\rangle \\ \langle\phi_3,\phi_1\rangle & \langle\phi_3,\phi_2\rangle & \langle\phi_3,\phi_3\rangle & \langle\phi_3,\phi_4\rangle \\ \langle\phi_4,\phi_1\rangle & \langle\phi_4,\phi_2\rangle & \langle\phi_4,\phi_3\rangle & \langle\phi_4,\phi_4\rangle \end{pmatrix} \begin{pmatrix} \frac{dX_1^{e_i}}{dt} \\ \frac{dX_{1,\xi}^{e_i}}{dt} \\ \frac{dX_2^{e_i}}{dt} \\ \frac{dX_{2,\xi}^{e_i}}{dt} \end{pmatrix}$$

which, after the computation of the integrals, can be written as:

$$\frac{\Delta z}{210} \begin{pmatrix} 78 & 22 & 27 & -13 \\ 22 & 8 & 13 & -6 \\ 27 & 13 & 78 & -22 \\ -13 & -6 & -22 & 8 \end{pmatrix} \begin{pmatrix} \frac{dX_1^{e_i}}{dt} \\ \frac{dX_{1,\xi}^{e_i}}{dt} \\ \frac{dX_2^{e_i}}{dt} \\ \frac{dX_{2,\xi}^{e_i}}{dt} \end{pmatrix}$$

The procedure for the assembly is very similar to the one presented in the case of Lagrangian elements. The main difference is that $X_{2,\xi}^{e_i} = X_{1,\xi}^{e_{i+1}}$ in addition to $X_2^{e_i} = X_1^{e_{i+1}}$ so that the overlap between the submatrices corresponding to the different finite elements is larger. The assembly leads to

$$
\mathbf{M} = \frac{\Delta z}{210}
\begin{pmatrix}
78 & 22 & 27 & -13 & & & & & & & & \\
22 & 8 & 13 & -6 & & & & & & & & \\
27 & 13 & 156 & 0 & 27 & -13 & & & & & & \\
-13 & -6 & 0 & 16 & 13 & -6 & & & & & & \\
& & 27 & 13 & 156 & 0 & 27 & -13 & & & & \\
& & -13 & -6 & 0 & 16 & 13 & -6 & & & & \\
& & & \ddots & \ddots & \ddots & \ddots & \ddots & \ddots & \ddots & & \\
& & & & \ddots & \ddots & \ddots & \ddots & \ddots & \ddots & \ddots & \\
& & & & & & 27 & 13 & 156 & 0 & 27 & -13 \\
& & & & & & -13 & -6 & 0 & 16 & 13 & -6 \\
& & & & & & & & 27 & 13 & 78 & -22 \\
& & & & & & & & -13 & -6 & -22 & 8
\end{pmatrix}
$$

To summarize, we have

$$
\int_{\Omega(z)} w_i \frac{\partial x}{\partial t} \mathrm{d}z \equiv \mathbf{M} \frac{\mathrm{d}\tilde{\mathbf{x}}}{\mathrm{d}t}, \quad \text{with} \quad i = 1, \ldots, 2N \tag{4.91}
$$

where the vector of dependent variables is $\tilde{\mathbf{x}} = [X_1, X_{1,\xi}, X_2, X_{2,\xi}, \ldots, X_{N-1}, X_{N-1,\xi}, X_N, X_{N,\xi}]^T$. The variable ordering naturally results from the previous developments, and corresponds to an arrangement according to the FEM nodes. Other arrangements could be considered as well. For instance, it would be possible to separate the solution values X_i from their derivatives $X_{i,\xi}$ (with $i = 1, 2, \ldots, N$), i.e., $\tilde{\mathbf{x}} = \left[X_1, X_2, \ldots, X_{N-1}, X_N, X_{1,\xi}, X_{2,\xi}, \ldots, X_{N-1,\xi}, X_{N,\xi}\right]^T$. Of course, the structure of the FEM matrices is then different and the construction of these matrices has to be adapted accordingly. In the sequel we will continue the method description using the first, natural, ordering of the variables.

4.7.2 First and Second Terms of the RHS Term of the Weighted Residual Solution

Again one element will contribute to four rows of the equation system resulting from the method of lines. We consider the first term, $\alpha_0 \int_{\Omega(z)^{e_i}} W_l \tilde{x} \mathrm{d}z$ and use the basis functions ϕ_i with $i = 1, 2, 3, 4$ as weighting functions to build

$$
\alpha_0 \frac{\Delta z}{2}
\begin{pmatrix}
\langle \phi_1, \phi_1 \rangle & \langle \phi_1, \phi_2 \rangle & \langle \phi_1, \phi_3 \rangle & \langle \phi_1, \phi_4 \rangle \\
\langle \phi_2, \phi_1 \rangle & \langle \phi_2, \phi_2 \rangle & \langle \phi_2, \phi_3 \rangle & \langle \phi_2, \phi_4 \rangle \\
\langle \phi_3, \phi_1 \rangle & \langle \phi_3, \phi_2 \rangle & \langle \phi_3, \phi_3 \rangle & \langle \phi_3, \phi_4 \rangle \\
\langle \phi_4, \phi_1 \rangle & \langle \phi_4, \phi_2 \rangle & \langle \phi_4, \phi_3 \rangle & \langle \phi_4, \phi_4 \rangle
\end{pmatrix}
\begin{pmatrix}
X_1^{e_i} \\
X_{1,\xi}^{e_i} \\
X_2^{e_i} \\
X_{2,\xi}^{e_i}
\end{pmatrix}
$$

The assembly of these elementary matrices to take account of the other finite elements results in the zeroth-order derivation matrix $\mathbf{D}_0$ which is exactly the same as the mass matrix and

$$\int\limits_{\Omega(z)} w_i \alpha_0 \frac{\partial x}{\partial z} \mathrm{d}z \equiv \alpha_0 \mathbf{D}_0 \tilde{\mathbf{x}}, \quad \text{with} \quad i = 1, \ldots, 2N \tag{4.92}$$

The second term, $\alpha_1 \int_{\Omega(z)^{e_i}} W_l \tilde{x}_z dz$, is given by

$$\alpha_1 \begin{pmatrix} \langle \phi_1, \phi_{1,\xi} \rangle & \langle \phi_1, \phi_{2,\xi} \rangle & \langle \phi_1, \phi_{3,\xi} \rangle & \langle \phi_1, \phi_{4,\xi} \rangle \\ \langle \phi_2, \phi_{1,\xi} \rangle & \langle \phi_2, \phi_{2,\xi} \rangle & \langle \phi_2, \phi_{3,\xi} \rangle & \langle \phi_2, \phi_{4,\xi} \rangle \\ \langle \phi_3, \phi_{1,\xi} \rangle & \langle \phi_3, \phi_{2,\xi} \rangle & \langle \phi_3, \phi_{3,\xi} \rangle & \langle \phi_3, \phi_{4,\xi} \rangle \\ \langle \phi_4, \phi_{1,\xi} \rangle & \langle \phi_4, \phi_{2,\xi} \rangle & \langle \phi_4, \phi_{3,\xi} \rangle & \langle \phi_4, \phi_{4,\xi} \rangle \end{pmatrix} \begin{pmatrix} X_1^{e_i} \\ X_{1,\xi}^{e_i} \\ X_2^{e_i} \\ X_{2,\xi}^{e_i} \end{pmatrix}$$

which, after the computation of the spatial integrals, results in

$$\alpha_1 \frac{1}{30} \begin{pmatrix} -15 & 6 & 15 & -6 \\ -6 & 0 & 6 & -2 \\ -15 & -6 & 15 & 6 \\ 6 & 2 & -6 & 0 \end{pmatrix} \begin{pmatrix} X_1^{e_i} \\ X_{1,\xi}^{e_i} \\ X_2^{e_i} \\ X_{2,\xi}^{e_i} \end{pmatrix}$$

The assembly leads to the first order differentiation matrix $\mathbf{D}_1$:

$$\mathbf{D}_1 = \frac{1}{30} \begin{pmatrix} -15 & 6 & 15 & -6 & & & & & & & & \\ -6 & 0 & 6 & -2 & & & & & & & & \\ -15 & -6 & 0 & 12 & 15 & -6 & & & & & & \\ 6 & 2 & -12 & 0 & 6 & -2 & & & & & & \\ & & -15 & -6 & 0 & 12 & 15 & -6 & & & & \\ & & 6 & 2 & -12 & 0 & 6 & -2 & & & & \\ & & & \ddots & \ddots & \ddots & \ddots & \ddots & \ddots & \ddots & & \\ & & & & \ddots & \ddots & \ddots & \ddots & \ddots & \ddots & \ddots & \\ & & & & & & -15 & -6 & 0 & 12 & 15 & -6 \\ & & & & & & 6 & 2 & -12 & 0 & 6 & -2 \\ & & & & & & & & -15 & -6 & 15 & 6 \\ & & & & & & & & 6 & 2 & -6 & 0 \end{pmatrix}$$

which allows the integral over the complete spatial domain to be computed as

$$\int_{\Omega(z)} w_i \alpha_1 \frac{\partial x}{\partial z} \mathrm{d}z \equiv \alpha_1 \mathbf{D}_1 \tilde{\mathbf{x}}, \quad \text{with} \quad i = 1, \ldots, 2N \tag{4.93}$$

4.7.3 *Third Term of the RHS Term of the Weighted Residual Solution*

When considering Langrangian (linear) elements it was required to use the so-called *weak form*. In this case, since Hermitian polynomials are twice differentiable this step is no longer necessary. The evaluation of $\alpha_2 \int_{\Omega(z)} w_i \tilde{x}_{zz} \mathrm{d}z$ on the finite element e_i yields

$$\alpha_2 \frac{2}{\Delta z} \begin{pmatrix} \langle\phi_1, \phi_{1,\xi\xi}\rangle & \langle\phi_1, \phi_{2,\xi\xi}\rangle & \langle\phi_1, \phi_{3,\xi\xi}\rangle & \langle\phi_1, \phi_{4,\xi\xi}\rangle \\ \langle\phi_2, \phi_{1,\xi\xi}\rangle & \langle\phi_2, \phi_{2,\xi\xi}\rangle & \langle\phi_2, \phi_{3,\xi\xi}\rangle & \langle\phi_2, \phi_{4,\xi\xi}\rangle \\ \langle\phi_3, \phi_{1,\xi\xi}\rangle & \langle\phi_3, \phi_{2,\xi\xi}\rangle & \langle\phi_3, \phi_{3,\xi\xi}\rangle & \langle\phi_3, \phi_{4,\xi\xi}\rangle \\ \langle\phi_4, \phi_{1,\xi\xi}\rangle & \langle\phi_4, \phi_{2,\xi\xi}\rangle & \langle\phi_4, \phi_{3,\xi\xi}\rangle & \langle\phi_4, \phi_{4,\xi\xi}\rangle \end{pmatrix} \begin{pmatrix} X_1^{e_i} \\ X_{1,\xi}^{e_i} \\ X_2^{e_i} \\ X_{2,\xi}^{e_i} \end{pmatrix}$$

which, after the computation of the spatial integrals and derivatives, results in:

$$\alpha_2 \frac{1}{15\Delta z} \begin{pmatrix} -18 & -33 & 18 & -3 \\ -3 & -8 & 3 & 2 \\ 18 & 3 & -18 & 33 \\ -3 & 2 & 3 & -8 \end{pmatrix} \begin{pmatrix} X_1^{e_i} \\ X_{1,\xi}^{e_i} \\ X_2^{e_i} \\ X_{2,\xi}^{e_i} \end{pmatrix}$$

The assembly of the elementary matrices leads to the second-order differentiation matrix $\mathbf{D}_2$

$$\mathbf{D}_2 = \frac{1}{15\Delta z} \begin{pmatrix} -18 & -33 & 18 & -3 & & & & & & \\ -3 & -8 & 3 & 2 & & & & & & \\ 18 & 3 & -36 & 0 & 18 & -3 & & & & \\ -3 & 2 & 0 & -16 & 3 & 2 & & & & \\ & & 18 & 3 & -36 & 0 & 18 & -3 & & \\ & & -3 & 2 & 0 & -16 & 3 & 2 & & \\ & & & \ddots & \ddots & \ddots & \ddots & \ddots & \ddots & \ddots \\ & & & & \ddots & \ddots & \ddots & \ddots & \ddots & \ddots & \ddots \\ & & & & & & 18 & 3 & -36 & 0 & 18 & -3 \\ & & & & & & -3 & 2 & 0 & -16 & 3 & 2 \\ & & & & & & & & 18 & 3 & -18 & 33 \\ & & & & & & & & -3 & 2 & 3 & -8 \end{pmatrix}$$

$$\int_{\Omega(z)} w_i \alpha_2 \frac{\partial^2 x}{\partial z^2} \mathrm{d}z \equiv \alpha_2 \mathbf{D}_2 \tilde{\mathbf{x}}, \quad \text{with} \quad i = 1, \ldots, 2N \tag{4.94}$$

4.7.4 Fourth Term of the RHS Term of the Weighted Residual Solution

The evaluation of the term $\int_{\Omega(z)} w_i f(\tilde{x}) dz$ on the finite element e_i is carried out in a similar way as for Lagrangian basis functions, i.e.,

$$\frac{\Delta z}{2} \begin{pmatrix} \langle \phi_1, f(\tilde{x}^{e_i}) \rangle \\ \langle \phi_2, f(\tilde{x}^{e_i}) \rangle \\ \langle \phi_3, f(\tilde{x}^{e_i}) \rangle \\ \langle \phi_4, f(\tilde{x}^{e_i}) \rangle \end{pmatrix} = \frac{\Delta z}{2} \begin{pmatrix} \langle \phi_1, f(X_1^{e_i} \phi_1 + X_{1,\xi}^{e_i} \phi_2 + X_2^{e_i} \phi_3 + X_{2,\xi}^{e_i} \phi_4) \rangle \\ \langle \phi_2, f(X_1^{e_i} \phi_1 + X_{1,\xi}^{e_i} \phi_2 + X_2^{e_i} \phi_3 + X_{2,\xi}^{e_i} \phi_4) \rangle \\ \langle \phi_3, f(X_1^{e_i} \phi_1 + X_{1,\xi}^{e_i} \phi_2 + X_2^{e_i} \phi_3 + X_{2,\xi}^{e_i} \phi_4) \rangle \\ \langle \phi_4, f(X_1^{e_i} \phi_1 + X_{1,\xi}^{e_i} \phi_2 + X_2^{e_i} \phi_3 + X_{2,\xi}^{e_i} \phi_4) \rangle \end{pmatrix}$$

The assembly over all the elements leads to[1]

$$\tilde{\mathbf{f}}_{NL}(\tilde{\mathbf{x}}) = \int_{\Omega(z)} w_i f(\tilde{x}) \mathrm{d}z \equiv \frac{\Delta z}{2} \begin{pmatrix} \langle \phi_1, f(\tilde{x}^{e_1}) \rangle \\ \langle \phi_2, f(\tilde{x}^{e_1})) \rangle \\ \langle \phi_3, f(\tilde{x}^{e_1}) \rangle + \langle \phi_1, f(\tilde{x}^{e_2}) \rangle \\ \langle \phi_4, f(\tilde{x}^{e_1}) \rangle + \langle \phi_2, f(\tilde{x}^{e_2}) \rangle \\ \langle \phi_3, f(\tilde{x}^{e_2}) \rangle + \langle \phi_1, f(\tilde{x}^{e_3}) \rangle \\ \langle \phi_4, f(\tilde{x}^{e_2}) \rangle + \langle \phi_2, f(\tilde{x}^{e_3}) \rangle \\ \vdots \\ \langle \phi_3, f(\tilde{x}^{e_{N-2}}) \rangle + \langle \phi_1, f(\tilde{x}^{e_{N-1}}) \rangle \\ \langle \phi_4, f(\tilde{x}^{e_{N-2}}) \rangle + \langle \phi_2, f(\tilde{x}^{e_{N-1}}) \rangle \\ \langle \phi_3, f(\tilde{x}^{e_{N-1}}) \rangle \\ \langle \phi_4, f(\tilde{x}^{e_{N-1}}) \rangle \end{pmatrix} \tag{4.95}$$

[1] The notation using the series expansion makes the expression of this vector too long to fit into one page, therefore we will use a more compact notation, i.e., $f(\tilde{x}^{e_i}) = f(X_i \phi_1 + X_{i,\xi} \phi_2 + X_{i+1} \phi_3 + X_{i+1,\xi} \phi_4)$

where the integrals can be evaluated using the FEM structure or numerical quadrature formulas.

Using Eqs. (4.91–4.95) we can rewrite Eq. (4.33), without the contribution of the boundary conditions, as:

$$\mathbf{M}\frac{d\tilde{\mathbf{x}}}{dt} = (\alpha_0\mathbf{D}_0 + \alpha_1\mathbf{D}_1 + \alpha_2\mathbf{D}_2)\tilde{\mathbf{x}} + \tilde{\mathbf{f}}_{NL}(\tilde{\mathbf{x}}) \tag{4.96}$$

Note that the number of ordinary differential equations resulting from the application of the finite element method with Hermitian elements is twice as much as with Lagrangian finite elements for the same number of nodes.

The last task to close the system is to include the contribution of the boundary conditions. This will be discussed in the following section.

4.7.5 Galerkin Method Over Hermitian Elements: Contribution of the Boundary Conditions

In this case the only nonzero function at z_L and z_R are, respectively, functions ϕ_1 of element e_1 and function ϕ_3 of element e_{N-1}.

4.7.5.1 Dirichlet Boundary Conditions

Recall the form of these boundary conditions:

$$x(z_L, t) = g_L(t); \quad x(z_R, t) = g_R(t) \tag{4.97}$$

The BCs will only affect the first and the $(2N-1)$th rows of system (4.96) with respectively, the term $-(x_1 - g_L)$ and the term $-(x_N - g_R)$. If we define the following vector,

$$\mathbf{g} = \begin{pmatrix} -(x_1 - g_L) \\ 0 \\ \vdots \\ 0 \\ -(x_N - g_R) \\ 0 \end{pmatrix} \tag{4.98}$$

the final system of equations can be written as:

$$\mathbf{M}\frac{\partial \tilde{\mathbf{x}}}{\partial t} = (\alpha_0 \mathbf{D}_0 + \alpha_1 \mathbf{D}_1 + \alpha_2 \mathbf{D}_2)\,\tilde{\mathbf{x}} + \mathbf{M}\tilde{f}_{\mathrm{NL}} + \mathbf{g} \tag{4.99}$$

4.7.5.2 Neumann Boundary Conditions

Neumann boundary conditions indicate the value of the gradient, i.e.,

$$\left.\frac{\partial \tilde{x}}{\partial z}\right|_{z_L} = -g_L(t); \quad \left.\frac{\partial \tilde{x}}{\partial z}\right|_{z_R} = g_R(t) \tag{4.100}$$

The former expressions can be rewritten in terms of variable ξ as:

$$-\left.\frac{\partial \tilde{x}}{\partial z}\right|_{z_L} - g_L(t) = -\left.\frac{\partial \tilde{x}}{\partial \xi}\right|_{z_L} - \frac{\Delta z}{2} g_L(t) \tag{4.101}$$

$$\left.\frac{\partial \tilde{x}}{\partial z}\right|_{z_R} - g_R(t) = \left.\frac{\partial \tilde{x}}{\partial \xi}\right|_{z_R} - \frac{\Delta z}{2} g_R(t) \tag{4.102}$$

Note that in contrast with the Lagrangian elements, spatial derivatives belong to the vector of dependent variables (i.e., coefficients of the basis functions) so that

$$-\left.\frac{\partial \tilde{x}}{\partial \xi}\right|_{z_L} - \frac{\Delta z}{2} g_L = -X_{1,\xi} - \frac{\Delta z}{2} g_L \tag{4.103}$$

$$\left.\frac{\partial \tilde{x}}{\partial \xi}\right|_{z_R} - \frac{\Delta z}{2} g_R = X_{N,\xi} - \frac{\Delta z}{2} g_R \tag{4.104}$$

and the vector $\mathbf{g}$ in Eq. (4.70) takes a simple form:

$$\mathbf{g} = \begin{pmatrix} -X_{1,\xi} - \frac{\Delta z}{2} g_L \\ 0 \\ \vdots \\ 0 \\ X_{N,\xi} - \frac{\Delta z}{2} g_R \\ 0 \end{pmatrix} \tag{4.105}$$

4.8 An Illustrative Example

Again Burgers' equation will be used to illustrate the methodology and MATLAB implementation of the finite element method with Hermitian elements.

As in the previous section, let us start with the ODE function. This function is listed in `burgerspdes_hermite` and we can note the following:

```
function xt = burgerspdes_hermite(t,x)
global mu
global wquad xquad
global n z0 zL D2 xx

% Second spatial derivative xzz computed through the
% second differentiation matrix
xzz = D2*x;

% Rename state variable x to pass it to integrands
% functions
xx = x;

% Spatial integration of the integrands in the nonlinear
% term of Burgers equation. Nonlinear term: fx = x*xz
y1 = feval('integrand1_hermite',xquad')*wquad';
y2 = feval('integrand2_hermite',xquad')*wquad';
y3 = feval('integrand3_hermite',xquad')*wquad';
y4 = feval('integrand4_hermite',xquad')*wquad';
fx = y1+y2+y3+y4;

% Boundary conditions
gL           = burgers_exact(z0,t);
gR           = burgers_exact(zL,t);
gv(1,1)      = x(1)-gL;
gv(2*n-1,1)  = x(2*n-1)-gR;
gv(2*n,1)    = 0;

% ODEs
xt = mu*xzz - fx - gv;
```

Function burgerspdes_hermite Function `burgerspdes_hermite` to define the ODE set for Burgers equation with Hermitian elements

- Four terms (`y1`, `y2`, `y3`, and `y4`) are used to construct the FEM approximation of the nonlinear function. Each one corresponds to the projection onto the basis functions ϕ_1, ϕ_2, ϕ_3, and ϕ_4, respectively.
- Instead of having `n` equations we have `2*n` equations since the spatial derivatives of the field at the discretization points are also expansion coefficients of the Hermitian basis functions.
- The last boundary condition is introduced in row `2*n-1`.
- The integrand functions have changed. As an example, we include the listing of the first one (`integran1_hermite`) where the MATLAB codes `trialfunctionsher2n` and `trialfunctionsher2np` compute, respectively, the basis functions (4.87–4.90) and their first derivatives.

```
function out = integrand1_hermite(xi)
% Computation of the first integrand on the nonlinear term
% for the Burgers' equation. This term corresponds with
%       y1 = phi_1*(xfem*xfem_z)

global xx n

% Dependent variable values and their derivatives with the
% FEM local nomenclature
x1    = xx(1:2:2*n-3);
x1_xi = xx(2:2:2*n-2);
x2    = xx(3:2:2*n-1);
x2_xi = xx(4:2:2*n);

% Preallocation of memmory for the output
out = zeros(2*n,length(xi));

for i = 1:length(xi)
   % Basis function computation
   [phi1, phi2, phi3, phi4]  = trialfunctionsher2n(xi(i));
   % First derivative of the basis function computation
   [phi1p,phi2p,phi3p,phi4p] = trialfunctionsher2np(xi(i));
   % Finite element approximation of the dependent
   % variable
   xfem = phi1*x1 + phi2*x1_xi + phi3*x2 + phi4*x2_xi;
   % Finite element approximation of the first spatial
   % derivative of the dependent variable
   xfem_z = phi1p*x1+phi2p*x1_xi+phi3p*x2+phi4p*x2_xi;
   out(1:2:2*n-3,i) = phi1*(xfem.*xfem_z);
end
```

Function integran1_hermite Function `integrand1_hermite` to compute the first integrand of the nonlinear term for Burgers equation with Hermitian elements

```
function [phi1,phi2,phi3,phi4] = trialfunctionsher2n(xi)
% Computation of the hermitian basis functions of the
% finite element method
phi1 = ((xi.^2-3).*xi+2)/4;
phi2 = (((xi-1).*xi-1).*xi+1)/4;
phi3 = ((-xi.^2+3).*xi+2)/4;
phi4 = (((xi+1).*xi-1).*xi-1)/4;
```

Function trialfunctionsher2n Function `trialfunctionsher2n` to compute the Hermitian basis functions

```
function [phi1p,phi2p,phi3p,phi4p] = trialfunctionsher2np(x)
% Computation of the derivative of the hermitian basis
% functions of the finite element method
phi1p = 3*(x^2-1)/4;
phi2p = ((3*x-2)*x-1)/4;
```

```
phi3p = -3*(x^2-1)/4;
phi4p = ((3*x+2)*x-1)/4;
```

Function trialfunctionsher2np Function `trialfunctionsher2np` to compute the first spatial derivative of the Hermitian basis functions

The main program calling function `burgerspdes_hermite` is listed in `main_burgers_FEM_hermite`.

```
close all
clear all
global mu
global wquad xquad
global n dz z0 zL D2

% Finite element spatial grid
z0  = 0;
zL  = 1;
n   = 101;
nel = n-1;
dz  = (zL - z0)/(n-1);
z   = linspace(z0,zL,n);

% Second order differentiation matrix
D2 = hermiteD2(dz,n);

% Problem parameters
mu = 0.001;

% Computation of the weigths and the abcissa to be
% employed in the numerical integration via the
% Gauss-Legendre quadrature
nquad      = 2;
beta       = 0.5./sqrt(1-(2*(1:nquad)).^(-2));
T          = diag(beta,1) +diag(beta,-1);
[V,D]      = eig(T);
xquad      = diag(D);
[xquad,i] = sort(xquad);
wquad      = 2*V(1,i).^2;

% Initial conditions
x  = zeros(1,n);
xz = zeros(1,n);
for i = 1:n
    x(i) = burgers_exact(z(i),0);
    xz(i) = derburgers_exact(z(i),0);
end
xx(1:2:2*n-1)=x;
xx(2:2:2*n)=dz*xz/2;

% Time instants at which the IVP solver will save the
% solution
dt   = 0.1;
time = (0:dt:1);
nt   = length(time);
```

```
% Time integration with the IVP solver
M       = hermitemass(dz,n); % Finite element mass matrix
options = odeset('Mass',M,'RelTol',1e-6,'AbsTol',1e-6);
[tout,xout] = ode15s(@burgerspdes_hermite,time,xx,options);

% Recovering the states from the ode15s output (xout)
x_direct = xout(:,1:2:2*n-1);

% Recovering the solution using the hermitian polynomials
figure
dzvis = 0.5;
zvis = -1:dzvis:1;
nvis = length(zvis);
% Basis functions
[phi1, phi2, phi3, phi4] = trialfunctionsher2n(zvis);
z_absc   = linspace(z0,zL,(n-1)*(nvis-1)+1);

for k=1:nel
    xx = xout(:,2*k-1)*phi1+xout(:,2*k)*phi2+...
         xout(:,2*k+1)*phi3+xout(:,2*k+2)*phi4;
    if k == 1
        n1 = 1;
        n2 = 5;
        ordo(:,n1:n2) = xx;
    else
        n1 = n2+1;
        n2 = n2+4;
        ordo(:,n1:n2) = xx(:,2:end);
    end
end

% Plot the results
figure
hold
ztheor=(0:.001:1);
ntheor=length(ztheor);
for k=1:length(tout)
    for i=1:ntheor
        yexact(k,i)= burgers_exact(ztheor(i),tout(k));
    end
end
plot(z,x_direct,'-k')
hold on
plot(ztheor,yexact,'r')
axis([0 1 0 1.4])
hold off



plot(z_absc,ordo,'-k')
hold on
plot(ztheor,yexact,'r')
axis([0 1 0 1.4])
hold off
```

Script main_burgers_FEM_hermite Main script calling function burgerspdes_hermite

Table 4.4 Comparison in terms of efficiency and accuracy of the FEM using Lagrange and Hermite polynomials for Burgers' equation

	N	Maximum error (%)	Mean error (%)	CPU time
Lagrange	101	36.80	0.42	1
	201	14.04	0.07	1.8
	401	3.49	0.02	8.7
Hermite	51	36.99	0.33	2.8
	101	5.67	0.04	3.5
	151	1.53	0.005	8.5

CPU time was normalized with respect the most efficient test (Lagrange polynomials with 101 discretization points)

We can notice the following differences with respect to the program with linear Lagrangian elements:

- Both the second-order differentiation matrix and the mass matrix are computed using functions `hermiteD2` and `hemitemass` which can be found in the companion software.
- Initial conditions are given for the dependent variables `x` and their spatial derivatives `xz`. Therefore we end up with `2*n` ODEs.
- The final solution can be computed directly from the output of `ode15s`, i.e. `xout` or using the Hermitian basis functions (which will give a more precise solution).

Accuracy and efficiency of the FEM using Lagrange and Hermite polynomials for Burgers' equation with $\mu = 0.001$ are summarized in Table 4.4. As expected, increasing the number of discretization points (N) improves the accuracy and increases the computational effort. Also, for the same N, Hermite polynomials allow a more accurate solution to be obtained at the price of a higher computational cost. However, it should be noted that, for comparable computation times ($N = 401$ and $N = 151$ for Lagrange and Hermite polynomials, respectively), FEM with Hermite polynomials appears as a more accurate scheme for this particular problem.

4.9 The Orthogonal Collocation Method

As mentioned before, orthogonal collocation belongs to the family of the WRM. In this technique, the weighting functions are chosen as Dirac impulses located at given collocation points. The main difference with respect to the classical collocation method is that, the collocation points are distributed in the spatial domain so as to maximize the accuracy of the results. The advantage with respect to other WRM is that the computation of spatial integrals is no longer required (in fact this computation is trivial and does not require any numerical procedure). The Hermitian elements that we have introduced in the previous sections are a popular choice in conjunction with the collocation method. Let us rewrite, for the sake of clarity, the expressions of the four Hermitian basis functions:

$$\phi_1(\xi) = \frac{1}{4}(1-\xi)^2(2+\xi)$$
$$\phi_2(\xi) = \frac{1}{4}(1-\xi^2)(1-\xi)$$
$$\phi_3(\xi) = \frac{1}{4}(1+\xi)^2(2-\xi)$$
$$\phi_4(\xi) = \frac{1}{4}(-1+\xi^2)(1+\xi)$$

The approximation of the field using the *truncated series* is of the form:

$$\tilde{x}(\xi,t) = X_1(t)\phi_1(\xi) + X_{1,\xi}(t)\phi_2(\xi) + X_2(t)\phi_3(\xi) + X_{2,\xi}(t)\phi_4(\xi) \tag{4.106}$$

As in the previous section, we will use the following generic PDE to illustrate the procedure

$$x_t = \alpha_0 x + \alpha_1 x_z + \alpha_2 x_{zz} + f(x)$$

The contribution of an Hermitian element e^i to the system of equations in the orthogonal collocation method is given by:

$$\begin{aligned} \int_{\Omega^{e_i}} \delta(z_{\text{col}})\tilde{x}_t dz = \alpha_0 \int_{\Omega^{e_i}} \delta(z_{\text{col}})\tilde{x} dz + \alpha_1 \int_{\Omega^{e_i}} \delta(z_{\text{col}})\tilde{x}_z dz \\ + \alpha_2 \int_{\Omega^{e_i}} \delta(z_{\text{col}})\tilde{x}_{zz} dz + \int_{\Omega^{e_i}} \delta(z_{\text{col}}) f(\tilde{x}) dz \end{aligned} \tag{4.107}$$

Since $\delta(z_{\text{col}})$ represents the Dirac delta, Eq. (4.107) can be rewritten as:

$$\tilde{x}_t(z_{\text{col}}) = \alpha_0 \tilde{x}(z_{\text{col}}) + \alpha_1 \tilde{x}_z(z_{\text{col}}) + \alpha_2 \tilde{x}_{zz}(z_{\text{col}}) + f(\tilde{x}(z_{\text{col}})) \tag{4.108}$$

The spatial domain is subdivided into N elements, and the definition of the Hermitian polynomials on these elements involve $2N+2$ unknow coefficients. Hence, two collocation points per element are required to provide $2N$ residual ODEs, which can be completed by two equations related to the boundary conditions.

A popular choice for the location of the two collocation points is the so-called Legendre points (i.e., they are the roots of the second-order Legendre polynomial $1/2(3z^2-1)$). Further discussion about this choice is given in Sect. 4.10. Here we will just use this result and proceed with the evaluation of the residual equations.

$$\xi_1 = -\frac{1}{\sqrt{3}}; \quad \xi_2 = \frac{1}{\sqrt{3}} \tag{4.109}$$

Let us now treat all the terms in Eq. (4.108) separately.

4.9.1 LHS Term of the Collocation Residual Equation

One element will contribute to two rows (one per collocation point) of the final system. For the first collocation point, denoted by ξ_1 in the transformed coordinates, we have

$$\tilde{x}_t(\xi_1) = \frac{\mathrm{d}X_1}{\mathrm{d}t}\phi_1(\xi_1) + \frac{\mathrm{d}X_{1,\xi}}{\mathrm{d}t}\phi_2(\xi_1) + \frac{\mathrm{d}X_2}{\mathrm{d}t}\phi_3(\xi_1) + \frac{\mathrm{d}X_{2,\xi}}{\mathrm{d}t}\phi_4(\xi_1)$$

where the series expansion (4.106) has been used. Likewise, we have for the second collocation point, ξ_2:

$$\tilde{x}_t(\xi_2) = \frac{\mathrm{d}X_1}{\mathrm{d}t}\phi_1(\xi_2) + \frac{\mathrm{d}X_{1,\xi}}{\mathrm{d}t}\phi_2(\xi_2) + \frac{\mathrm{d}X_2}{\mathrm{d}t}\phi_3(\xi_2) + \frac{\mathrm{d}X_{2,\xi}}{\mathrm{d}t}\phi_4(\xi_2)$$

In matrix form, these two terms can be rewritten as:

$$\tilde{x}_t\left(z_{\mathrm{col}}\right) \Rightarrow \begin{pmatrix} \phi_1\left(\xi_1\right) & \phi_2\left(\xi_1\right) & \phi_3\left(\xi_1\right) & \phi_4\left(\xi_1\right) \\ \phi_1\left(\xi_2\right) & \phi_2\left(\xi_2\right) & \phi_3\left(\xi_2\right) & \phi_4\left(\xi_2\right) \end{pmatrix} \begin{pmatrix} \dfrac{\mathrm{d}X_1}{\mathrm{d}t} \\ \dfrac{\mathrm{d}X_{1,\xi}}{\mathrm{d}t} \\ \dfrac{\mathrm{d}X_2}{\mathrm{d}t} \\ \dfrac{\mathrm{d}X_{2,\xi}}{\mathrm{d}t} \end{pmatrix}$$

The assembly procedure is simpler as compared to the Galerkin method. In the latter, the coefficients in the basis functions correspond to the values of the solution at the edges of the finite elements. As these points are shared by two elements, the elementary matrices overlap. In the orthogonal collocation, the collocation points belong only to one element and, therefore, there is no overlapping between two elementary matrices. Hence, the assembly results in:

$$\mathbf{M}\tilde{x}_t = \begin{pmatrix} a & b & c & d & & & & & & \\ c & -d & a & -b & & & & & & \\ & & a & b & c & d & & & & \\ & & c & -d & a & -b & & & & \\ & & & & \cdots & \cdots & \cdots & & & \\ & & & & \cdots & \cdots & \cdots & & & \\ & & & & & & a & b & c & d \\ & & & & & & c & -d & a & -b \\ & & & & & & & & a & b \; c \; d \\ & & & & & & & & c & -d \; a \; -b \end{pmatrix} \begin{pmatrix} X_1 \\ X_{1,\xi} \\ X_2 \\ X_{2,\xi} \\ \vdots \\ X_{N-1} \\ X_{N-1,\xi} \\ X_N \\ X_{N,\xi} \end{pmatrix}_t \tag{4.110}$$

where

$$\begin{aligned}\phi_1(\xi_1)=\phi_3(\xi_2)=a=\frac{3\sqrt{3}+4}{6\sqrt{3}};\quad \phi_2(\xi_1)=-\phi_4(\xi_2)=b=\frac{\sqrt{3}+1}{6\sqrt{3}};\\ \phi_3(\xi_1)=\phi_1(\xi_2)=c=\frac{3\sqrt{3}-4}{6\sqrt{3}};\quad \phi_4(\xi_1)=-\phi_2(\xi_2)=d=\frac{-\sqrt{3}+1}{6\sqrt{3}}\end{aligned} \tag{4.111}$$

Equation (4.110) constitutes a system with $2N$ equations and $2N+2$ unknowns. The missing equations are provided by the boundary conditions. The equations for the boundary conditions are added to the previous ones as algebraic equations, arbitrarily located at the first and last rows of the final system. It follows that the mass matrix **M** is:

$$\mathbf{M}=\begin{pmatrix} 0 & 0 & 0 & 0 \\ a & b & c & d \\ c & -d & a & -b \\ & & a & b & c & d \\ & & c & -d & a & -b \\ & & & & \cdots & \cdots & \cdots & \cdots \\ & & & & \cdots & \cdots & \cdots & \cdots \\ & & & & & & a & b & c & d \\ & & & & & & c & -d & a & -b \\ & & & & & & & & a & b & c & d \\ & & & & & & & & c & -d & a & -b \\ & & & & & & & & 0 & 0 & 0 & 0 \end{pmatrix} \tag{4.112}$$

4.9.2 First Three Terms of Collocation Residual Equation

As in the LHS term, each element will contribute, for these three terms, to two rows of the final system.

(a) Since $\tilde{x}=X_1\phi_1+X_{1,\xi}\phi_2+X_2\phi_3+X_{2,\xi}\phi_4$—see Eq. (4.106) we have $\alpha_0\tilde{x}=\alpha_0\left(X_1\phi_1+X_{1,\xi}\phi_2+X_2\phi_3+X_{2,\xi}\phi_4\right)$. Evaluating this term in the collocation points, ξ_1 and ξ_2 results in:

$$\begin{aligned}\alpha_0\tilde{x}(\xi_1)&=\alpha_0\left(X_1\phi_1(\xi_1)+X_{1,\xi}\phi_2(\xi_1)+X_2\phi_3(\xi_1)+X_{2,\xi}\phi_4(\xi_1)\right)\\ \alpha_0\tilde{x}(\xi_2)&=\alpha_0\left(X_1\phi_1(\xi_2)+X_{1,\xi}\phi_2(\xi_2)+X_2\phi_3(\xi_2)+X_{2,\xi}\phi_4(\xi_2)\right),\end{aligned}$$

respectively. In matrix from, this gives:

$$\alpha_0\tilde{x}\left(z_{\mathrm{col}}\right)\Rightarrow\alpha_0\begin{pmatrix}\phi_1(\xi_1) & \phi_2(\xi_1) & \phi_3(\xi_1) & \phi_4(\xi_1)\\ \phi_1(\xi_2) & \phi_2(\xi_2) & \phi_3(\xi_2) & \phi_4(\xi_2)\end{pmatrix}\begin{pmatrix}X_1\\ X_{1,\xi}\\ X_2\\ X_{2,\xi}\end{pmatrix}$$

Following the same procedure as for the LHS term, the assembly leads to the following matrix:

$$\alpha_0 \mathbf{D_0} \tilde{x} = \alpha_0 \begin{pmatrix} 0 & 0 & 0 & 0 & & & & & & \\ a & b & c & d & & & & & & \\ c & -d & a & -b & & & & & & \\ & & a & b & c & d & & & & \\ & & c & -d & a & -b & & & & \\ & & & & \cdots & \cdots & \cdots & \cdots & & \\ & & & & \cdots & \cdots & \cdots & \cdots & & \\ & & & & a & b & c & d & & \\ & & & & c & -d & a & -b & & \\ & & & & & & a & b & c & d \\ & & & & & & c & -d & a & -b \\ & & & & & & 0 & 0 & 0 & 0 \end{pmatrix} \tilde{x} \tag{4.113}$$

Note once more that $\mathbf{D}_0 = \mathbf{M}$.

(b) The first step in the treatment of the term $\alpha_1 \tilde{x}_z (z_{\mathrm{col}})$ is to transform the spatial derivative in the coordinates z, this is $\tilde{x}_z$, into the spatial derivative in the transformed coordinates $\tilde{x}_\xi$

$$\tilde{x}_z = \frac{\partial \tilde{x}}{\partial \xi} \frac{\mathrm{d}\xi}{\mathrm{d}z} = \frac{\partial \tilde{x}}{\partial \xi} \frac{2}{\Delta z}$$

Also, for the collocation points, ξ_1 and ξ_2, we have

$$\alpha_1 \tilde{x}_\xi (\xi_1) = \alpha_1 \left(X_1 \phi_{1,\xi} (\xi_1) + X_{1,\xi} \phi_{2,\xi} (\xi_1) + X_2 \phi_{3,\xi} (\xi_1) + X_{2,\xi} \phi_{4,\xi} (\xi_1) \right)$$
$$\alpha_1 \tilde{x}_\xi (\xi_2) = \alpha_1 \left(X_1 \phi_{1,\xi} (\xi_2) + X_{1,\xi} \phi_{2,\xi} (\xi_2) + X_2 \phi_{3,\xi} (\xi_2) + X_{2,\xi} \phi_{4,\xi} (\xi_2) \right)$$

or in matrix form:

$$\alpha_1 \tilde{x}_\xi (z_{\mathrm{col}}) \frac{2}{\Delta z} = \alpha_1 \frac{2}{\Delta z} \begin{pmatrix} \phi_{1\xi} (\xi_1) & \phi_{2\xi} (\xi_1) & \phi_{3\xi} (\xi_1) & \phi_{4\xi} (\xi_1) \\ \phi_{1\xi} (\xi_2) & \phi_{2\xi} (\xi_2) & \phi_{3\xi} (\xi_2) & \phi_{4\xi} (\xi_2) \end{pmatrix} \begin{pmatrix} X_1 \\ X_{1,\xi} \\ X_2 \\ X_{2,\xi} \end{pmatrix}$$

The assembly leads to

$$\alpha_1 \mathbf{D_1} \tilde{x} = \alpha_1 \frac{1}{\Delta z} \begin{pmatrix} 0 & 0 & 0 & 0 & & & & & & \\ -1 & e & 1 & -e & & & & & & \\ -1 & -e & 1 & e & & & & & & \\ & & -1 & e & 1 & -e & & & & \\ & & -1 & -e & 1 & e & & & & \\ & & & & \cdots & \cdots & \cdots & \cdots & & \\ & & & & \cdots & \cdots & \cdots & \cdots & & \\ & & & & -1 & e & 1 & -e & & \\ & & & & -1 & -e & 1 & e & & \\ & & & & & & -1 & e & 1 & -e \\ & & & & & & -1 & -e & 1 & e \\ & & & & & & 0 & 0 & 0 & 0 \end{pmatrix} \tilde{x} \tag{4.114}$$

with $e = \frac{1}{\sqrt{3}}$.

(c) For the term $\alpha_2 \tilde{x}_{zz}(z_{col})$ we also have to use the transformed coordinates:

$$\tilde{x}_{zz} = \frac{\partial^2 \tilde{x}}{\partial \xi^2}\left(\frac{2}{\Delta z}\right)^2$$

and the expression for this term in matrix form becomes:

$$\alpha_2 \tilde{x}_{\xi\xi}(z_{col})\left(\frac{2}{\Delta z}\right)^2 = \alpha_2\left(\frac{2}{\Delta z}\right)^2\begin{pmatrix} \phi_{1,\xi\xi}(\xi_1) & \phi_{2,\xi\xi}(\xi_1) & \phi_{3,\xi\xi}(\xi_1) & \phi_{4,\xi\xi}(\xi_1) \\ \phi_{1,\xi\xi}(\xi_2) & \phi_{2,\xi\xi}(\xi_2) & \phi_{3,\xi\xi}(\xi_2) & \phi_{4,\xi\xi}(\xi_2) \end{pmatrix}\begin{pmatrix} X_1 \\ X_{1,\xi} \\ X_2 \\ X_{2,\xi} \end{pmatrix}$$

The assembly leads to

$$\alpha_2 \mathbf{D}_2 \tilde{x} = \alpha_2 \frac{2}{\Delta z^2}\begin{pmatrix} 0 & 0 & 0 & 0 & & & & & & & & \\ -f & -g & f & h & & & & & & & & \\ f & -h & -f & g & & & & & & & & \\ & & -f & -g & f & h & & & & & & \\ & & f & -h & -f & g & & & & & & \\ & & & & \cdots & \cdots & \cdots & \cdots & & & & \\ & & & & \cdots & \cdots & \cdots & \cdots & & & & \\ & & & & & & -f & -g & f & h & & \\ & & & & & & f & -h & -f & g & & \\ & & & & & & & & -f & -g & f & h \\ & & & & & & & & f & -h & -f & g \\ & & & & & & & & 0 & 0 & 0 & 0 \end{pmatrix}\tilde{x} \tag{4.115}$$

with $f = \sqrt{3}$; $g = 1 + \sqrt{3}$; $h = 1 - \sqrt{3}$.

4.9.3 Fourth Term of the RHS of the Collocation Residual Equation

Each element will contribute to two rows of the final system. The term $f(\tilde{x}(z_{col}))$ becomes

$$\begin{pmatrix} f(\tilde{x}(\xi_1)) \\ f(\tilde{x}(\xi_2)) \end{pmatrix}$$

the assembly leads to:

$$\begin{pmatrix} 0 \\ f\left(x^{e_1}\left(\xi_1\right)\right) \\ f\left(x^{e_1}\left(\xi_2\right)\right) + f\left(x^{e_2}\left(\xi_1\right)\right) \\ \vdots \\ f\left(x^{e_{N-2}}\left(\xi_2\right)\right) + f\left(x^{e_{N-1}}\left(\xi_1\right)\right) \\ f\left(x^{e_{N-1}}\left(\xi_2\right)\right) \\ 0 \end{pmatrix} \tag{4.116}$$

4.9.4 Contribution of the Boundary Conditions

As previously stated, the boundary conditions are introduced in the RHS members of the first and last rows of the global system. Let us use Burgers' equation to illustrate how to introduce the boundary conditions in the system.

The MATLAB function to implement the resulting system of ODEs is presented in `burgerspdes_ortcol`.

```
function xt = burgerspdes_ortcol(t,u)
global mu
global n z0 zL D2 u1

u1 = u;

% Linear part
ut = mu*D2*u;

% Nonlinear part
y1 = feval('integrand_ortcol',-1/sqrt(3));
y2 = feval('integrand_ortcol',1/sqrt(3));

% Time derivative
ut(2:2:2*n-2,1) = ut(2:2:2*n-2,1)-y1;
ut(3:2:2*n-1,1) = ut(3:2:2*n-1,1)-y2;

% Boundary conditions and construction of xt
xt(1,1)         = u(1)- burgers_exact(z0,t);
xt(2:2*n-1,1) = ut(2:2*n-1,1);
xt(2*n,1)       = u(2*n-1) - burgers_exact(zL,t);
```

Function burgerspdes_ortcol Function to implement Burgers PDEs with the orthogonal collocation method

Nonlinear terms are computed using the MATLAB function `integrand_ortcol`. Finally, boundary conditions are implemented as in Eq. (4.98).

```
function out = integrand_ortcol(x)

global u1 n dz

% Basis function
[N1 N2 N3 N4]     = trialfunctionsher_1(x);
[N1p N2p N3p N4p] = der1trialfunctionsher_1(x);

% Output
out(1:n-1,1) = (2/dz)*(N1*u1(1:2:2*n-3) +...
                      N2*u1(2:2:2*n-2) +...
                      N3*u1(3:2:2*n-1) +...
                      N4*u1(4:2:2*n)).*...
                      (N1p*u1(1:2:2*n-3) +...
                      N2p*u1(2:2:2*n-2) +...
                      N3p*u1(3:2:2*n-1) +...
                      N4p*u1(4:2:2*n));
```

Function integrand_ortcol Function to compute the basis functions in the orthogonal collocation method.

4.9.5 A Brief Benchmark

As an indication, the solution of Burgers equation, with the usual parameters (101 grid points and $\mu = 0.001$) and four different methods, gives the statistics shown in Table 4.5, where the error is computed as:

$$err = \left\| \bar{x}_{\text{exact}}(\bar{\xi}, 1) - \bar{x}_{\text{num}}(\bar{\xi}, 1) \right\|$$

In the finite difference solution, the first and second spatial derivatives are computed using a two point upwind and a three point centered method, respectively.

4.10 Chebyshev Collocation

In the previous section, we have introduced the basic principle of the orthogonal collocation method using the finite element method with Hermite polynomial basis functions. Of course, all the weighted residual methods, including Galerkin and orthogonal collocation, can be used on several interconnected spatial elements (which is the essence of FEM methods) or on the global spatial domain, when the solution is sufficiently smooth and does not require the discretization of the spatial domain. We now consider a particularly efficient method in this context: Chebyshev collocation. An excellent treatment of this method and very elegant MATLAB codes can be found in the book by Trefethen [7]. In the following, we present some basic information and a code, available in the companion software, largely inspired from [7].

As already stressed, the selection of interpolation functions and the location of the collocation points have a significant effect on the accuracy of the numerical solution. The theory of interpolation tells us that:

Table 4.5 Comparison between the different techniques presented in this chapter

Method	*err*	CPU Time
Finite differences	0.329	1
Finite elements (Lagrangian)	0.172	3
Orthogonal collocation	0.085	7
Finite elements (Hermitian)	0.012	11

CPU time is normalized with respect to the fasted method (finite differences)

- Interpolation with trigonometric polynomials in equidistributed points (i.e., finite linear combinations of cos (nz) and sin (nz) functions, n being natural numbers) suffers from the Gibbs phenomenon (spurious oscillations near jumps of the function to interpolate) as already illustrated in Fig. 4.4.
- Interpolation with conventional polynomials in equally spaced points (i.e., finite combinations of z^n terms, with n natural) suffers from the Runge phenomenon (interpolating polynomials with increasing orders will develop oscillations near the boundaries of the spatial domain).

To avoid these undesirable effects, we have already explored the possibility to discretize the spatial domain into small spatial elements on which low-order interpolation polynomials can be used (the FEM). Another option is to use specific, nonuniform, distribution of the interpolation points. Whereas the density of $N+1$ evenly spaced points on the spatial domain $[-1, +1]$ is given by:

$$d(z) = \frac{N}{2}$$

the density of Legendre or Chebyshev points is

$$d(z) = \frac{N}{\pi\sqrt{1-z^2}}$$

i.e., the points are more densely located near the boundaries.

For instance, the Chebyshev points are the projection on the interval $[-1, +1]$ of equally spaced points on the unit circle . On an interval $[a, b]$, these points are given by

$$z_j = \frac{a+b}{2} + \frac{b-a}{2}\cos\left(\frac{j\pi}{N}\right); \quad j = 0, 1, 2, \ldots, N$$

their distribution for $N = 11$ is represented in Fig. 4.10.

Another choice is given by the Legendre points (or Gauss-Legendre points), which are the zeroes of Legendre polynomial. So-called extreme points can also be used, i.e., the Gauss-Lobatto-Legendre points are the extrema of the Legendre polynomials, and the Gauss-Lobatto-Chebyshev points are the extrema of the Chebyshev polynomials.

These specific choices lead to a vanishing interpolation error for an increasing number of interpolation points, if the function is Lipschitz, i.e., if the function is smooth (i.e., it is limited in how fast it can change).

The above-mentioned Runge phenomenon was originally discovered by Carl Runge in 1901 [2]. The occurrence of this phenomenon can be conveniently illustrated with the

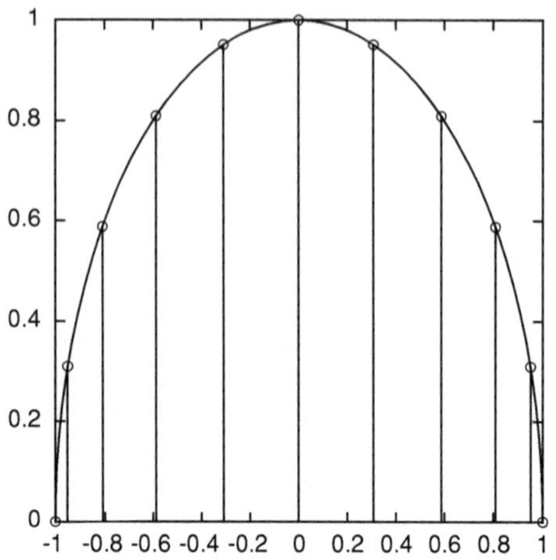

Fig. 4.10 $N = 11$ Chebyshev collocation points on the interval $[-1 + 1]$

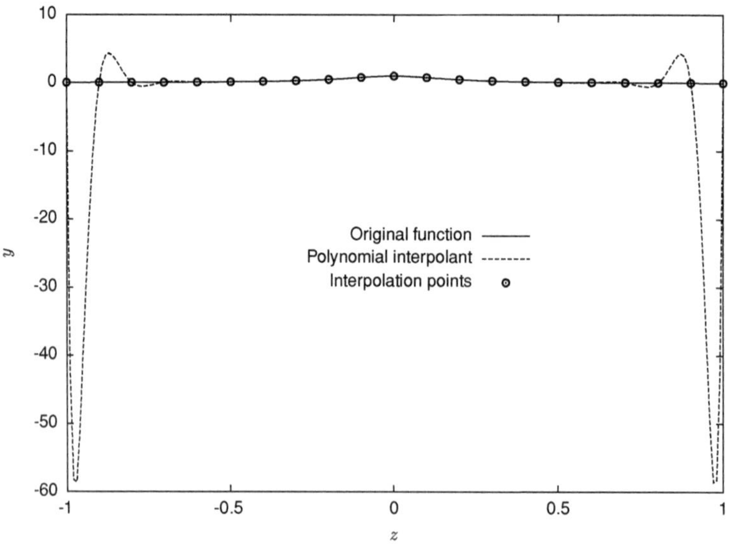

Fig. 4.11 Interpolation of the Runge function using 21 equidistant points

nowadays so-called Runge function $\frac{1}{1+25z^2}$ on the interval $[-1, +1]$. When interpolated with equidistant points, the interpolation polynomial (computed in Fig. 4.11 with the MATLAB function `polyfit`) gives rise to large oscillations at the boundaries. When using Chebyshev points, the interpolating polynomial converges for an increasing number of points (see Fig. 4.12). In both Figs. 4.11 and 4.12, $N = 21$ has been used.

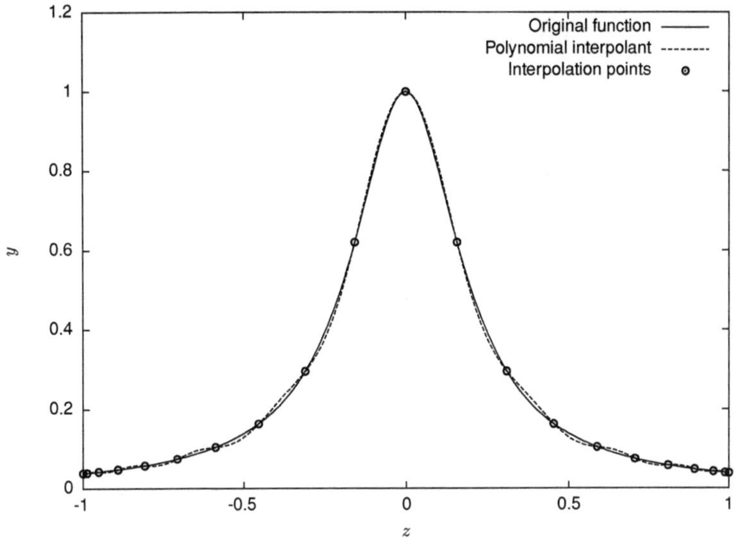

Fig. 4.12 Interpolation of the Runge function using 21 Chebychev points

The first Chebyshev polynomials on the interval $[-1, +1]$ are given by

$$T_0(z) = \frac{1}{2}\left(\xi^0 + \xi^{-0}\right) = 1$$
$$T_1(z) = \frac{1}{2}\left(\xi^1 + \xi^{-1}\right) = z$$
$$T_2(z) = \frac{1}{2}\left(\xi^2 + \xi^{-2}\right) = \frac{1}{2}\left(\xi^1 + \xi^{-1}\right)^2 - 1 = 2z^2 - 1$$
$$T_3(z) = \frac{1}{2}\left(\xi^3 + \xi^{-3}\right) = \frac{1}{2}\left(\xi^1 + \xi^{-1}\right)^3 - \frac{3}{2}\left(\xi^1 + \xi^{-1}\right) = 4z^3 - 3z$$

with

$$\xi = \exp(i\theta); \qquad z = Re(\xi) = \frac{1}{2}\left(\xi^1 + \xi^{-1}\right) = \cos(\theta)$$

or more generally

$$T_n(z) = Re(\xi^n) = \frac{1}{2}\left(\xi^n + \xi^{-n}\right) = \cos(n\theta)$$

Figure 4.13 shows the graphical interpretation of the variables ξ, z and θ.

Using Chebyshev collocation, it is possible to define differentiation matrices so that

$$\tilde{x}_z = \mathbf{D}_1\tilde{x} \tag{4.117}$$

and

$$\tilde{x}_{zz} = \mathbf{D}_1\left(\mathbf{D}_1\tilde{x}\right) \tag{4.118}$$

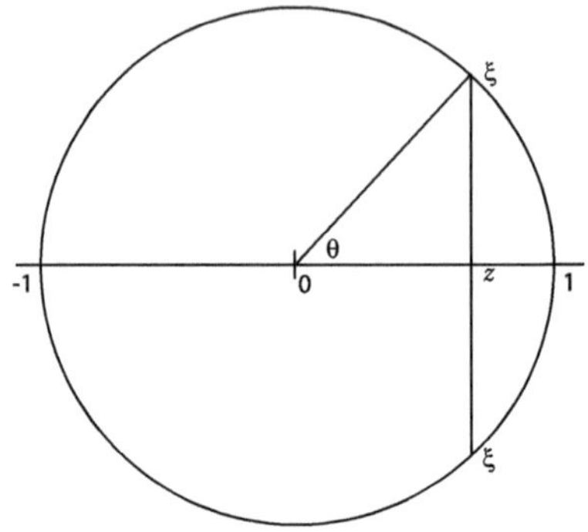

Fig. 4.13 Graphical interpretation of symbols z, ξ and θ

The computation of the differentiation matrix can be achieved using Fast Fourier Transform (FFT) so as to reduce the computation order to $O(N \log(N))$ instead of $O(N^2)$. Very elegant and compact codes are provided by Trefethen [7] which have been adapted in the companion software (see function `chebyshev_spectral_D1`).

```
function [D,z] = chebyshev_spectral_D1(z0,zL,n)
% function chebishev_spectral_D1 returns the differentiation
% matrix for computing the first derivative, x_z  , of a variable
% x over the spatial domain z0 < x < zL from a spectral method on
% a clustered grid.
% This code is a slight adapatation of a code taken from
% (Trefethen, 2000)
%
% argument list
%   z0     left value of the spatial independent variable (input)
%   zL     right value of the spatial independent variable (input)
%   n      number of spatial grid points, including the end points
%          (input)
%   D      differentiation matrix (output)
%   z      Chebishev points (output)

% compute the spatial grid
L = zL-z0;
n = n-1;                              % zi, i = 0,...,n-1
z = cos(pi*(n:-1:0)/n)';              % Chebishev points on [-1, 1]

% discretization matrix
Z = repmat(z,1,n+1);    % create a (nxn) square matrix by
                        % replicating each Chebishev point in a
                        % line of the matrix
dZ = Z-Z';      % create a matrix whose elements represent the
                % distance to the diagonal elements
c = [2; ones(n-1,1); 2].*(-1).^(0:n)';
D = (c*(1./c)')./(dZ+(eye(n+1)));     % compute off-diagonal entries
D = D - diag(sum(D'));                % adjust diagonal entries
z = z0+(z+1)*(zL-z0)/2;               % convert to interval [z0, zL]
D = 2/(zL-z0)*D;
```

Function chebyshev_spectral_D1 Function to numerically compute the differentiation matrix $\mathbf{D}_1$ using the Chebyshev collocation points

We now illustrate the use of this function for the solution of a simple PDE model describing an algae bloom.

Plankton blooms that appear as patches on the surface of water can be modeled by Beltrami [8]:

$$\frac{\partial x}{\partial t} = \mu \frac{\partial^2 x}{\partial z^2} + rx\left(1 - \frac{x}{K}\right) \tag{4.119}$$

where μ is the diffusion coefficient, r represents the maximum specific growth rate and K the carrying capacity (logistic growth). The boundary conditions model the unfavorable conditions outside the island-like patch

$$x(z_0, t) = x(z_L, t) = 0 \tag{4.120}$$

There is a minimal size for a patch if the population is not to die out: $z_L > \pi\sqrt{(\mu/r)}$. Here, twice this critical size is considered. Finally, the initial conditions are taken as

$$x(z, 0) = x_0 \tag{4.121}$$

The MATLAB implementation of this problem is not very different from the other application examples presented so far in this chapter. The main difference is the computation of the first-order differentiation matrix $\mathbf{D}_1$ used in Eq. (4.117). In this case $\mathbf{D}_1$ is computed using function `chebyshev_spectral_D1`.

The second-order differentiation matrix is obtained by stage-wise differentiation as in Eq. (4.118), i.e., $\mathbf{D}_2 = \mathbf{D}_1\mathbf{D}_1$. After the application of the collocation method, the following system of ODEs is obtained:

$$\frac{\mathrm{d}\tilde{x}}{\mathrm{d}t} = \mu\tilde{x}_{zz} + r\tilde{x}\left(1 - \frac{\tilde{x}}{K}\right) \tag{4.122}$$

where $\tilde{x}$ is the discretized version of the dependent variable x, i.e., $\tilde{x} = [x_1, x_2, \ldots, x_N]^T$ and x_{zz} is computed as $\tilde{x}_{zz} = \mathbf{D}_2\tilde{x}$.

The ODE system (4.122) is implemented in function `algae_bloom_pde`:

```
function xt = algae_bloom_pde(t,x)

% set global variables
global mu r K kd;
global z0 zL z n D1;

% boundary conditions at z = 0
x(1) = 0;

% boundary conditions at z = L
x(n) = 0;

% spatial derivatives (stagewise differentiation)
xz  = D1*x;
xzz = D1*xz;
```

```
% temporal derivatives
xt    = mu*xzz + r*x.*(1-x/K);
xt(1) = 0;
xt(n) = 0;
```

Function algae_bloom_pde MATLAB implementation of the ODE system (4.122).

The code begins with the definition of the global variables which are passed to this function and/or will be passed to other programs. After this, the values of the dependent variable x in the first and the last points are imposed according to boundary conditions (4.120). The next step is the computation of the second-order derivative of $\tilde{x}$. Finally, the ODE system is constructed and the time derivatives in the first and last points are zeroed due to the nature of the BCs. Note the close resemblance between Eq. (4.122) and its implementation.

Figures 4.14 and 4.15 show the numerical results obtained with an extremely small number of nodes. Only the values of the solution at the four Chebyshev points is displayed in the figures, using the linear interpolation provided by the MATLAB function plot. A polynomial interpolation should be used to get a better picture of the solution profile. Chebyshev collocation appears therefore as a very powerful method to solve parabolic PDEs.

4.11 The Proper Orthogonal Decomposition

The application of the classical numerical techniques for PDEs, such as finite differences or the FEM previously described may lead to a large system of ODEs. This is particularly critical when considering 2D or 3D spatial domains, where the number of equations necessary to obtain a satisfactory result may increase to tens or hundreds of thousands. Fortunately, it is possible to derive *reduced order models* (ROMs), which are much cheaper from a computational point of view. Basically, model reduction techniques neglect solution features with very fast dynamics, as compared to other features which occur in the dynamical range of interest. In some sense, this idea is similar to the concept of data compression used nowadays to store images or music (jpg, mp3) where non relevant features (information that cannot be captured by the human eye or the ear) are neglected.

Many efforts have been spent in the development of techniques for the reduction of linear models. Amongst them one can find the *Balance and Truncate* [9] or the *Hankel approximation* [10].With regard to nonlinear distributed parameter systems, many techniques for model reduction are based on the Galerkin method with globally defined basis functions and have been applied to the simulation of a wide range of systems during the last six decades.

In this section, we will focus on one of the most efficient order reduction techniques, *the proper orthogonal technique* (POD). This method, first proposed in [11], arose in the context of turbulence simulation. Since then, this technique has been used in many different fields including atmospheric modeling [12], chemical systems [13–15] , fluid

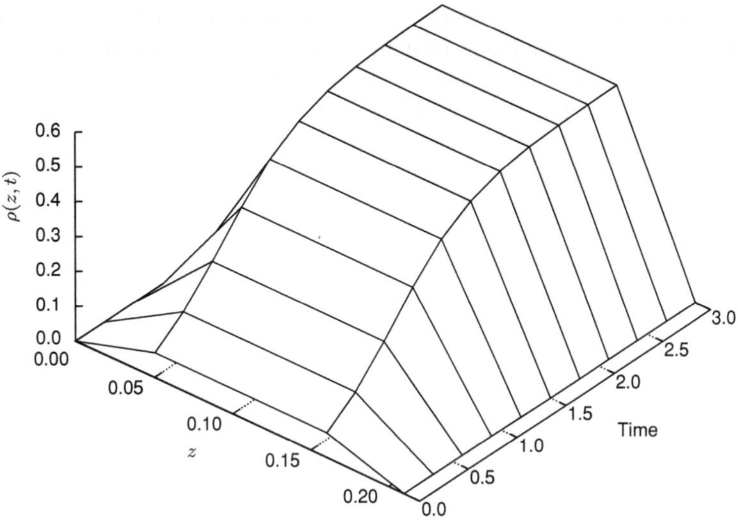

Fig. 4.14 Solution of the Algae bloom problem with a large diameter of the algae patch using the Chebyshev collocation method

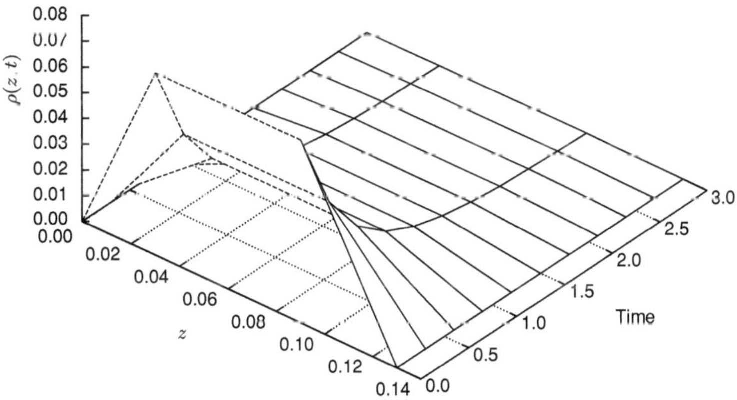

Fig. 4.15 Solution of the Algae bloom problem with a small diameter of the algae patch using the Chebyshev collocation method

dynamics [16–18], biological systems [19] or thermal treatment of canned food [20] among others.

As in the other methods presented in this chapter, the first step is to expand the solution of the PDE in a truncated series of the form of Eq. (4.13), i.e.,

$$x(z,t) \approx \tilde{x}(z,t) = \sum_{i=1}^{N} m_i(t)\phi_i(z)$$

The main difference between order reduction techniques lies in the selection of basis functions ($\phi_i(z)$) in the Galerkin method. In the POD, such basis functions are obtained from experimental data (either in silico or in vitro). The experiments provide us with a set of measurements of the time evolution and spatial distribution of the relevant fields (temperature, compound concentration, population density, etc.). For example, consider a tubular reactor where 10 temperature sensors are distributed along the reactor length and the sensors send the temperature measurements every minute. If we keep measuring for 1 h, we could sort all the temperature data in a matrix with 10 rows (one per sensor) and 60 columns (one per measurement time). This matrix is then used in the POD technique to obtain the set of basis functions in an optimal way. The measurement obtained for a given field at a given time and different spatial locations is called *snapshot*.

The main idea behind the POD technique is to find the set of basis functions which minimize the distance from the experimental data to the subspace defined by the basis functions. This is equivalent to maximizing the projection of the data set $x(z,t)$ onto the basis functions. Mathematically this can be expressed as [21]:

$$\max_{\phi} J : J = \left\{ \left| \int_{\Omega} x\phi \mathrm{d}z \right|^2 \right\} - \lambda \left(\int_{\Omega} \phi\phi \mathrm{d}z - 1 \right) \tag{4.123}$$

where $|\cdot|$ indicates the modulus and $\{\cdot\}$ is an averaging operation. It can be shown that finding the maximum of J in (4.123) is equivalent to solving the following eigenvalue problem [21, 22]:

$$\lambda_i \phi_i(z) = \int_{\Omega} \mathscr{K}(z, z') \phi_i(z') \mathrm{d}z' \tag{4.124}$$

where the kernel $\mathscr{K}(z, z')$ is constructed as a two point correlation kernel of the form:

$$\mathscr{K} = \frac{1}{k} \sum_{i=1}^{k} x_i x_i^T. \tag{4.125}$$

where $x_i \in \mathbb{R}^N$ is the vector of values of the field x at a finite number N of spatial points and at a given time t_i (snapshot). Since the kernel $\mathscr{K}$ is real symmetric, its eigenvalues λ_i are real numbers [1] and may be arranged so that $|\lambda_i| \geq |\lambda_j|$ for $i < j$ [1]. Furthermore, it can be shown that $\lambda_p \to 0$ as $p \to \infty$ [23, 24].

Since the basis functions in the (POD) method are obtained by solving an optimization problem, they form a set of empirical basis functions which are optimal with respect to other possible expansions. This set is optimal in the sense that for a given number of basis functions, it captures most of the relevant dynamic behavior of the original distributed system in the range of initial conditions, parameters, inputs, and/or perturbations of the experimental data [25].

It should be stressed that the eigenvalues λ_i can be used as an a priori measurement of the accuracy of the approximation. Indeed, the total energy captured by the full set of PODs is computed through the eigenvalues as $E = \sum_{i=1}^{N} \lambda_i$. Thus the percentage of energy captured by a given number p of PODs is:

$$E(\%) = \frac{\sum_{i=1}^{p} \lambda_i}{\sum_{i=1}^{N} \lambda_i} 100. \tag{4.126}$$

The more the energy captured, the better the quality of the approximation.

4.11.1 The Method of Snapshots

When the number of discretization points (or measurement points) of variable x in Eq. (4.125) is large, i.e., N is large, solving Eq. (4.124) can be computationally involved. In order to avoid this problem and save computation time, a useful alternative, proposed by Sirovich [11] and known as the *method of snapshots* or *indirect method*, is briefly discussed. In this method, each eigenfunction is expressed in terms of the original data as:

$$\phi_j = \sum_{i=1}^{k} w_i^j x_i, \tag{4.127}$$

where w_i^j are the weights to be computed. To this purpose, a new matrix is defined as:

$$\mathscr{S}_{ij} = \frac{1}{k} \int_{\Omega} x_i x_j \mathrm{d}\xi. \tag{4.128}$$

Introducing Eqs. (4.127) and (4.128) in the eigenvalue problem (4.124), results in:

$$\mathscr{S}\mathscr{W}_j = \lambda_j \mathscr{W}_j, \tag{4.129}$$

where the eigenvectors $\mathscr{W}_j$ have as elements the weights in Eq. (4.127) so that $\mathscr{W}_j = [w_1^j, w_2^j, \ldots, w_k^j]^T$.

Both the direct and the snapshot methods are implemented in function `matpod.m` available in the companion software. Let us illustrate now the use of this function. Consider a 1D process where 11 temperature sensors are equidistributed along the process spatial length $L = 1$ m. Sensors are therefore located at positions $s_p = [0, 0.1, 0.2, 0.3, 0.4, 0.5, 0.6, 0.7, 0.8, 0.9, 1]$ m. Consider also that the sensors have registered five measurements at times $t_m = [0, 0.5, 1.0, 1.5, 2.0]$ s. Table 4.6 collects in a matrix form all the data taken by the sensors

In the following, this 11×5 matrix will be denoted by V and it will be used to describe the different steps for the derivation of the POD basis, in function `matpod.m`. For the sake of clarity, function `matpod.m` was simplified here with respect to the one in the companion software which can handle several fields (temperature, compounds concentration, bacteria population, etc) at the same time. By considering only one field (in this case temperature), the code becomes much simpler to describe.

Table 4.6 Experimental data taken by the temperature sensors at different spatial locations and time instants

s_p	t=0	t=0.5	t=1.0	t=1.5	t=2.0
0.0	1.000	0.657	0.367	0.205	0.115
0.1	1.000	0.654	0.365	0.204	0.114
0.2	1.000	0.643	0.360	0.201	0.112
0.3	1.000	0.624	0.349	0.195	0.109
0.4	1.000	0.598	0.334	0.187	0.104
0.5	1.000	0.565	0.316	0.177	0.099
0.6	1.000	0.526	0.294	0.164	0.092
0.7	1.000	0.480	0.268	0.150	0.084
0.8	1.000	0.429	0.239	0.134	0.075
0.9	1.000	0.373	0.208	0.116	0.065
1.0	1.000	0.312	0.174	0.097	0.054

The code of `matpod.m` begins, as any other MATLAB function, by defining input and output parameters.

```
function [pods] = matpod(V , M , mth , ener)
switch (nargin)
    case {2}
        mth     = 'd';
        ener    = 99.99;
    case {3}
        ener    = 99.99;
end
```

Mandatory input parameters are the matrix containing the experimental data V and the mass matrix resulting from the finite element method $\mathbf{M}$. This matrix will be used to perform the spatial integrals involved in the method although any other quadrature formula could be exploited as well. Of course, the spatial grid used to obtain the experimental measurements of V must coincide with the grid underlying the definition of matrix $\mathbf{M}$. The other two input parameters correspond to the selection of the method, *direct* (`'d'`) or *indirect* (`'i'`), and the energy level to be captured. These two input arguments are optional. If the number of input arguments `nargin` is 2 then default values for `mth` and `ener` are used. On the other hand if `nargin`= 3 then `ener` is set to its default value.

The paths to obtain the POD basis in the direct and indirect methods are different. The MATLAB function `switch-case` is used to distinguish between both techniques. The first step is to construct the matrices used in the eigenvalue problem. For instance, in the case of the indirect method, each element of this matrix is computed through Eq. (4.128). As shown previously, the mass matrix of the FEM can be used to compute spatial integrals, so that, matrix $\mathscr{S}$ in Eq. (4.128) can be numerically computed as `S = 1/L*X'*M*X` where `X` is the matrix of snapshots. In the direct method, the kernel $\mathscr{K}$ is constructed as in Eq. (4.125). The following piece of code is used to compute matrix $\mathscr{S}$ in the indirect method and $\mathscr{K}$ in the direct method.

```
switch (mth)
    case {'i'}
        S = 1/L*X*M*X;
    case {'d'}
        K   = 1/L*X*X';
        S   = K*M;
end
```

Now MATLAB function eig is used to compute the eigenfunctions of the relevant matrices.

```
% Computation of Eigenfunctions and eigenvalues
[W , lambda_nn] = eig(full(S));
switch (mth)
    case {'i'}
        phi_nn   = X*W;
    case {'d'}
        phi_nn   = W;
end
```

In the indirect method, such eigenfunctions are the weights used in Eq. (4.127) to compute the POD basis. In matrix form, Eq. (4.127) can be written as $\phi_j = XW_j$. In the direct method, the eigenfunctions coincide with the POD basis.

The next part of the code selects the most representative POD basis according to the energy criterion. The total energy is computed as the sum of all eigenvalues while the energy of a given number j of POD basis is computed as the sum of the eigenvalues corresponding with such POD basis. When the value of energy selected by the user is reached, the loop stops.

```
% Selection of the more representative POD
tot_ener = sum(diag(lambda_nn));
for j = 1 : L
    par_ener(j) = sum( diag( lambda_nn(1:j , 1:j) ) );
    cap_ener     = par_ener(j)/tot_ener*100;
    if (cap_ener >= ener)
        phinn         = phi_nn(: , 1:j);
        pods.lambda = lambda_nn(1:j , 1:j);
        break
    end
end

% Eigenfunctions normalization
for ii = 1 : size(P(i).phinn,2)
    cc.par(ii)      = phinn(:,ii)'*MM*phinn(:,ii);
    pods.phi(:,ii)  = 1/sqrt(cc.par(ii))*phinn(:,ii);
end
```

In order to illustrate the procedure, advantages, and weaknesses of the POD approach, several examples will be considered next.

4.11.2 Example: The Heat Equation

The heat equation has already been presented at the beginning of this chapter. Let us, for the sake of clarity, rewrite the equation for this problem:

$$\frac{\partial T}{\partial t} = \alpha \frac{\partial^2 T}{\partial z^2}, \tag{4.130}$$

with $\alpha = 0.1$ and the length of the spatial domain is $L = 1$. Homogeneous Neumann boundary conditions are considered:

$$\left.\frac{\partial T}{\partial n}\right|_{z=0} = \left.\frac{\partial T}{\partial n}\right|_{z=L} = 0. \tag{4.131}$$

and initial conditions are given by:

$$T(z, 0) = 5\left(\frac{z^2}{2} - \frac{z^4}{4}\right) + 1;$$

The analytical solution for this simple example has already been obtained using the method of separation of variables, see Eqs. (4.9) and (4.11). It will be used to check the accuracy of the POD technique.

As pointed out before, the first step to obtain the PODs basis is the construction of a representative set of data (snapshots). In this example such a set will be obtained from the analytical solution. To this purpose the time domain $t = [0, 4]$ is divided into three parts. This separation is based on the dynamic behavior of the system:

- First interval $t = [0, 0.5]$: the temperature changes relatively fast so that the time interval (δt) between two consecutive measurements should be short enough. In this case $\delta t = 0.05$, i.e., 11 time measurements.
- Second interval $t = [0.5, 2]$: the temperature approaches steady-state so that the measurement interval may increase $\delta t = 0.1$, i.e., 14 time measurements.
- Third interval $t = [2, 4]$: the temperature has reached steady state. A couple of measurements are enough ($\delta t = 0.5$).

The basis functions are now computed following the methodology described in the previous section. Figure 4.16 shows the shape of the four more representative basis functions, i.e., those associated with the largest eigenvalues. Note that the number of spatial oscillations in the basis functions decreases as they are more representative (i.e., more energy is captured). In general, the number of spatial oscillations of the basis function can be used to determine their contribution to the dynamic behavior of the system in a qualitative manner.

The value of the eigenvalues and the energy captured by the POD basis functions is represented in Table 4.7. Note that the value of the first eigenvalue is orders of magnitude larger than the second one, the second is orders of magnitude larger than the third and so on. In this case, two basis functions are able to capture more than the 99.99 % of the energy. This is because we are only considering diffusion in 1D in this example. When

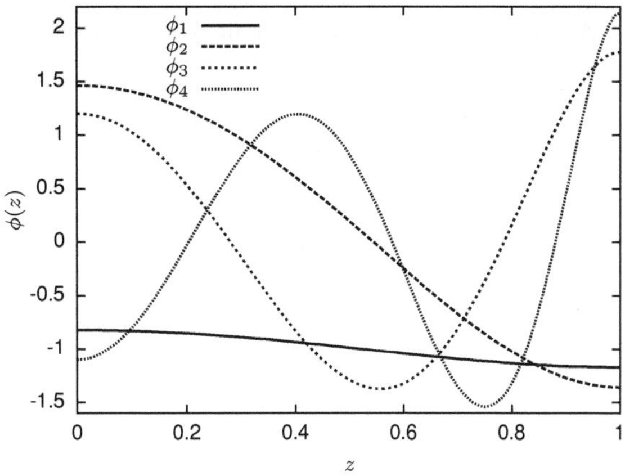

Fig. 4.16 Representation of the four more representative POD basis functions for the heat diffusion problem

Table 4.7 Value of the eigenvalues for the heat diffusion problem and energy captured by them

	λ	Energy (%)
First	2.5453	99.3507
Second	1.6621×10^{-2}	0.6488
Third	1.3914×10^{-5}	5.4310×10^{-4}
Fourth	1.2166×10^{-7}	4.7489×10^{-6}

source terms or more spatial dimensions are considered, the value of the eigenvalues will be closer and more POD basis will be necessary to approximate the solution with a given level of accuracy.

Let us now project Eq. (4.130) over a given basis function. Mathematically this is performed by multiplying Eq. (4.130) by the POD basis functions and integrating the result over the spatial domain, i.e.,

$$\int_{\Omega} \phi_i \frac{\partial T}{\partial t} \mathrm{d}z = \int_{\Omega} \phi_i \alpha \frac{\partial^2 T}{\partial z^2} \mathrm{d}z = \alpha \int_{\Omega} \phi_i \frac{\partial^2 T}{\partial z^2} \mathrm{d}z$$

Green's theorem allows us to express the RHS of the previous relation in terms of the boundary integral, i.e.,

$$\alpha \int_{\Omega} \phi_i \frac{\partial^2 T}{\partial z^2} \mathrm{d}z = \alpha \int_{\Gamma} \phi_i \frac{\partial T}{\partial n} \mathrm{d}z - \alpha \int_{\Omega} \frac{\mathrm{d}\phi_i}{\mathrm{d}z} \frac{\partial T}{\partial z} \mathrm{d}z$$

It is easy to see that, due to the boundary conditions of this problem—Eq. (4.131) the first term of the RHS is zero so that

$$\int_{\Omega} \phi_i \frac{\partial T}{\partial t} \mathrm{d}z = -\alpha \int_{\Omega} \frac{\mathrm{d}\phi_i}{\mathrm{d}z} \frac{\partial T}{\partial z} \mathrm{d}z$$

Approximating the temperature field by a truncated Fourier series, we have

$$\int_{\Omega} \phi_i \frac{\partial\left(\sum_{j=1}^{p} m_j\phi_j\right)}{\partial t}\mathrm{d}z = -\alpha \int_{\Omega} \frac{\mathrm{d}\phi_i}{\mathrm{d}z}\frac{\partial\left(\sum_{j=1}^{p} m_j\phi_j\right)}{\partial z}\mathrm{d}z$$

Since the POD basis functions (ϕ_i) do not depend on time and the coefficients of the expansion (m_i) do not depend on the spatial coordinates, the previous equation can be rewritten as:

$$\sum_{j=1}^{p}\frac{\mathrm{d}m_j}{\mathrm{d}t}\int_{\Omega}\phi_i\phi_j\mathrm{d}z = -\alpha\sum_{j=1}^{p} m_j\int_{\Omega}\frac{\mathrm{d}\phi_i}{\mathrm{d}z}\frac{\mathrm{d}\phi_j}{\mathrm{d}z}\mathrm{d}z$$

The basis functions are orthonormal, so that the integral $\int_{\Omega}\phi_i\phi_j\mathrm{d}z = 1$ if $i = j$ and $\int_{\Omega}\phi_i\phi_j\mathrm{d}z = 0$ otherwise, and

$$\frac{\mathrm{d}m_i}{\mathrm{d}t} = -\alpha\sum_{j=1}^{p} m_j\int_{\Omega}\frac{\mathrm{d}\phi_i}{\mathrm{d}z}\frac{\mathrm{d}\phi_j}{\mathrm{d}z}\mathrm{d}z$$

The expressions resulting from the projection of the POD basis functions ϕ_i, with $i = 1, 2, \ldots, p$, can be collected onto the following matrix form.

$$\frac{\mathbf{dm}}{\mathrm{d}t} = \alpha\mathscr{A}\mathbf{m} \tag{4.132}$$

where $\mathbf{m} = [m_1, \ldots, m_p]^T$ and $\mathscr{A}$ is the projection of the Laplacian operator. That is, the elements of matrix $\mathscr{A}$ are computed as:

$$\mathscr{A}_{i,j} = \int_{\Omega}\frac{\mathrm{d}\phi_i}{\mathrm{d}z}\frac{\mathrm{d}\phi_j}{\mathrm{d}z}dz \tag{4.133}$$

The solution of this integral can be obtained in two different ways:

- Using a numerical approach like a quadrature formula or the FEM approach. In the latter case matrix $\mathbf{D}_2^{\text{int}}$—see Eq. (4.55) is used:

$$\int_{\Omega}\frac{\mathrm{d}\phi_i}{\mathrm{d}z}\frac{\mathrm{d}\phi_j}{\mathrm{d}z}dz \approx \phi_i\mathbf{D}_2^{\text{int}}\phi_j$$

- Fitting the POD data to polynomials of given order which are easy to integrate and differentiate (see [20] for details).

The reduced-order model, i.e., Eq. (4.132), is implemented in function `pde_dif_pod`. Note the close resemblance of the ODE system—Eq. (4.132)- $m_t = \alpha\mathscr{A}\mathbf{m}$ and the programming of the ODE system `m_t = kappa*A*m;`.

```
function m_t = pde_dif_pod(t, m, A, kappa)

% PDE
m_t = kappa*A*m;
```

Function pde_dif_pod Function to implement Eq. (4.132)

Function pde_dif_pod is called from the main program main_diffusion_rom

```
% The diffusion problem solved using the fem

clear all
clc

% Spatial coordinates
nd = 101;                    % Discretization points
z  = linspace(0,1,nd)';

% Parameters
kappa = 0.1;

% FEM matrices
[MM, D2_int] = matfem (z, 'neu', 'neu', 'MM', 'DM');

% Load the POD data
load fig_pod_diff
neig = 5;
phi  = pods.phi(:,1:neig);

% Initial conditions
x0 = 5*(z.^2/2 - z.^4/4) + 1;
m0 = phi'*MM*x0;

% Projection of the laplacian operator
A = -phi'*D2_int*phi;

% Time span
tlist = 0:0.1:4;

% ODE integration
options  = odeset('RelTol', 1e-6, 'AbsTol', 1e-6);
[tout,m] = ode15s(@pde_dif_pod,tlist,m0,options,A,kappa);

% Recovery the field
x_num = phi*m';

% Plot the results
mesh(z, tout, x_num')
```

Script main_diffusion_rom Main script to call function pde_dif_pod

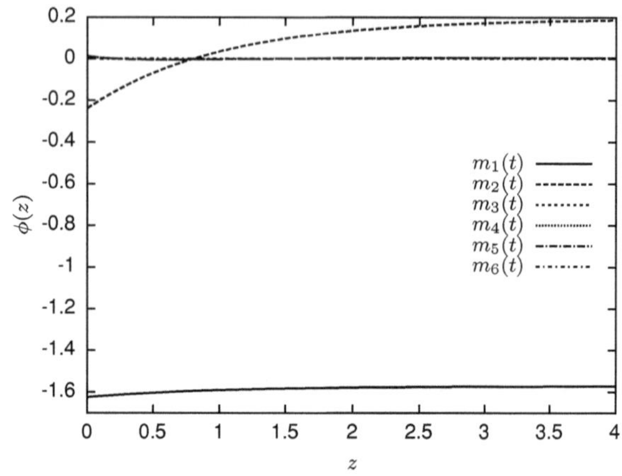

Fig. 4.17 Evolution of the first six coefficients, computed using the POD method, for the heat diffusion problem

The structure of this script is very similar to those already presented in this chapter.

- First, the spatial coordinates and the relevant parameters of the system are defined.
- Then, the mass and second-order differentiation matrices for the FEM are computed. These matrices will be used to perform the projection (which requires to compute spatial integrals) and to construct the discrete version of the Laplacian operator.
- The next step is to load the POD basis functions previously computed. Initial conditions are first defined for all the spatial domain, $x_0 = 5 * (z.^2/2 - z.^4/4) + 1$, and then they are projected onto the basis functions, using the mass matrix of the FEM, in order to obtain the initial conditions for the time dependent coefficients.
- Next, the projection of the spatial operator is numerically computed using the second-order differentiation matrix `A=-phi'*D2_int*phi`.
- Then, the output times are defined and the ODE system is solved using the MATLAB ODE suite in order to obtain the time evolution of the coefficients used in the series expansion.
- Finally, the temperature field is recovered by multiplying the POD basis by the time-dependent coefficients and the solution is plotted.

The time evolution of the first six coefficients is represented in Fig. 4.17. The plot shows that, as expected, the first three coefficients are much more important than the remaining ones, which rapidly converge to zero. Hence, using $p = 3$ should be enough to accurately represent the solution $T(z, t)$, see Fig. 4.18.

A quantitative comparison between the accuracy of the FEM with Lagrangian elements and the POD technique is presented in Table 4.8.[2] Three POD basis functions are enough to obtain a level of accuracy comparable to the FEM with 10 finite elements. Four basis functions allow to reach a level of accuracy corresponding to 30 elements.

[2] The accuracy is computed as the error between the numerical technique and the analytical solution.

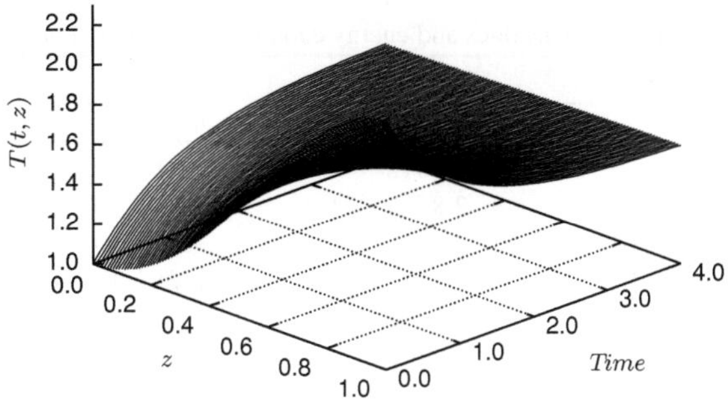

Fig. 4.18 Numerical solution to the Fourier equation computed with the POD method ($p = 3$)

Table 4.8 Errors obtained with the numerical techniques POD and FEM as compared with the analytical solution for the diffusion equation and for different number of ODEs

Number of ODEs	FEM			ROM		
	11	21	31	3	4	5
Maximum error	0.131	0.033	0.014	0.12	0.016	0.0024
Mean error	0.048	0.012	0.0053	0.015	0.0021	0.00057

4.11.3 Example: The Brusselator

Let us now consider another classical example already presented in Chap. 3, *the Brusselator*. The system is described by the following set of two PDEs:

$$\begin{aligned} u_t &= \alpha u_{zz} + f(u, v), \quad f(u, v) = a - (b + 1)u + u^2 v \\ v_t &= \alpha v_{zz} + g(u, v), \quad g(u, v) \;= bu - u^2 v \end{aligned} \tag{4.134}$$

with $0 \leq z \leq 1$ and $t \geq 0$. Note that, in contrast with the heat equation, this system has a nonlinear term $u^2 v$ in both equations. Dirichlet conditions are considered at both sides of the spatial domain:

$$u(0, t) = u(1, t) = 1; \quad v(0, t) = v(1, t) = 3 \tag{4.135}$$

and initial conditions are given by:

$$u(z, 0) = 1 + \sin(2\pi z); \quad v(z, 0) = 3 \tag{4.136}$$

with parameter values: $a = 1, b = 3, \alpha = 0.02$.
The solution (u, v) of the problem is approximated by a truncated Fourier series of the form:

Table 4.9 Value of the first eigenvalues and energy captured for the brusselator system

	State ***u***		State ***v***	
	λ	Energy (%)	λ	Energy (%)
First	1.11	94.64	9.99	98.96
Second	3.36×10^{-2}	2.85	7.27×10^{-2}	0.72
Third	2.87×10^{-2}	2.43	3.06×10^{-2}	0.31
Fourth	7.83×10^{-4}	6.47×10^{-2}	1.48×10^{-3}	1.47×10^{-2}
Fifth	7.18×10^{-5}	6.10×10^{-3}	7.87×10^{-5}	7.79×10^{-4}

$$u \approx \sum_{i=1}^{p_u} m_{u,i}(t)\phi_{u,i}(z)$$
$$v \approx \sum_{i=1}^{p_v} m_{v,i}(t)\phi_{v,i}v(z) \tag{4.137}$$

The POD basis functions ϕ_u and ϕ_v are computed following the procedure previously described. Once these basis are available, they are used to project the original PDE system and, in this way, obtain the time-dependent coefficients m_u and m_v. The procedure is described below.

The first step to obtain the POD basis is to collect a set of data representative of the dynamic behavior of the system. For this particular example, the analytical solution is unknown, therefore we will use the FEM solution to obtain the snapshots and to check the accuracy of the POD method. The time interval between two consecutive snapshots is $\Delta t = 0.1$ and the last snapshot is taken at $t = 20$ units of time.

Since two dependent variables, u and v, are considered in this example, two different sets of PODs, ϕ_u and ϕ_v should be computed. To this purpose, function `matpod.m`, included in the companion software, is used.

The value of the first five eigenvalues for states u and v as well as the associated energy are presented in Table 4.9. In order to capture 99.99 % of the energy, four basis functions per state variable are enough.

The reduced-order model is derived by projecting the first and second equations onto (4.134) over the POD basis ϕ_u and ϕ_v, respectively. Let us start with the projection of the first equation in (4.134) onto a given POD basis function $\phi_{u,i}$.

$$\int_\Omega \phi_{u,i} u_t \mathrm{d}z = \int_\Omega \phi_{u,i} \alpha u_{zz} \mathrm{d}z + \int_\Omega \phi_{u,i} f(u, v) \mathrm{d}z, \tag{4.138}$$

As in the heat equation example, u_t is approximated using the truncated Fourier series—see (4.137). Taking into account that POD basis functions are orthogonal, it is easy to see that the RHS term of the previous equation can be expressed as:

$$\int_\Omega \phi_{u,i} u_t \mathrm{d}z = \frac{\mathrm{d}m_{u,i}}{\mathrm{d}t} \tag{4.139}$$

The second term on the RHS of (4.138) is numerically computed. As mentioned before, different options are available including the use of a quadrature formula or the mass matrix of the FEM **M**. For simplicity, this term will be denoted by F_i, i.e..

$$F_i = \int_{\Omega} \phi_{u,i} f(u, v) \mathrm{d}z \tag{4.140}$$

Green's theorem is used to deal with the remaining term in (4.138):

$$\alpha \int_{\Omega} \phi_{u,i} \frac{\partial^2 u}{\partial z^2} \mathrm{d}z = \alpha \int_{\Gamma} \phi_{u,i} \frac{\partial u}{\partial z} \mathrm{d}z - \alpha \int_{\Omega} \frac{\mathrm{d}\phi_{u,i}}{\mathrm{d}z} \frac{\partial u}{\partial z} \mathrm{d}z \tag{4.141}$$

The first term on the RHS of the previous equation represents the contribution of the boundary conditions. Note that Neumann and Robin boundary conditions enter naturally in this formulation since they fix the value of the field spatial derivative $\partial u/\partial z$. When Dirichlet boundary conditions are considered, as it is the case in this example, the "trick" described in Sect. 3.10—see Eq. (3.181) can be used. Note that boundary conditions in (4.135) can be expressed as

$$0 \left.\frac{\partial u}{\partial z}\right|_{z=0} = 1 - u(0, t), \quad 0 \left.\frac{\partial u}{\partial z}\right|_{z=L} = 1 - u(L, t)$$

The "trick" consists of substituting the zero-factor in the previous expression by a small epsilon factor, i.e.,

$$\varepsilon \left.\frac{\partial u}{\partial z}\right|_{z=0} = 1 - u(0, t), \quad \varepsilon \left.\frac{\partial u}{\partial z}\right|_{z=L} = 1 - u(L, t)$$

or

$$\left.\frac{\partial u}{\partial z}\right|_{z=0} = \frac{1}{\varepsilon}(1 - u(0, t)), \quad \left.\frac{\partial u}{\partial z}\right|_{z=L} = \frac{1}{\varepsilon}(1 - u(L, t)) \tag{4.142}$$

which can be easily introduced, as Robin boundary conditions, into the formulation. It must be mentioned that the lower the value of ε, the better the approximation. However, extremely low values of ε will result in numerical problems when solving the system of equations.

Using the Fourier expansion on the second term on the RHS of Eq. (4.141) and since the coefficients $m_{u,i}$ do not depend on the spatial coordinates, we obtain

$$\alpha \int_{\Omega} \frac{\mathrm{d}\phi_{u,i}}{\mathrm{d}z} \frac{\partial u}{\partial z} \mathrm{d}z = \alpha \int_{\Omega} \frac{\mathrm{d}\phi_{u,i}}{\mathrm{d}z} \frac{\partial \sum_{i=j}^{p_u} m_{u,j} \phi_{u,j}}{\partial z} \mathrm{d}z = \alpha \sum_{i=j}^{p_u} m_{u,j} \int_{\Omega} \frac{\mathrm{d}\phi_{u,i}}{\mathrm{d}z} \frac{\mathrm{d}\phi_{u,j}}{\mathrm{d}z} \mathrm{d}z \tag{4.143}$$

Substituting expressions (4.139–4.143) into (4.138) results in

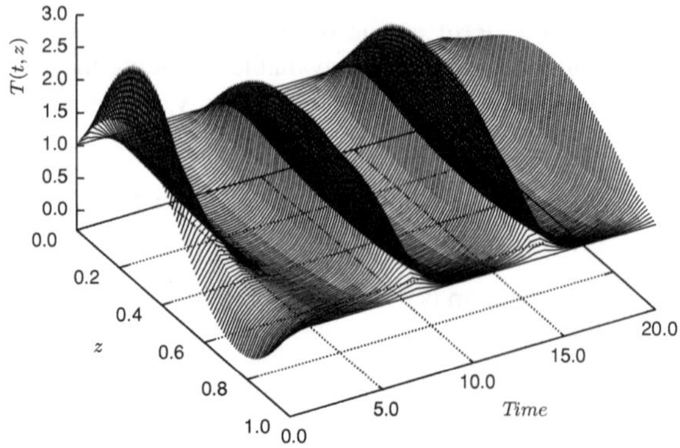

Fig. 4.19 Numerical solution (u) of the Brusselator problem (4.134)–(4.136) using the POD technique with 6 and 4 POD basis for the states u and v, respectively

$$\frac{\mathrm{d}m_{u,i}}{\mathrm{d}t} = \frac{1}{\varepsilon}\left(\phi_{u,i}(L)(1-u(L,t)) - \phi_{u,i}(0)(1-u(0,t))\right) - \alpha \sum_{i=j}^{p_u} m_{u,j} \int_{\Omega} \frac{\mathrm{d}\phi_{u,i}}{\mathrm{d}z}\frac{\mathrm{d}\phi_{u,j}}{\mathrm{d}z}\mathrm{d}z + F_i \tag{4.144}$$

or, in a more compact form:

$$\frac{\mathrm{d}\mathbf{m}_u}{\mathrm{d}t} = \alpha \mathscr{A}^u \mathbf{m}_u + F + \frac{1}{\varepsilon}\left(\phi_u(L)(1-u(L,t)) - \phi_u(0)(1-u(0,t))\right) \tag{4.145}$$

where $\mathbf{m}_u = [m_{u,1}, m_{u,2}, \ldots, m_{u,p_u}]^T$, $\phi_u = [\phi_{u,1}, \phi_{u,2}, \ldots, \phi_{u,p_u}]$, $F = [F_1, F_2, \ldots, F_{p_u}]^T$ and

$$\mathscr{A}^u_{i,j} = -\int_{\Omega} \frac{\mathrm{d}\phi_{u,i}}{\mathrm{d}z}\frac{\mathrm{d}\phi_{u,j}}{\mathrm{d}z}\mathrm{d}z \tag{4.146}$$

The same procedure is followed with the second equation in (4.134) to obtain:

$$\frac{\mathrm{d}\mathbf{m}_v}{\mathrm{d}t} = \alpha \mathscr{A}^v \mathbf{m}_v + G + \frac{1}{\varepsilon}\left(\phi_v(L)(3-v(L,t)) - \phi_v(0)(3-v(0,t))\right) \tag{4.147}$$

We can note that matrices $\mathscr{A}^u$ and $\mathscr{A}^v$ do not depend on time—see Eq. (4.146), therefore they can be computed beforehand. This is not the case for the projection of the nonlinear term, F^u, therefore, such projection must be performed in the ODE file defining the ODE system.

The solution obtained with the reduced-order model is represented in Figs. 4.19 and 4.20. One can qualitatively see that the dynamic behavior is the same as the one predicted by the finite differences scheme of Chap. 3 (see Figs. 3.11 and 3.12).

A more quantitative idea of the accuracy of the method can be obtained by using the mean and maximum relative errors between the POD technique and the FEM with 500 finite elements (see Table 4.10). Three different cases are considered:

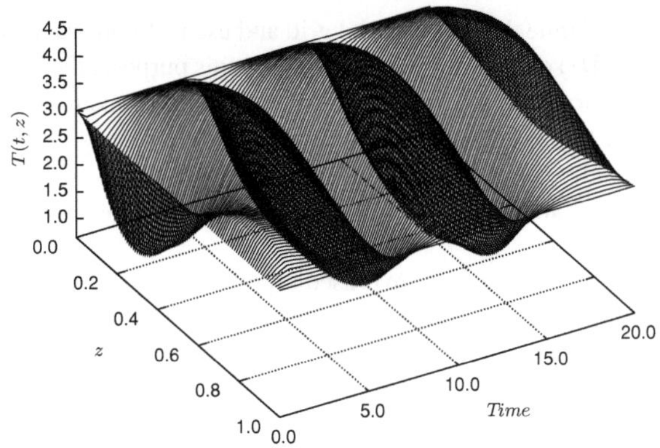

Fig. 4.20 Numerical solution (v) of the Brusselator problem (4.134)–(4.136) using the POD technique with 6 and 4 POD basis for the states u and v, respectively

Table 4.10 Relative errors between the FEM solution (501 discretization points) and the POD technique with different number of basis functions for the Brusselator problem

	State ***u***		State ***v***	
	Max. error (%)	Mean error(%)	Max. error (%)	Mean error (%)
Case 1	16.29	0.86	5.78	0.44
Case 2	12.68	0.64	5.93	0.42
Case 3	9.29	0.48	6.38	0.37

- Case 1 where $p_u = p_v = 4$
- Case 2 where $p_u = 5$ and $p_v = 4$
- Case 3 where $p_u = 6$ and $p_v = 4$

where p_u and p_v are the number of POD basis functions used in the projection for the state variables u and v, respectively. In all the considered cases, the mean relative errors are below 1 %.

4.12 On the Use of SCILAB and OCTAVE

In this section, the numerical solution of Burgers' Eq. (4.77) using the finite element method with linear Lagrangian elements is developed. The Gauss-Legendre quadrature is chosen to perform spatial integration and the procedure described in (4.4) is used to implement the boundary conditions.

Several SCILAB functions were written to implement the different parts of the method. Let us begin the description with the main file `main_burgers_FEM.sci`. The following points can be highlighted:

- The program begins by defining the global variables and the set of functions will be used in the solution procedure.
- Then, as it would be the case in the MATLAB implementation, the problem parameters are defined (in this case $\mu = 0.01$).

- The next step is to define the FEM spatial grid and use it to construct the mass (**M**) and the second-order ($\mathbf{D}_2$) differentiation matrices. To this purpose, the SCILAB functions `lagrlinmass.sci` and `lagrlinD2.sci` are used. Function `lagrlinmass.sci` is described below while function `lagrlinD2.sci` is only included in the companion software since its code is similar to function `lagrlinmass.sci`. The inverse of the mass matrix is also computed since it will be used in function `burgers_pdes.sci`.
- Since the Gauss-Legendre quadrature is used to perform the required spatial integration, the weights and the abscissa required for this procedure are computed in the next part of the code. Such code is similar to its MATLAB counterpart, the main difference is the use of function `spec` to compute the eigenvalues instead of `eig` and function `gsort` instead of `sort` to order the values of `xquad`.

```
nquad      = 2;
pbeta      = 0.5./sqrt(1-(2*(1:nquad)).^(-2));
T          = diag(pbeta,1) +diag(pbeta,-1);
[V,D]      = spec(T);
xquad      = diag(D);
[xquad,i]  = gsort(xquad);
wquad      = 2*(V(1,i) .^2);
```

  Two points (`nquad = 2`) are used in the quadrature. Function `diag` constructs a diagonal matrix using the values of the first argument (`pbeta`). These values are placed in one of the diagonals of the resulting matrix according to the value of the second argument (0 for the the main diagonal; 1 and -1 for the first upper and lower diagonals, respectively; 2 and -2 for the second upper and lower diagonals and so on).
- Before the call to the IVP solver, initial conditions and integration time are defined.
- As in the example of Chap. 3, the ODEPACK package is used to numerically compute the solution of the ODE system. The BDF method is chosen by the first input parameter `"stiff"`

```
yout = ode("stiff",x0,tl(1),tl,list(burgers_pdes));
```

- Finally, the numerical and analytical solutions are plotted. In order to compute the exact solution, function `burgers_exact.sci` is used. This function is included in the companion software.

```
clear

// Display mode
mode(-1);

// Global variables (shared by other files)
global("mu")
global("wquad","xquad")
global("n","dz","z0","zL","D2")
global("iM")
```

```
// Load the subroutines
exec('lagrlinD2.sci');
exec('lagrlinmass.sci');
exec('burgers_exact.sci');
exec('burgers_pdes.sci');
exec('integrand1.sci');
exec('integrand2.sci');
exec('trialfunctionslagrlin.sci');

// Problem parameters
mu = 0.01;

// Finite element spatial grid
z0  = 0;
zL  = 1;
n   = 201;
nel = n-1;
dz  = (zL-z0)/(n-1);
z   = (z0:dz:zL)';

// FEM mass matrix
ne   = 1;
M    = lagrlinmass(dz,n,ne);
iM   = inv(M);

// Second order differentiation matrix
D2 = lagrlinD2(dz,n);

// Computation of the weigths and the abscissa used in the
// numerical integration via the Gauss-Legendre quadrature
nquad      = 2;
pbeta      = 0.5./sqrt(1-(2*(1:nquad)).^(-2));
T          = diag(pbeta,1) +diag(pbeta,-1);
[V,D]      = spec(T);
xquad      = diag(D);
[xquad,i]  = gsort(xquad);
wquad      = 2*(V(1,i) .^2);

// Initial conditions
x0 = zeros(n,1);
for ii = 1:n
  x0(ii) = burgers_exact(z(ii),0);
end;

// Time instants at which the IVP solver will save the solution
dt = 0.1;
tl = 0:dt:1;
nt = length(tl);

// Time integration with the IVP solver
tic
yout = ode("stiff",x0,tl(1),tl,list(burgers_pdes));

// read the stopwatch timer
tcpu = toc();

// Plot the solution
plot(z,yout)
```

```
set(gca(),"auto_clear","off")
yexact = zeros(n,length(tl));
for k = 1:length(tl)
    for i = 1:n
        yexact(i,k) = burgers_exact(z(i),tl(k));
    end;
end;
plot(z,yexact,'*')
```

Script main_burgers_FEM.sci Main program for the implementation of the burgers equation (4.77) with the finite element method in SCILAB. This program calls the functions

Function `lagrlinmass.sci` is used to compute the mass matrix of the finite element method according to Eq. (4.44). Note that the only elements different from zero in this matrix are those of the main diagonal (`d0` in the code) as well as the first upper and lower (`d1` and `dm1` in the code) diagonals. The operator `.*.` is used to compute the Kronecker tensor product of two matrices. Since most of the elements of the mass matrix are zero, the command `sparse` is used to store only the elements different from zero.

```
// Display mode
mode(-1);

function [M] = lagrlinmass(h,n,ne)

// Output variables initialisation (not found in input variables)
M=[];

// Computation of the mass matrix of the finite element method
// with linear Lagrangian elements

// Main diagonal of the mass matrix
d0 = diag(ones(1,ne).*.[2 4*ones(1,n-2) 2], 0);

// First upper diagonal of the mass matrix
d1 = diag([ones(1,ne-1).*.[ones(1,n-1) 0] ones(1,n-1)],1);

// First lower diagonal of the mass matrix
dm1 = diag([ones(1,ne-1).*.[ones(1,n-1) 0] ones(1,n-1)],-1);

// Mass matrix
M = sparse((h/6)*(d0 + d1 + dm1));

endfunction
```

Function lagrlinmass.sci SCILAB function to construct the finite element mass matrix from a given uniform spatial grid.

The ODE system resulting from the FEM methodology:

$$\mathbf{M}\frac{d\tilde{\mathbf{x}}}{dt} = (\mu\mathbf{D}_2)\,\tilde{\mathbf{x}} + \tilde{f}_{NL} + \mathbf{g} \tag{4.148}$$

is implemented in function `burgers_pdes.sci`. To this purpose, the second-order derivative of the dependent variable $x(z, t)$ is numerically computed using matrix $\mathbf{D}_2$. After this, the finite element version of the nonlinear term $(x \cdot x_z)$, i.e.,

$$f_{NL} = y1 + y2; \quad \text{with} \quad yi = \int \phi_i (x x_z) \mathrm{d}z$$

is computed using the Gauss-Legendre quadrature. Boundary conditions are defined as in (4.69) and the system of ODEs is constructed. The mass matrix is not used as such by the ODE solver, but an explicit formulation in the spirit of (4.85) is preferred where the inverse of the matrix $\mathbf{M}$ is computed.

```
// Display mode
mode(-1);

function [xt] = burgers_pdes(t,x)

// Output variables initialisation (not found in input variables)
xt=[];

// Global variables
global("mu")
global("wquad","xquad")
global("n","z0","zL","D2","xx")
global("iM")

// Second spatial derivative xzz computed through the second
// differentiation matrix
xzz = D2*x;

// Rename state variable x to pass it to integrands functions
xx = x;

// Spatial integration of the integrands in the nonlinear term of
// Burgers equation. The nonlinear term is fx = x*xz
y1 = integrand1(xquad')*wquad';
y2 = integrand2(xquad')*wquad';
fx = y1+y2;

// Boundary conditions
gL      = burgers_exact(z0,t);
gR      = burgers_exact(zL,t);
gv(1,1) = x(1)-gL;
gv(n,1) = x(n)-gR;

// System of ordinary differential equations
xt = mu*xzz - fx - gv;

// Multiplication by the inverse of the mass matrix
xt = iM*xt;

endfunction
```

Function burgers_pdes.sci Implementation of the ODE system resulting from the application of the finite element methodology to the Burgers PDE system.

Functions `integrand1.sci` and `trialfunctionslagrlin.sci` will not be described here as they are very similar to their MATLAB counterparts. The interested reader can find them in the companion software.

4.13 Summary

In this chapter, we have introduced weighted residual methods, which represent the solution as a series of basis functions whose coefficients are determined so as to make the PDE (and BC) residuals as small as possible (in some average sense). In particular, we have focused attention on the Galerkin and collocation methods, and have discussed global and local basis functions, the latter leading to the famous Finite Element Method—FEM. Together with finite difference methods, FEM is one of the most popular methods for solving IBVP. Various basis functions can be used in conjunction with the FEM, especially Lagrange or Hermite polynomials. The construction of the FEM matrices is presented in detail for these two types of polynomials, when using a Galerkin method, and for Hermite polynomials, when using an orthogonal collocation approach. It is also possible to compute, using experimental or simulation data, optimal basis functions, i.e., the ones that are able to capture most of the solution features with a limited number of functions. This is the essence of the proper orthogonal decomposition (POD). Several examples are studied including the heat equation, and the Brusselator.

References

1. Courant R, Hilbert D (1937) Methods of mathematical physics. Wiley, New York, USA
2. Runge C (1901) Über empirische funktionen und die interpolation zwischen äquidistanten ordinaten. Zeitschrift für Mathematik und Physik 46:224–243
3. Lanczos C (1938) Trigonometric interpolation of empirical and analytic functions. J Math Phys 17:123–199
4. Finlayson BA (1971) Packed bed reactor analysis by orthogonal collocation. Chem Eng Sci 26:1081–1091
5. de Boor C, Swartz B (1973) Collocation at Gaussian points. SIAM J Numer Anal 10:582–606
6. Reddy JN (1993) Introduction to the finite element method, 2nd edn. McGraw Hill, New York
7. Trefethen LN (2000) Spectral methods in matlab. SIAM.
8. Beltrami E (1998) Mathematics for dynamic modeling. Academic Press, Massachusetts
9. Tombs MS, Postlethwaite I (1987) Truncated balanced realization of stable, non-minimal state-space systems. Int J Control 46:1319–1330
10. Lemouel A, Neirac F, Maisi N (1994) Heat-transfer equation–modeling by rational Hankel approximation methods. Revue Generale de Thermique 33(389):336–343
11. Sirovich L (1987) Turbulence and the dynamics of coherent structures. Part I: Coherent structures. Quaterly Appl Math 45(3):561–571.
12. Lorenz EN (1960) Energy and numerical weather prediction. Tellus 12(4):364–373
13. Alonso AA, Frouzakis CE, Kevrekidis IG (2004) Optimal sensor placement for state reconstruction of distributed process systems. AIChE J 50(7):1438–1452
14. García MR, Vilas C, Santos LO, Alonso AA (2012) A robust multi-model predictive controller for distributed parameter systems. J Process Control 22(1):60–71

15. Vilas C, Vande A (2011) Wouwer. combination of multi-model predictive control and the wave theory for the control of simulated moving bed plants. Chem Eng Sci 66:632–641
16. Berkooz G, Holmes P, Lumley L (1993) The proper orthogonal decomposition in the analysis of turbulent flows. Ann Rev Fluid Mech 25:539–575
17. Holmes P, Lumley JL, Berkooz G (1996) Turbulence, coherent structures, dynamical systems and symmetry. Cambridge University Press, Cambridge
18. García MR, Vilas C, Banga JR, Alonso AA (2007) Optimal field reconstruction of distributed process systems from partial measurements. Ind Eng Chem Res 46(2):530–539
19. Vilas C, García MR, Banga JR, Alonso AA (2008) Robust feed-back control of travelling waves in a class of reaction-diffusion distributed biological systems. Physica D: Nonlinear Phenomena 237(18):2353–2364
20. Balsa-Canto E, Alonso AA, Banga JR (2002) A novel, efficient and reliable method for thermal process design and optimization. Part II: applications. J Food Eng 52(3):235–247
21. Holmes PJ, Lumley JL, Berkooz G, Mattingly JC, Wittenberg RW (1997) Low-dimensional models of coherent structures in turbulence. Phys Rep 287(4):338–384
22. Ravindran SS (2000) A reduced-order approach for optimal control of fluids using proper orthogonal decomposition. Int. J Numer Meth Fluids 34(5):425–448
23. Smoller J (1994) Shock waves and Reaction-Diffusion equations, 2nd edn. Springer, New york
24. Reddy BD (1998) Introductory functional analysis: with applications to boundary value problems and finite elements. Springer, New York
25. Balsa-Canto E, Alonso AA, Banga JR (2004) Reduced-order models for nonlinear distributed process systems and their application in dynamic optimization. Ind Eng Chem Res 43(13):3353–3363
26. Buchanan GR (1995) Schaum's outlines of theory and problems of finite element analysis. McGraw Hill, New York
27. Courant R, Hilbert D (1937) Methods of mathematical physics. Wiley, New York
28. Lapidus L, Pinder GF (1999) Numerical solution of partial differential equations in science and engineering. Wiley, New York
29. Pozrikidis C (2005) Introduction to finite and spectral element methods using Matlab. Chapman & Hall/CRC, Boca Raton
30. Reddy JN (1993) Introduction to the finite element method, 2nd edn. McGraw Hill, New York
31. Sirovich L (1987) Turbulence and the dynamics of coherent structures. Part I: Coherent structures. Quaterly of Appl Math 45(3):561–571.
32. Trefethen LN (2000) Spectral methods in Matlab. SIAM

Chapter 5
How to Handle Steep Moving Fronts?

As we have seen in previous chapters, some IBVPs have solutions which can develop steep spatial gradients and in some cases, steep fronts traveling through the spatial domain. For instance, Burgers equation, which was our first PDE example in Chap. 1.

$$x_t = -xx_z + \mu x_{zz} \tag{5.1}$$

has a solution

$$x_a(z,t) = \frac{0.1e^a + 0.5e^b + e^c}{e^a + e^b + e^c} \tag{5.2}$$

with

$$a = \frac{0.05}{\mu}(z - 0.5 + 4.95t); \quad b = \frac{0.25}{\mu}(z - 0.5 + 0.75t); \quad c = \frac{0.5}{\mu}(z - 0.375)$$

which displays *front sharpening*, i.e., the solution front becomes sharper and sharper as it travels from left to right (see Fig. 5.1).

The numerical solution of such problems is delicate as classical discretization schemes will usually suffer from spurious oscillations (as already outlined in Chap. 3, when studying the convection-diffusion-reaction equation). The objective of this chapter is to introduce a few methods that can be useful in dealing with problems with steep moving fronts. We will first briefly discuss the concept of conservation laws and describe an analytical solution procedure called the method of characteristics, which can be applied to a certain class of problems (hyperbolic conservation laws). After the review of these theoretical concepts, which will allow us to have a better grasp of the origin of moving fronts and nonlinear waves, we will focus attention on several numerical schemes which can provide good numerical solutions, depending on the problem characteristics, among which:

- Upwind finite difference schemes (which have already been touched upon in Chap. 3)
- Alternating methods

A. Vande Wouwer et al., *Simulation of ODE/PDE Models with MATLAB®, OCTAVE and SCILAB*, DOI: 10.1007/978-3-319-06790-2_5,

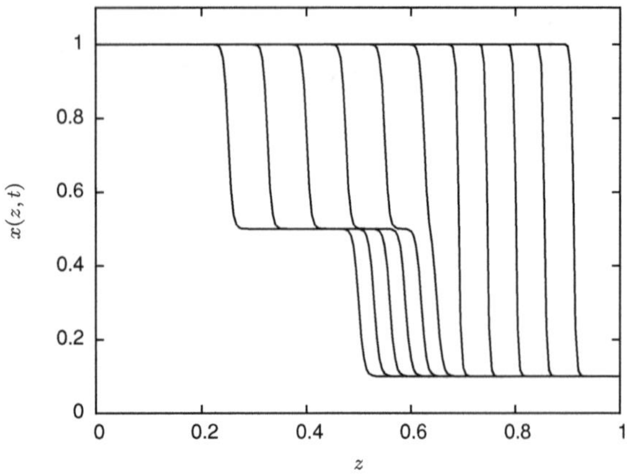

Fig. 5.1 Front sharpening effect in Burgers equation (with a viscosity $\mu = 0.001$)

- Finite volume methods, slope limiters, and centered schemes
- Moving grid methods.

The first category of methods are finite difference schemes on dedicated stencils (i.e., stencils oriented in the upwind direction). They can achieve satisfactory results for problems with mild fronts. The second group of methods use a *divide and conquer* approach by decomposing the original problem into subproblems that can be more easily handled by one of the previously introduced techniques (mixing the results of the method of characteristics with finite difference techniques). The third group is based on finite volume methods, i.e., they make use of cell averaging and significantly depart from the finite difference methods as they are nonlinear in essence. Finally, moving or adaptive grid methods use a spatial discretization grid with a nonuniform density, which is adapted so as to concentrate the grid points in regions of high spatial activity. The grid points can be moved continuously or only at discrete times, following some front tracking algorithm.

5.1 Conservation Laws

The conservation of mass, energy or momentum are fundamental principles that are at the root of the derivation of many PDEs. Let $x(z, t)$ be, for instance, the density of cars on a highway. The number of cars in the road segment $[z_L, z_R]$ is then given by

$$\int_{z_L}^{z_R} x(z, t)\mathrm{d}z$$

and the variation of this number of cars can be expressed as

$$\frac{\mathrm{d}}{\mathrm{d}t}\int_{z_L}^{z_R} x(z,t)\mathrm{d}z = \int_{z_L}^{z_R} \frac{\mathrm{d}x(z,t)}{\mathrm{d}t}\mathrm{d}z = f(x(z_L,t)) - f(x(z_R,t))$$

where the function $f(x)$ represents the flux of cars entering and exiting the highway. This expression can be rewritten in an integral form using the divergence theorem

$$\int_{z_L}^{z_R} \left[\frac{\mathrm{d}x(z,t)}{\mathrm{d}t} + \frac{\mathrm{d}f(x(z,t))}{\mathrm{d}z}\right] dz = 0$$

As this expression is valid whatever the road segment $[z_L, z_R]$ under consideration, the integrand must be exactly zero

$$\frac{\mathrm{d}x(z,t)}{\mathrm{d}t} + \frac{\mathrm{d}f(x(z,t))}{\mathrm{d}z} = 0$$

This latter PDE is in the form of a conservation law. It can be generalized for the case where there are sources or sinks in the spatial domain. For the car traffic example, this would correspond to highway entrances or exits:

$$\frac{\mathrm{d}x(z,t)}{\mathrm{d}t} + \frac{\mathrm{d}f(x(z,t))}{\mathrm{d}z} = S(x(z,t))$$

Let us consider some particular examples:
Example 1 car traffic with a velocity $v(x(z,t))$ depending on the car density (with reasonable drivers, one can imagine that the velocity of the cars decreases with increasing car density on the road). In this case, the flux $f(x) = v(x)x$ and

$$x_t + (v(x)x)_z = 0$$

If the car velocity is constant (independent of the car density), then the equation reduces to a simple advection equation

$$x_t + vx_z = 0$$

Example 2 Burgers equation. Up to now, we have written Burgers equation in the so-called *advection form* $x_t = -xx_z + \mu x_{zz}$. It is possible to express Burgers equation as a conservation law with $f(x) = 0.5x^2 - \mu x_z$:

$$x_t + (0.5x^2 - \mu x_z)_z = 0.$$

Actually, Burgers equation expresses the conservation of momentum. It is a simplified version of the Navier–Stokes equation, where it is assumed that the pressure is uniform and there is no external force. The inviscid Burgers equation corresponds to a zero kinematic viscosity $\mu = 0$, and reduces to a hyperbolic conservation law

$$x_t + (0.5x^2)_z = 0$$

Inviscid Burgers equation can be derived directly when considering a gas with free, noninteracting, particles.

Example 3 Wave equation. This equation models sound waves, water waves, and light waves. It is a second-order linear PDE with $f(x) = -c^2 x_z$

$$x_{tt} - (c^2 x_z)_z = 0.$$

5.2 The Methods of Characteristics and of Vanishing Viscosity

The method of characteristics [1] is an analytical procedure which allows the derivation of exact solutions for hyperbolic conservation laws, i.e., conservation laws where only first-order spatial derivatives appear, or in other words where there is no viscous, dissipative, terms.

The idea behind the method is to change coordinates from (z, t) to a new coordinate system (ξ, t) where the PDE becomes an ODE along certain curves (the characteristics) in the (z, t) plane. Each characteristic curve is associated to a particular value of ξ.

To introduce this method we consider the advection equation with constant velocity

$$x_t + v x_z = 0; \qquad x(z, 0) = x_0(z)$$

The total derivative of x with respect to time can be decomposed into

$$\frac{\mathrm{d}x}{\mathrm{d}t} = x_z \frac{\mathrm{d}z}{\mathrm{d}t} + x_t$$

If we introduce the characteristic equation

$$\frac{\mathrm{d}z}{\mathrm{d}t} = v$$

with the initial condition $z(0) = \xi$ (which represents a set of initial spatial locations), then the original PDE reduces to a simple ODE

$$\frac{\mathrm{d}x}{\mathrm{d}t} = 0; \qquad x(0) = x_0(\xi)$$

i.e., the solution $x(z, t)$ is constant along the characteristic curves $z = vt + \xi$, and can be expressed as $x(z, t) = x_0(\xi) = x_0(z - vt)$. Thus, the initial solution profile travels through the spatial domain at a constant velocity without change in shape. This is shown in Fig. 5.2 for $x_0(z) = \frac{e^{-100(z-0.1)}}{1+e^{-100(z-0.1)}}$ and $v = 2$.

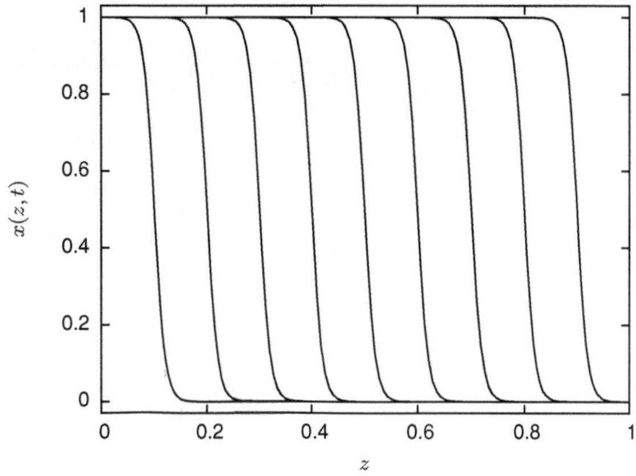

Fig. 5.2 Solution of the advection equation with $x_0(z) = e^{-100(z-0.1)}/1 + e^{-100(z-0.1)}$ and $v = 2$ at times $t = 0, 0.05, \ldots, 4$

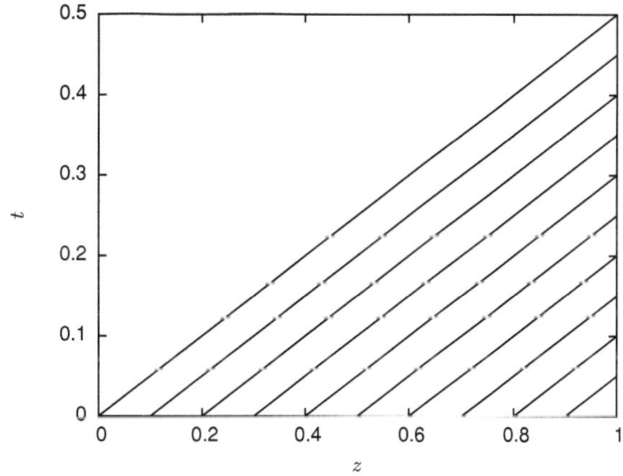

Fig. 5.3 Characteristic curves for the advection equation with $v = 2$

In a time-space diagram, such as Fig. 5.3, the characteristic curves can be rewritten as $t = (z - \xi)/2$. They are straight lines with slope $1/v$ (i.e., with slope $1/2$ in the considered example).

Figure 5.4 summarizes all the results by showing a 3D representation of the solution along with the characteristic lines. Clearly a point of the solution profile, which corresponds to a particular value of the dependent variable, moves along a characteristic line.

Let us now consider an advection equation with space-dependent velocity $v(z) = \beta z$:

$$x_t + v(z)x_z = 0; \qquad x(z, 0) = x_0(z)$$

The characteristic equation is now

$$\frac{\mathrm{d}z}{\mathrm{d}t} = v(z) = \beta z$$

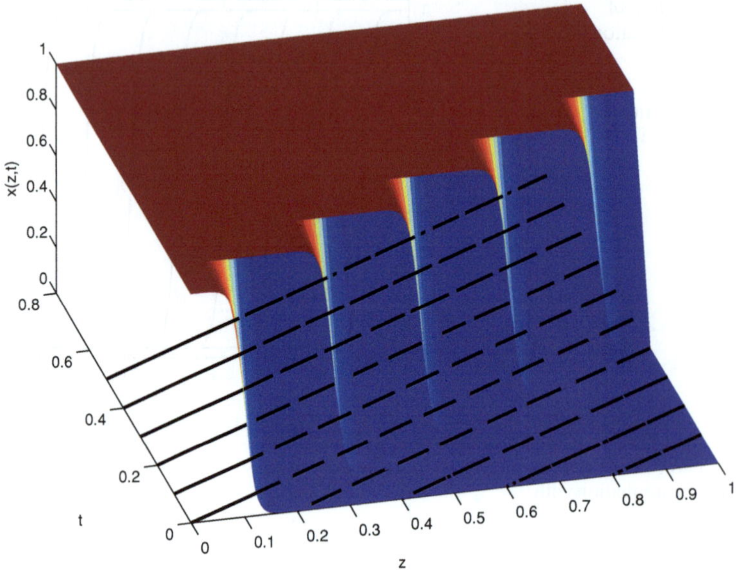

Fig. 5.4 3D representation of the solution of the advection equation with $v = 2$. *Black lines* in the plane (z, t) correspond with the characteristic curves

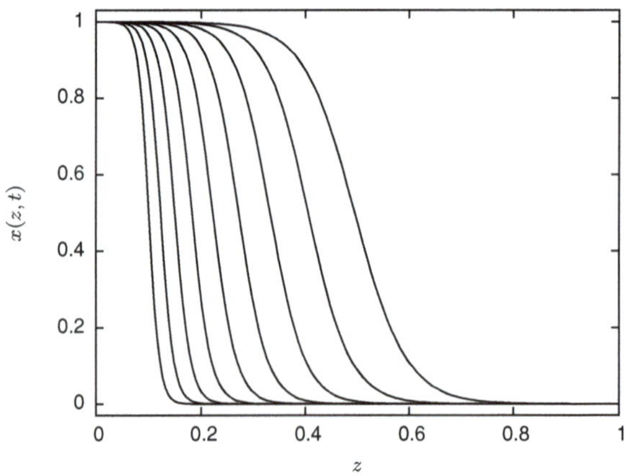

Fig. 5.5 Solution of the advection equation with $x_0(z) = e^{-100(z-0.1)}/1 + e^{-100(z-0.1)}$ and with $v = 2z$ at times $t = 0, 0.1, \ldots, 0.8$

with the initial condition $z(0) = \xi$, and solution $z = \xi e^{\beta t}$.

Again the solution is constant along the characteristic curves, which are no longer straight lines, and $x(z,t) = x_0(\xi) = x_0(ze^{-\beta t})$. The solution and characteristic curves are shown in Figs. 5.5 and 5.6.

Finally, we consider the inviscid Burgers equation

$$x_t + (0.5x^2)_z = 0; \qquad x(z,0) = x_0(z)$$

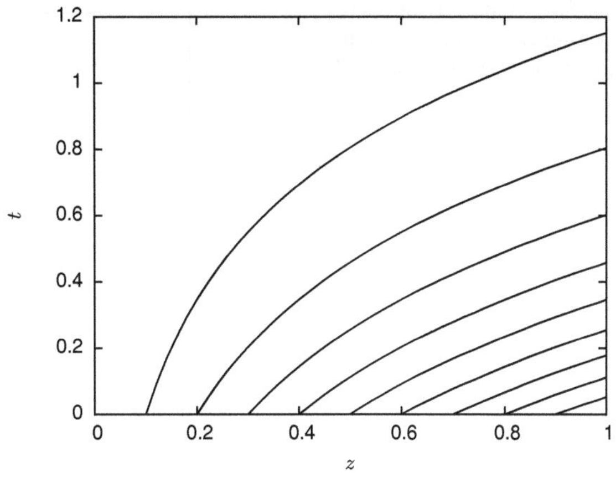

Fig. 5.6 Characteristic curves for the advection equation with $v = 2z$

The characteristic equation is now

$$\frac{\mathrm{d}z}{\mathrm{d}t} = x(z, t)$$

The solution is constant along these characteristic lines, as the total derivative of $x(z, t)$ is

$$\frac{\mathrm{d}x}{\mathrm{d}t} = x_z \frac{\mathrm{d}z}{\mathrm{d}t} + x_t = x_z \frac{\mathrm{d}z}{\mathrm{d}t} - (0.5x^2)_z = x_z \frac{\mathrm{d}z}{\mathrm{d}t} - x_z x = x_z \left(\frac{\mathrm{d}z}{\mathrm{d}t} - x \right) = 0$$

Considering these latter two equations,

$$\frac{\mathrm{d}z}{\mathrm{d}t} = x$$

$$\frac{\mathrm{d}x}{\mathrm{d}t} = 0$$

with the initial condition $z(0) = \xi$, then the solution is given by $z = x_0(\xi)t + \xi$.

The characteristic curves are straight lines as for the advection equation with constant velocity, but these lines are no longer parallel as they have different slopes $1/x_0(\xi)$. Depending on the initial condition, they can therefore intersect, as shown in Fig. 5.7, which is drawn with an initial condition computed using Eq. (5.2).

Before the characteristics intersect, the solution to the PDE problem is given by $x(z, t) = x_0(z - xt)$, which is in an implicit form and is therefore difficult to use. When the characteristics intersect, the method is no longer valid as it would lead to nonphysical, multivalued, functions. In fact, the solution parts with larger initial conditions move faster than the ones with smaller initial conditions, and a so-called *wave breaking phenomenon* occurs. The *breaking time* is easy to predict.

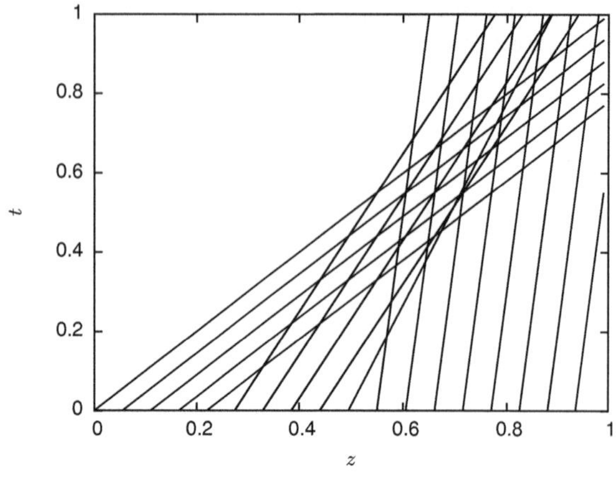

Fig. 5.7 Characteristic curves for Burgers equation with an initial condition based on (5.2)

If we consider two characteristic lines starting from z_i and z_j, they will intersect at time t_b when

$$z_i + x_0(z_i)t_b = z_j + x_0(z_j)t_b$$

so that

$$t_b = \frac{z_i - z_j}{x_0(z_j) - x_0(z_i)} = -\frac{\Delta z}{\Delta x_0}; \qquad \lim_{\Delta z \to 0}\left(-\frac{\Delta z}{\Delta x_0}\right) = -\frac{1}{x_{0,z}(z_i)}$$

In reality, there are not only two characteristic lines but a whole bunch of them, and the breaking time corresponds to the smallest intersection time, i.e., $t_b = \min_z \left(-1/x_{0,z}(z)\right)$. After the breaking time, discontinuities, also called *shocks*, may appear in the solution. To handle this phenomenon, an approach is to introduce in the equation a small dispersion term μx_{zz}, which counteracts the steepening effect of the nonlinear term. The presence of this dispersion term is natural in many realistic applications. In some sense, mathematical idealization leads to the inviscid Burgers equation, and the existence of solution displaying shocks. As $\mu \to 0$, we can expect that the solution to our modified PDE problem will approach the solution of the original inviscid problem. This approach is called the *vanishing viscosity method*, and has been proposed first by Oleinik [2] in one space dimension and extended by Kruzkov [3] in several space dimensions. This method ensures the uniqueness of the solution, and the shocks constructed by the vanishing viscosity method are physical shocks (in particular they satisfy the so-called *entropy condition* which states in mathematical terms that the entropy of the solution is a nondecreasing quantity, as thermodynamics tells us - more on the entropy condition can be found for instance in the nice book by LeVeque [4]).

5.3 Transformation-Based Methods

The consideration of a dispersion term in Burgers equation leads to the equation already introduced, i.e.,

$$x_t + (0.5x^2 - \mu x_z)_z = 0$$

which can be solved analytically using specific transformations. The earliest method is the *Cole-Hopf transformation* [5]. The idea is to introduce a transformation which eliminates the nonlinearity. We consider the function

$$\varphi(z,t) = \exp\left(-\frac{1}{2\mu}\int x dz\right)$$

so that

$$x(z,t) = -2\mu\frac{\varphi_z}{\varphi}$$

Substituting in Burgers equation, and evaluating term by term

$$x_t = -\frac{2\mu(\varphi_{z,t}\varphi - \varphi_z\varphi_t)}{\varphi^2}; \qquad x x_z = \frac{4\mu^2\varphi_z(\varphi_{zz}\varphi - \varphi_z^2)}{\varphi^3};$$

$$\mu x_{zz} = -\frac{2\mu^2(2\varphi_z^3 - 3\varphi\varphi_{zz}\varphi_z + \varphi^2\varphi_{zzz})}{\varphi^3}$$

gives

$$-\varphi\varphi_{z,t} + \varphi_z(\varphi_t - \mu\varphi_{zz}) + \mu\varphi\varphi_{zzz} = 0 \Rightarrow$$

$$\varphi_z(\varphi_t - \mu\varphi_{zz}) = \varphi\varphi_{z,t} - \mu\varphi\varphi_{zzz} = \varphi(\varphi_{z,t} - \mu\varphi_{zzz}) = \varphi(\varphi_t - \mu\varphi_{zz})_z$$

If $\varphi(z,t)$ is solution to the heat equation $\varphi_t - \mu\varphi_{zz} = 0$, then $x(z,t)$ is solution of Burgers equation. The original problem is thus transformed into

$$\varphi_t = \mu\varphi_{zz}; \qquad \varphi(z,t=0) = \exp\left(-\frac{1}{2\mu}\int_0^z x(\xi,t=0)d\xi\right) = \varphi_0(z)$$

This latter problem is easy to solve using Fourier transform (in the previous chapter we have already seen how to solve the heat equation using Fourier series). Let us denote by F the Fourier transform applied to φ considered as a function of the space variable:

$$F(\varphi(z,t)) = \hat{\varphi}(\omega,t) = \frac{1}{\sqrt{2\pi}}\int_{-\infty}^{\infty}\varphi(z,t)\exp(-i\omega z)dz$$

The heat equation can be transformed to

$$F(\varphi_t) = \frac{\mathrm{d}F(\varphi)}{\mathrm{d}t} = \hat{\varphi}_t$$
$$F(\mu\varphi_{zz}) = \mu(i\omega)^2 F(\varphi) = -\mu\omega^2\hat{\varphi}$$

so that the PDE problem is transformed into an ODE problem

$$\hat{\varphi}_t = -\mu\omega^2\hat{\varphi}$$
$$\hat{\varphi}(\omega, t=0) = \hat{\varphi}_0(\omega)$$

whose solution is simply $\hat{\varphi}(\omega, t) = \hat{\varphi}_0(\omega)\exp(-\mu\omega^2 t)$.

The inverse transform then gives the sought solution

$$\varphi(z,t) = F^{-1}(\hat{\varphi}(\omega,t)) = F^{-1}(\hat{\varphi}_0(\omega)\exp(-\mu\omega^2 t)) = \varphi_0(z) * F^{-1}(\exp(-\mu\omega^2 t))$$
$$= \frac{1}{\sqrt{4\pi\mu t}}\int\limits_{-\infty}^{+\infty} \varphi_0(s)\exp\left(-\frac{(z-s)^2}{4\mu t}\right)\mathrm{d}s$$

where $*$ denotes the convolution product, and

$$F^{-1}(\exp(-\mu\omega^2 t)) = g(z,t) = \frac{1}{\sqrt{2\pi}}\frac{1}{\sqrt{2\mu t}}\exp\left(-\frac{z^2}{4\mu t}\right)$$

Note that this latter function, $g(z,t)$, also called a *heat kernel*, is the solution of the heat equation when considering an initial heat point source (an initial condition defined as a Dirac delta function). The solution of the heat equation to an arbitrary initial condition is the convolution product of this initial condition with the heat kernel.

Now that we have the solution to the heat equation, we can return to the solution of Burgers equation applying the transformation in the reverse way

$$x(z,t) = \frac{\displaystyle\int_{-\infty}^{+\infty} \frac{z-s}{t}\varphi_0(s)\exp\left(-\frac{(z-s)^2}{4\mu t}\right)\mathrm{d}s}{\displaystyle\int_{-\infty}^{+\infty} \varphi_0(s)\exp\left(-\frac{(z-s)^2}{4\mu t}\right)\mathrm{d}s}$$

This approach is elegant, and not restricted to Burgers equation (see [6] for the extension of this approach to a class of nonlinear parabolic and hyperbolic equations, and [7] for the derivation of solution to a seventh-order Korteweg-de-Vries equation). The integrals appearing in the exact solution are however not trivial to evaluate. Numerical schemes have been proposed, for instance by Sakai and Kimura [8]. Other transformations have been introduced, such as the *tanh method* [9] or the *exp-*

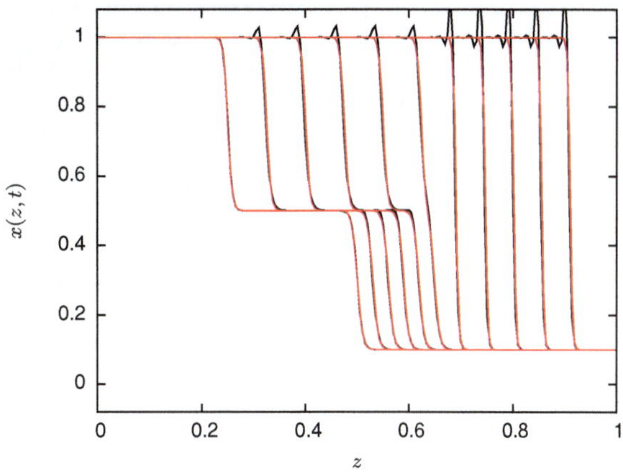

Fig. 5.8 Numerical solution of Burgers equation ($\mu = 0.001$) at times $t = 0, 0.1, \ldots, 1$ with $N = 200$ and a 3-point centered FD scheme for the first spatial derivative and a 5-point centered FD scheme for the second spatial derivative. *Red curves* exact solution. *Black curves* numerical solution

function method [10], which have allowed several exact solutions to be obtained [11]. These methods are particularly convenient for an implementation using a symbolic manipulation language [12].

Here, we will not follow this direction, which provides very interesting insight in the solution of nonlinear wave problems, but rather turn back to general purpose numerical techniques, and first have a look at what classical numerical methods can give.

5.4 Upwind Finite Difference and Finite Volume Schemes

We consider the viscid Burgers equation in the form

$$x_t = -x x_z + \mu x_{zz}$$

with a viscosity $\mu = 0.001$ and an initial condition as in Fig. 5.1, and we first apply several finite difference schemes as introduced in Chap. 3. It is important to recall that first-order derivative terms are better approximated using upwind schemes, whereas centered schemes are appropriate for second-order spatial derivatives. Indeed, if we use a 3-point centered FD scheme for the first spatial derivative and a 5-point centered FD scheme for the second spatial derivative (on a grid with 201 nodes) then we get the results of Fig. 5.8. The front sharpening effect is relatively well captured, but spurious oscillations appear at the front edges. If a 2-point upwind scheme is applied to the convective term, the picture changes to Fig. 5.9. Oscillations have disappeared, but significant *numerical dissipation and dispersion* has been introduced, i.e., the fronts are smoothed off artificially, and eventually the speed of the front wave is not well-estimated leading to a significant phase shift of the fronts.

The wrong estimation of the wave velocity can be alleviated by considering the equation in conservation form, i.e.,

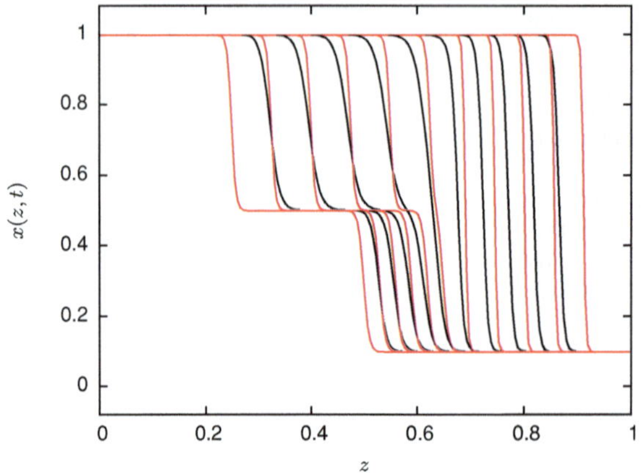

Fig. 5.9 Numerical solution of Burgers equation ($\mu = 0.001$) at times $t = 0, 0.1, \ldots, 1$ with $N = 200$ and 2-point upwind FDs for first-order spatial derivatives and 5-point centered FDs for second-order spatial derivatives. *Red curves* exact solution. *Black curves* numerical solution

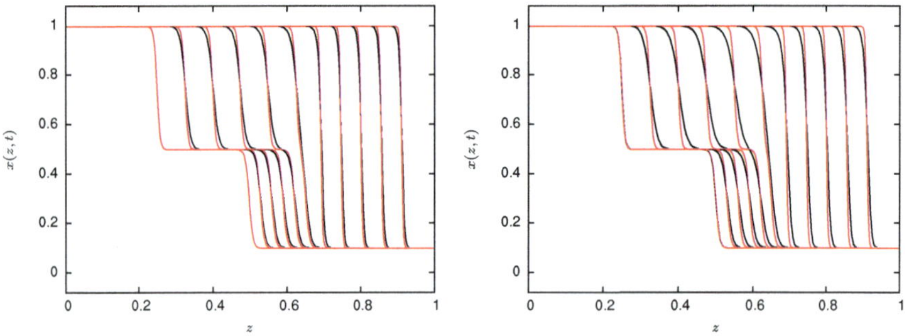

Fig. 5.10 Numerical solution of Burgers equation ($\mu = 0.001$) in conservation form at times $t = 0, 0.1, \ldots, 1$ with $N = 200$ (*right*) and $N = 400$ (*left*), 2-point upwind FDs for first-order spatial derivatives and 5-point centered FDs for second-order spatial derivatives. *Red curves* exact solution. *Black curves* numerical solution

$$x_t = (-0.5x^2)_z + \mu x_{zz}$$

and applying numerical discretization directly to the corresponding terms. The much improved results are shown in Fig. 5.10.

The use of numerical schemes written in conservation form, i.e.,

$$\frac{\mathrm{d}x_i}{\mathrm{d}t} = -\frac{F(x_{i+p}, \ldots, x_{i-q}) - F(x_{i+p-1}, \ldots, x_{i-q-1})}{\Delta z}$$

where F represents a numerical approximation of the flux function, appears to be of paramount importance to obtain a satisfactory solution, with the correct speed.

The formulation of the discretization scheme in conservation form appears naturally when deriving a basic *finite volume scheme*. The idea is to compute cell averages by integrating the PDE over a cell extending between intermediate points $z_{i-1/2}$ and $z_{i+1/2}$ (our grid points z_i are the centers of each of these cells).

$$\int_{z_{i-1/2}}^{z_{i+1/2}} x_t \mathrm{d}z = \Big[-0.5x^2 + \mu x_z\Big]_{z_{i-1/2}}^{z_{i+1/2}} = [-f]_{z_{i-1/2}}^{z_{i+1/2}}$$

If x_t is continuous on the interval $[z_{i-1/2}, z_{i+1/2}]$ then the mean-value theorem allows us to write

$$\int_{z_{i-1/2}}^{z_{i+1/2}} x_t \mathrm{d}z = x_t(\xi_i, t)\Delta z = x_t(z_i, t)\Delta z + O(\Delta z^2)$$

where ξ_i is a point in the interval $\left[z_{i-1/2}, z_{i+1/2}\right]$, which is approximately taken as the mid-point z_i. Thus,

$$x_t(z_i, t) = \frac{[-f]_{z_{i-1/2}}^{z_{i+1/2}}}{\Delta z} = \frac{f(z_{i-1/2}) - f(z_{i+1/2})}{\Delta z} + O(\Delta z)$$

To evaluate the terms at the cell interfaces $z_{i-1/2}$ and $z_{i+1/2}$ we can use averages based on the values in the grid points z_{i-1}, z_i, z_{i+1}, as well as a centered approximation of derivatives, e.g.,

$$x_z(z_{i+1/2}, t) = \frac{x(z_{i+1}) - x(z_i)}{\Delta z} + O(\Delta z^2)$$
$$x_z(z_{i-1/2}, t) = \frac{x(z_i) - x(z_{i-1})}{\Delta z} + O(\Delta z^2)$$

so that

$$\begin{aligned}
x_t(z_i, t) &= 0.5\frac{x^2(z_{i-1/2}) - x^2(z_{i+1/2})}{\Delta z} + \mu\frac{x_z(z_{i+1/2}) - x_z(z_{i-1/2})}{\Delta z} + O(\Delta z)\\
&= 0.5\frac{x^2(z_{i-1}) + x^2(z_i) - x^2(z_i) - x^2(z_{i+1})}{2\Delta z}\\
&\quad + \frac{\mu}{\Delta z}\left(\frac{x(z_{i+1}) - x(z_i)}{\Delta z} - \frac{x(z_i) - x(z_{i-1})}{\Delta z}\right) + O(\Delta z)\\
&= 0.5\frac{x^2(z_{i-1}) - x^2(z_{i+1})}{2\Delta z} + \mu\frac{x(z_{i+1}) - 2x(z_i) + x(z_{i-1})}{\Delta z^2} + O(\Delta z)
\end{aligned}$$

This scheme is quite simple, the main difference with the previous FD schemes being that the nonlinear flux function is evaluated at intermediate points. This scheme can be applied for $i = 1, \ldots, N-1$, which corresponds to entire cells. At the boundaries, $i = 0$, and $i = N$, the BCs inspired from the exact solution can be applied. Figure 5.11 shows results with $N = 400$ points.

As apparent in Figs. 5.10 and 5.11, numerical dissipation can be decreased using additional grid points (and the finite volume scheme performs a bit better in this

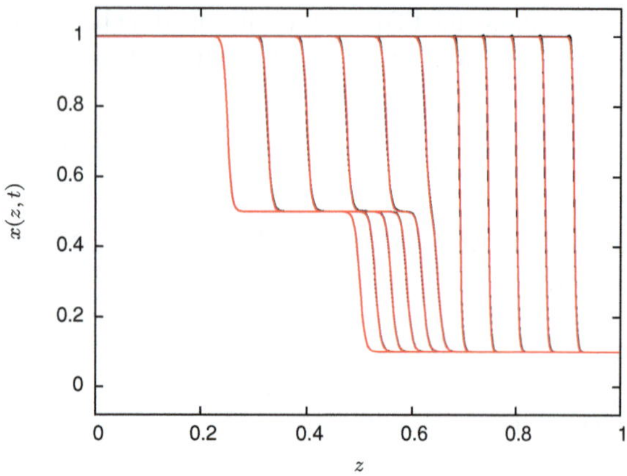

Fig. 5.11 Numerical solution of Burgers equation ($\mu = 0.001$) at times $t = 0, 0.1, \ldots, 1$ with $N = 400$, and a first-order accurate finite volume scheme. *Red curves* exact solution. *Black curves* numerical solution

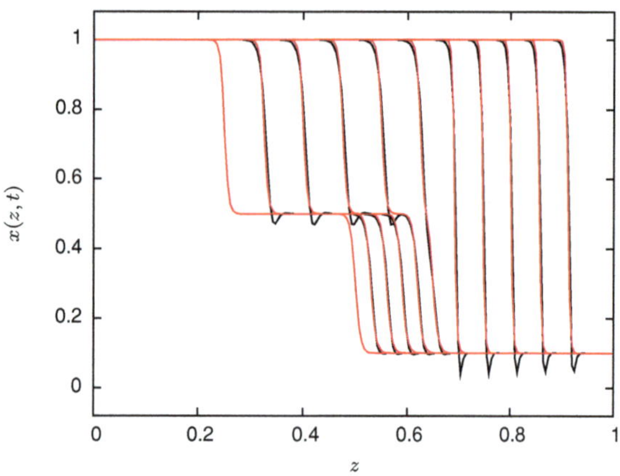

Fig. 5.12 Numerical solution of Burgers equation ($\mu = 0.001$) at times $t = 0, 0.1, \ldots, 1$ with $N = 200$, 3-point upwind FDs for first-order spatial derivatives, and 5-point centered FDs for second-order spatial derivatives. *Red curves* exact solution. *Black curves* numerical solution

respect). Alternatively higher order schemes can be used. Figure 5.12 shows results with a 3-point upwind finite difference scheme and $N = 200$. Numerical dissipation is much smaller than with a 2-point upwind scheme, but spurious oscillations appear.

Better results can be obtained with a 5-point biased-upwind FD scheme, as shown in Fig. 5.13.

This approach can cope with smaller viscosities but will be limited by the presence of spurious oscillations. If we consider for instance a viscosity $\mu = 0.0000001$, which is a much harder problem, Fig. 5.14 shows the results with $N = 700$.

5.5 A Divide and Conquer Approach

A general approach when tackling a difficult problem, is whenever possible to decompose it into a sequence of easier tasks. This is the underlying idea of the so-called *operator splitting methods*.

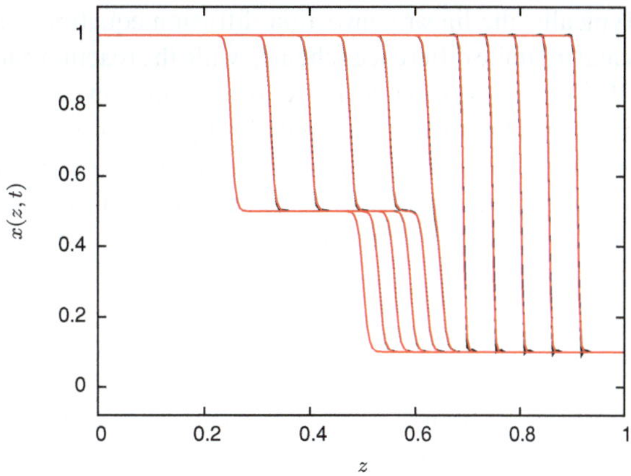

Fig. 5.13 Numerical solution of Burgers equation ($\mu = 0.001$) at times $t = 0, 0.1, \ldots, 1$ with $N = 200$, 5-point upwind FDs for first-order spatial derivatives, and 5-point centered FDs for second-order spatial derivatives. *Red curves* exact solution. *Black curves* numerical solution

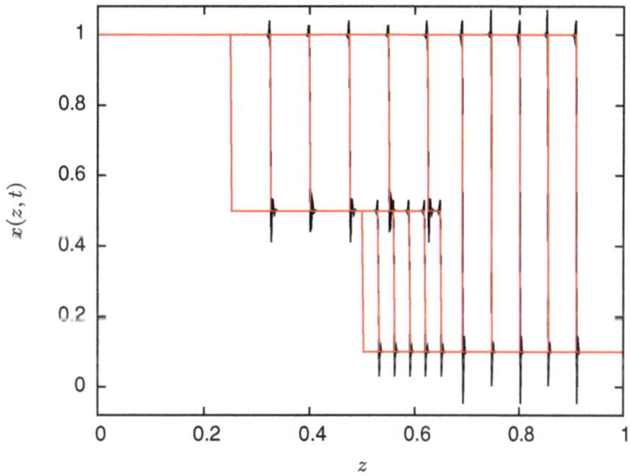

Fig. 5.14 Numerical solution of Burgers equation ($\mu = 0.0000001$) at times $t = 0, 0.1, \ldots, 1$ with $N = 700$, 5-point upwind FDs for first-order spatial derivatives and 5-point centered FDs for second-order spatial derivatives. *Red curves* exact solution. *Black curves* numerical solution

The rationale behind operator splitting, also called *time splitting* methods or *fractional step* methods, is to split the original convection-reaction-diffusion PDE studied in Chap. 3.

$$x_t = -v x_z + D x_{zz} + r(x)$$

into different subproblems which are solved sequentially within each time step, in order to take advantage of solution techniques that are adapted to each of the different subproblems [13]. This is true for the spatial discretization techniques, but also for time integration. In general, splitting algorithms can be classified into two- and three-step procedures. Traditional two-step schemes involve the isolation of the reaction term, yielding a linear convection-diffusion PDE, and a system of nonlinear ODEs:

1. $x_t = -v x_z + D x_{zz}$
2. $x_t = r(x)$.

Typically, the linear convection-diffusion equation is solved using the MOL with a standard finite difference scheme, while the reaction part is solved with an appropriate ODE integrator. Chemical reaction systems can be very stiff so that an implicit solver might be required for the second subproblem. However, for convection-dominated problems, this strategy can be computationally expensive due to the fine grid required for solving the convection-diffusion PDE. As an alternative, [14] has proposed to consider first a linear diffusion equation, followed by a nonlinear convection-reaction equation,

1. $x_t = Dx_{zz}$
2. $x_t = -vx_z + r(x)$.

The heat equation is an easy problem that can be solved by a variety of methods (Fourier series, Fourier transform, finite differences, finite elements), but the nonlinear hyperbolic convection-reaction PDE can be significantly more difficult to solve.

Three-step splitting algorithms separate all phenomena, i.e., diffusion, convection and reaction, yielding two linear PDEs, and a nonlinear system of ODEs

1. $x_t = Dx_{zz}$
2. $x_t = -vx_z$
3. $x_t = r(x)$.

Each of these steps can be solved by a dedicated technique.

A question naturally arises on how to select a particular sequence and on its potential influence on the solution. In [13, 15], necessary conditions for commutativity are given:

- Convection commutes with diffusion, if the velocity v and the diffusion coefficient D do not explicitly depend on the spatial coordinate z.
- Convection commutes with reaction, if the velocity v and the reaction $r(x)$ do not explicitly depend on the spatial coordinate z.
- Diffusion commutes with reaction, if the reaction $r(x)$ is linear in x, and independent of the spatial coordinate z.

In general, the first two necessary conditions are satisfied. The third condition is, however, often not fulfilled, which may lead to different solutions for different sequences. However, the correct sequence can be found by comparison with a low-order finite differences method.

As a particular implementation of this latter approach, we will consider the sequencing method proposed by Renou et al. [16]. This method uses the above-mentioned 3-step philosophy, in the following order:

1. $x_t = -vx_z$
2. $x_t = r(x)$
3. $x_t = Dx_{zz}$.

At time $t = 0$, the convection PDE is solved for a time step Δt with as initial condition the original profile $x_0(z)$, yielding a profile $x^*(z)$. This latter profile is then used as initial condition for solving the reaction ODE over time step Δt, resulting in a profile $x^{**}(z)$. This new profile is again used as an initial condition, but now for the diffusion PDE. This provides the final solution $x^{***}(z)$ for the current time interval. This solution is used as starting point to repeat the three-step solution procedure on the next time step from Δt to $2\Delta t$, and so on.

Back to numerical methods, the first step, i.e., the solution of the advection equation can be easily determined using the method of characteristics described in Sect. 5.2. With a constant velocity v, the initial profile is simply moved through the spatial domain of a distance $v\Delta t$ without change of shape. This corresponds to a transportation lag. The second step is the solution of a system of ordinary differential equations. If the reaction rates are linear, this system could even be solved analytically. If they are nonlinear, then the full range of solvers is available from a simple explicit Euler method to, for instance, an implicit Rosenbrock method for stiff ODEs. The last step is the solution of the heat equation which can be achieved using Fourier transformation, by convolution of the initial condition with the heat kernel. In [16] another approach is preferred, which is based on the discretization of the heat equation using a finite difference matrix $\mathbf{D}_2$.

$$\dot{\mathbf{x}} = D\mathbf{x}_{zz} = D(\mathbf{D}_2\mathbf{x}); \qquad \mathbf{x}(t_0) = \mathbf{x}^{**}$$

and an analytic solution of the resulting linear system of ODEs

$$\mathbf{x}(t_0 + \Delta t) = \exp\left(D(\mathbf{D}_2\mathbf{x})\Delta t\right)\mathbf{x}^{**}$$

Note that the transition matrix $\phi(t_0, t_0 + \Delta t) = \phi(\Delta t) = \exp\left(D(\mathbf{D}_2\mathbf{x})\Delta t\right)$ does not depend on t_0, and once a discretization matrix $\mathbf{D}_2$ and a time step size Δt have been chosen, can be computed once and for all. Hence the solution of the diffusion equation on a time step, can be simply implemented by a vector-matrix multiplication

$$\mathbf{x}(t_0 + \Delta t) = \phi\mathbf{x}^{**}$$

To illustrate the use of this solution procedure, we consider a fixed bed bioreactor [17], for instance an anaerobic digester for waste water treatment. Bacterial population forming a biomass X $[g/L]$ is attached to the reactor bed of length L $[m]$, and grows on a limiting substrate S $[g/L]$ that is fed at the reactor inlet:

$$\frac{\partial S}{\partial t} = -v\frac{\partial S}{\partial z} + D\frac{\partial^2 S}{\partial z^2} - k\mu(S,X)X$$

$$\frac{\partial X}{\partial t} = -\mu(S,X)X - k_d X$$

where the specific growth rate $[1/h]$ involves substrate activation as well as both biomass and substrate inhibition at higher concentration:

$$\mu(S,X) = \mu_0 \frac{S}{K_X X + S + S^2/K_S}$$

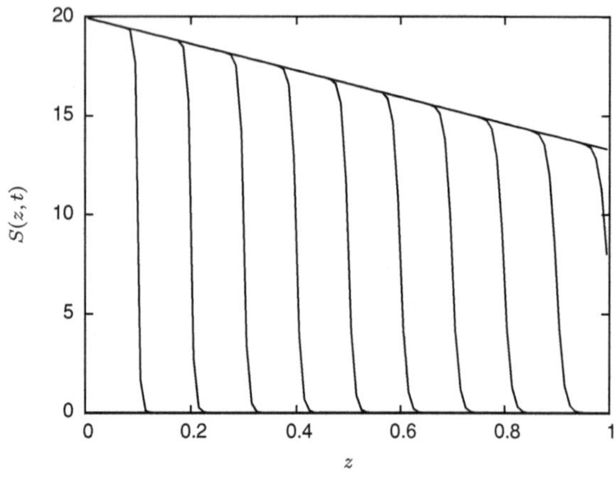

Fig. 5.15 Numerical solution of the fixed bed bioreactor example at times $t = 0, 0.1, 0.2, \ldots$ for $v = 1$ m/h, $D = 0.0001\ m^2/h$, with $N = 100$, 3-point centered FDs for second-order spatial derivatives, and `ode45`

In this expression, K_X is the biomass inhibition constant and K_S the substrate inhibition constant. The biomass growth is partly counterbalanced by a decay phenomenon (mortality and detachment from the bed) with rate k_d.

The boundary and initial conditions are given by

$$D\frac{\partial S}{\partial z}(0,t) = v\,(S(0,t) - S_{in})$$

$$D\frac{\partial S}{\partial z}(L,t) = 0$$

$$S(z,0) = S_0(z) = S_{in} = 20\ \ g/l$$

$$X(z,0) = X_0(z) = 300\ \ g/l$$

Figure 5.15 shows the evolution of the substrate profile at times $t = 0, 0.1, \ldots$ corresponding to a Peclet number $Pe = vL/D = 10000$, i.e., a convection-dominated problem. The procedure uses 100 grid points, a 3-point centered scheme for finite difference approximation, and the explicit solver `ode45`. The sequencing method is extremely powerful, providing accurate results with very modest computational load. A more detailed evaluation of the method, including additional tests and comparisons with alternative methods can be found in [18]. Codes for various examples, including the fixed-bioreactor, are available in the companion software.

5.6 Finite Volume Methods and Slope Limiters

As we have seen in the previous sections, the solution of conservation laws, such as Burgers equation

$$x_t = (-0.5x^2)_z + \mu x_{zz}$$

is very challenging as the solution can develop a front sharpening phenomenon, and classical methods are unable to resolve the steep slopes without excessive numerical dissipation or spurious oscillations.

The question that comes to mind is therefore: is it possible to preserve the monotonic character of the solution, i.e., avoid the appearance of spurious local extrema and the amplification of existing local maxima/minima, without introducing too much artificial dissipation?

The starting point to answer this question is an important result established by Lax in 1973 [19] who showed that for a scalar conservation law

$$\frac{\partial x}{\partial t} + \frac{\partial f(x)}{\partial z} = 0$$

the total variation of any physically possible solution

$$TV = \int \left\| \frac{\partial x}{\partial z} \right\| dz$$

does not increase with time.

If we consider a discrete-space solution

$$TV(x) = \sum_i \|x_{i+1} - x_i\|$$

a numerical scheme is said *total variation diminishing* if the TV of the numerical solution is not growing with time, i.e.,

$$TV(x^{k+1}) \leqslant TV(x^k)$$

If a numerical scheme preserving the solution monotonicity is called *monotone*, then [20] showed that a monotone scheme is TVD, and a TVD scheme is monotonicity preserving.

In fact, the simple 2-point upwind finite difference and finite volume schemes of Sect. 5.4 are first-order TVD schemes which can be used to solve, for instance, Burgers equation but introduces ample numerical dissipation. The development of higher order TVD schemes is based on the finite volume discretization that we will now derive in more details. For this, we come back to the scalar conservation law

$$x_t + f_z = 0$$

and integrate it over time between t_0 and t to obtain

$$x(z,t) = x(z,t_0) - \int_{t_0}^{t} f_z \mathrm{d}t$$

We now integrate this latter expression over a finite volume (usually called cell) $[z_{i-1/2}, z_{i+1/2}]$ and define the cell average value $\bar{x}_i$ as

$$\begin{aligned}
\bar{x}_i(t) &= \frac{1}{\Delta z}\int_{z_i-\frac{\Delta z}{2}}^{z_i+\frac{\Delta z}{2}} x(z,t)\mathrm{d}z = \frac{1}{\Delta z}\int_{z_i-\frac{\Delta z}{2}}^{z_i+\frac{\Delta z}{2}} \left[x(z,t_0) - \int_{t_0}^{t} f_z \mathrm{d}t\right]\mathrm{d}z \\
&= \bar{x}_i(t_0) - \frac{1}{\Delta z}\int_{t_0}^{t}\left[f\left(z_i + \frac{\Delta z}{2}\right) - f\left(z_i - \frac{\Delta z}{2}\right)\right]\mathrm{d}t
\end{aligned}$$

Differentiating with respect to time gives

$$\frac{\partial \bar{x}_i(t)}{\partial t} = \frac{f(z_{i-1/2}) - f(z_{i+1/2})}{\Delta z}$$

where the values of the flux at the cell interfaces have to be refined in order to achieve a higher accuracy (this is a major difference with finite difference schemes where higher order approximation is achieved by increasing the stencil of the formula).

Note that in Sect. 5.4, we have used $x(z_i)$, which is the value of the dependent variable in the mid-point of the cell $[z_{i-1/2}, z_{i+1/2}]$, and can be viewed as a first-order approximation of a cell-average value

$$x(z_i) = \frac{1}{\Delta z}\int_{z_i-\frac{\Delta z}{2}}^{z_i+\frac{\Delta z}{2}} x(z)\mathrm{d}z + O(\Delta z) = \bar{x}_i(t) + O(\Delta z)$$

To get a more detailed picture of the error term (represented by the big O), we replace the dependent variable by its Taylor series development

$$x(z) = x(z_i) + (z - z_i)\left.\frac{\mathrm{d}x}{\mathrm{d}z}\right|_{z_i} + \frac{(z-z_i)^2}{2!}\left.\frac{\mathrm{d}^2 x}{\mathrm{d}z^2}\right|_{z_i} + \cdots$$

and obtain

$$\begin{aligned}
\bar{x}_i &= \frac{1}{\Delta z}\int_{z_i-\frac{\Delta z}{2}}^{z_i+\frac{\Delta z}{2}} x(z)\mathrm{d}z = \frac{1}{\Delta z}\int_{z_i-\frac{\Delta z}{2}}^{z_i+\frac{\Delta z}{2}}\left[x(z_i) + (z-z_i)\left.\frac{\mathrm{d}x}{\mathrm{d}z}\right|_{z_i} + \frac{(z-z_i)^2}{2!}\left.\frac{\mathrm{d}^2x}{\mathrm{d}z^2}\right|_{z_i} + \cdots\right]\mathrm{d}z \\
&= x(z_i) + \frac{\Delta z^2}{24}\left.\frac{\mathrm{d}^2 x}{\mathrm{d}z^2}\right|_{z_i} + \cdots
\end{aligned}$$

We see more clearly the distinction between $\bar{x}_i$, the cell-average value, and $x(z_i)$, the mid-cell value (the two are exactly the same if the solution profile is a straight line, but they will in general differ depending on the solution curvature). Let us now eliminate $x(z_i)$ in the Taylor series

$$x(z) = \bar{x}_i + (z - z_i) \left.\frac{dx}{dz}\right|_{z_i} + \left(\frac{(z - z_i)^2}{2!} - \frac{\Delta z^2}{24}\right) \left.\frac{d^2x}{dz^2}\right|_{z_i} + \cdots$$

and replace the derivatives in this latter equation by well-known finite difference formulas

$$\left.\frac{dx}{dz}\right|_{z_i} = \frac{x(z_{i+1}) - x(z_{i-1})}{2\Delta z} - \frac{\Delta z^2}{3!} \left.\frac{d^3x}{dz^3}\right|_{z_i} + \cdots$$

$$\left.\frac{d^2x}{dz^2}\right|_{z_i} = \frac{x(z_{i+1}) - 2x(z_i) + x(z_{i-1})}{\Delta z^2} - \frac{\Delta z^2}{12} \left.\frac{d^4x}{dz^4}\right|_{z_i} + \cdots$$

that we would like to also express in terms of cell-average values

$$\bar{x}_{i-1} = x(z_{i-1}) + \frac{\Delta z^2}{24} \left.\frac{d^2x}{dz^2}\right|_{z_{i-1}} + \cdots$$

$$\bar{x}_i = x(z_i) + \frac{\Delta z^2}{24} \left.\frac{d^2x}{dz^2}\right|_{z_i} + \cdots$$

$$\bar{x}_{i+1} = x(z_{i+1}) + \frac{\Delta z^2}{24} \left.\frac{d^2x}{dz^2}\right|_{z_{i+1}} + \cdots$$

The first formula thus becomes

$$\left.\frac{dx}{dz}\right|_{z_i} = \frac{x(z_{i+1}) - x(z_{i-1})}{2\Delta z} - \frac{\Delta z^2}{3!} \left.\frac{d^3x}{dz^3}\right|_{z_i} + \cdots$$

$$= \frac{\bar{x}_{i+1} - \bar{x}_{i-1}}{2\Delta z} - \frac{\Delta z^2}{48} \left(\left.\frac{d^2x}{dz^2}\right|_{z_{i+1}} - \left.\frac{d^2x}{dz^2}\right|_{z_{i-1}} \right) - \frac{\Delta z^2}{3!} \left.\frac{d^3x}{dz^3}\right|_{z_i} + \cdots$$

whereas the second one can be written as

$$\left.\frac{d^2x}{dz^2}\right|_{z_i} = \frac{x(z_{i+1}) - 2x(z_i) + x(z_{i-1})}{\Delta z^2} - \frac{\Delta z^2}{12} \left.\frac{d^4x}{dz^4}\right|_{z_i} + \cdots$$

$$= \frac{\bar{x}_{i+1} - 2\bar{x}_i + \bar{x}_{i-1}}{\Delta z^2} - \frac{1}{24} \left(\left.\frac{d^2x}{dz^2}\right|_{z_{i+1}} - 2 \left.\frac{d^2x}{dz^2}\right|_{z_i} + \left.\frac{d^2x}{dz^2}\right|_{z_{i-1}} \right) - \frac{\Delta z^2}{12} \left.\frac{d^4x}{dz^4}\right|_{z_i} + \cdots$$

These latter expressions can be simplified by considering the Taylor series developments

$$\left.\frac{d^2x}{dz^2}\right|_{z_{i-1}} = \left.\frac{d^2x}{dz^2}\right|_{z_i} - \Delta z \left.\frac{d^3x}{dz^3}\right|_{z_i} + \frac{\Delta z^2}{2!}\left.\frac{d^4x}{dz^4}\right|_{z_i}$$

$$\left.\frac{d^2x}{dz^2}\right|_{z_{i+1}} = \left.\frac{d^2x}{dz^2}\right|_{z_i} + \Delta z \left.\frac{d^3x}{dz^3}\right|_{z_i} + \frac{\Delta z^2}{2!}\left.\frac{d^4x}{dz^4}\right|_{z_i}$$

giving

$$\begin{aligned}\left.\frac{dx}{dz}\right|_{z_i} &= \frac{\bar{x}(z_{i+1}) - \bar{x}(z_{i-1})}{2\Delta z} - \frac{5\Delta z^2}{24}\left.\frac{d^3x}{dz^3}\right|_{z_i} + \cdots = \frac{\bar{x}(z_{i+1}) - \bar{x}(z_{i-1})}{2\Delta z} + O(\Delta z^2)\\ \left.\frac{d^2x}{dz^2}\right|_{z_i} &= \frac{\bar{x}(z_{i+1}) - 2\bar{x}(z_i) + \bar{x}(z_{i-1})}{\Delta z^2} - \frac{\Delta z^2}{8}\left.\frac{d^4x}{dz^4}\right|_{z_i} + \cdots\\ &= \frac{\bar{x}(z_{i+1}) - 2\bar{x}(z_i) + \bar{x}(z_{i-1})}{\Delta z^2} + O(\Delta z^2)\end{aligned}$$

second-order accurate formulas based on the cell-average values.

If we now come back to our initial Taylor series development of the solution, we can make use of these results to write

$$x(z) = \bar{x}_i + (z - z_i)\frac{\bar{x}_{i+1} - \bar{x}_{i-1}}{2\Delta z} + \left(\frac{(z-z_i)^2}{2!} - \frac{\Delta z^2}{24}\right)\frac{\bar{x}(z_i+1) - 2\bar{x}(z_i) + \bar{x}(z_i-1)}{\Delta z^2} + \cdots$$

From this expression we can deduce a first-order approximation

$$x(z) = \bar{x}_i$$

a second-order approximation

$$x(z) = \bar{x}_i + (z - z_i)\frac{\bar{x}_{i+1} - \bar{x}_{i-1}}{2\Delta z}$$

and a third-order approximation

$$x(z) = \bar{x}_i + (z - z_i)\frac{\bar{x}_{i+1} - \bar{x}_{i-1}}{2\Delta z} + \left(\frac{(z-z_i)^2}{2!} - \frac{\Delta z^2}{24}\right)\frac{\bar{x}(z_i+1) - 2\bar{x}(z_i) + \bar{x}(z_i-1)}{\Delta z^2}$$

These results can be summarized into a single formula whose accuracy can be varied thanks to an adjustable coefficient κ

$$x(z) = \bar{x}_i + (z - z_i)\frac{\bar{x}_{i+1} - \bar{x}_{i-1}}{2\Delta z} + \kappa\left(\frac{(z-z_i)^2}{2!} - \frac{\Delta z^2}{24}\right)\frac{\bar{x}(z_{i+1}) - 2\bar{x}(z_i) + \bar{x}(z_{i-1})}{\Delta z^2}$$

We can now compute the value of $x(z)$ in any point of the cell $\left[z_i - \frac{\Delta z}{2}, z_i + \frac{\Delta z}{2}\right]$, and in particular at the cell interfaces

$$\begin{aligned}
x(z_{i-1/2}) &= \bar{x}_i - \frac{\bar{x}_{i+1} - \bar{x}_{i-1}}{4} + \kappa\frac{\bar{x}_{i+1} - 2\bar{x}_i + \bar{x}_{i-1}}{12} \\
&= \bar{x}_i - \left(\frac{1}{4} + \frac{\kappa}{12}\right)(\bar{x}_i - \bar{x}_{i-1}) - \left(\frac{1}{4} - \frac{\kappa}{12}\right)(\bar{x}_{i+1} - \bar{x}_i) \\
x(z_{i+1/2}) &= \bar{x}_i + \frac{\bar{x}_{i+1} - \bar{x}_{i-1}}{4} + \kappa\frac{\bar{x}_{i+1} - 2\bar{x}_i + \bar{x}_{i-1}}{12} \\
&= \bar{x}_i + \left(\frac{1}{4} - \frac{\kappa}{12}\right)(\bar{x}_i - \bar{x}_{i-1}) + \left(\frac{1}{4} + \frac{\kappa}{12}\right)(\bar{x}_{i+1} - \bar{x}_i)
\end{aligned}$$

These values could be used to evaluate the flux functions in the discretized equation

$$\frac{\partial \bar{x}_i(t)}{\partial t} = \frac{f(z_{i-1/2}) - f(z_{i+1/2})}{\Delta z}$$

However, before proceeding with this computation, we have to observe a few important points:

- If we consider the neighboring intervals $\left[z_{i-1} - \frac{\Delta z}{2}, z_{i-1} + \frac{\Delta z}{2}\right]$ and $\left[z_{i+1} - \frac{\Delta z}{2}, z_{i+1} + \frac{\Delta z}{2}\right]$, similar formulas can also be derived at the cell interfaces. We will therefore distinguish the values of $x^+(z_{i-1/2})$ computed from the interval $\left[z_i - \frac{\Delta z}{2}, z_i + \frac{\Delta z}{2}\right]$ and $x^-(z_{i-1/2})$ computed from the interval $\left[z_{i-1} - \frac{\Delta z}{2}, z_{i-1} + \frac{\Delta z}{2}\right]$. In total, we have 4 formulas for the values at the two interface points of cell i

$$\begin{aligned}
x^-(z_{i-1/2}) &= \bar{x}_{i-1} + \left(\frac{1}{4} - \frac{\kappa}{12}\right)(\bar{x}_{i-1} - \bar{x}_{i-2}) + \left(\frac{1}{4} + \frac{\kappa}{12}\right)(\bar{x}_i - \bar{x}_{i-1}) \\
x^+(z_{i-1/2}) &= \bar{x}_i - \left(\frac{1}{4} + \frac{\kappa}{12}\right)(\bar{x}_i - \bar{x}_{i-1}) - \left(\frac{1}{4} - \frac{\kappa}{12}\right)(\bar{x}_{i+1} - \bar{x}_i) \\
x^-(z_{i+1/2}) &= \bar{x}_i + \left(\frac{1}{4} - \frac{\kappa}{12}\right)(\bar{x}_i - \bar{x}_{i-1}) + \left(\frac{1}{4} + \frac{\kappa}{12}\right)(\bar{x}_{i+1} - \bar{x}_i) \\
x^+(z_{i+1/2}) &= \bar{x}_{i+1} - \left(\frac{1}{4} + \frac{\kappa}{12}\right)(\bar{x}_{i+1} - \bar{x}_i) - \left(\frac{1}{4} - \frac{\kappa}{12}\right)(\bar{x}_{i+2} - \bar{x}_{i+1})
\end{aligned}$$

- The coefficient κ can be selected so as to obtain upwind formulas (which are formulas we are interested in, as we consider problem with steep moving fronts). If we select $\kappa = -3$, then the above expressions become

$$x^-(z_{i-1/2}) = \bar{x}_{i-1} + \frac{1}{2}(\bar{x}_{i-1} - \bar{x}_{i-2})$$
$$x^+(z_{i-1/2}) = \bar{x}_i - \frac{1}{2}(\bar{x}_{i+1} - \bar{x}_i)$$
$$x^-(z_{i+1/2}) = \bar{x}_i + \frac{1}{2}(\bar{x}_i - \bar{x}_{i-1})$$
$$x^+(z_{i+1/2}) = \bar{x}_{i+1} - \frac{1}{2}(\bar{x}_{i+2} - \bar{x}_{i+1})$$

The choice of the "+" or "−" values will depend on the direction of flow. To assess the flow direction, the sign of $\frac{\partial f}{\partial x}$ has to be evaluated. Indeed

$$\frac{\partial x}{\partial t} = -\frac{\partial f(x)}{\partial z} = -\frac{\mathrm{d}f}{\mathrm{d}x}\frac{\partial x}{\partial z}$$

If $\frac{\mathrm{d}f}{\mathrm{d}x} > 0$, then we will choose

$$x^-(z_{i-1/2}) = \bar{x}_{i-1} + \frac{1}{2}(\bar{x}_{i-1} - \bar{x}_{i-2})$$

$$x^-(z_{i+1/2}) = \bar{x}_i + \frac{1}{2}(\bar{x}_i - \bar{x}_{i-1})$$

Otherwise, if $\frac{\mathrm{d}f}{\mathrm{d}x} < 0$

$$x^+(z_{i-1/2}) = \bar{x}_i - \frac{1}{2}(\bar{x}_{i+1} - \bar{x}_i)$$

$$x^+(z_{i+1/2}) = \bar{x}_{i+1} - \frac{1}{2}(\bar{x}_{i+2} - \bar{x}_{i+1})$$

Figure 5.16 shows, for the case $\frac{\mathrm{d}f}{\mathrm{d}x} > 0$, the above second-order approximation and compares it with the first-order approximation $x^-(z_{i-1/2}) = \bar{x}_{i-1}$ and $x^-(z_{i+1/2}) = \bar{x}_i$. Clearly, the second-order approximation is prone to oscillations, and to alleviate this problem, a slope limiter $\phi(r)$ is introduced

$$x^-(z_{i-1/2}) = \bar{x}_{i-1} + \frac{\phi(r_i)}{2}(\bar{x}_{i-1} - \bar{x}_{i-2})$$
$$x^-(z_{i+1/2}) = \bar{x}_i + \frac{\phi(r_i)}{2}(\bar{x}_i - \bar{x}_{i-1})$$

This slope limiter senses the relative variation in the solution slope

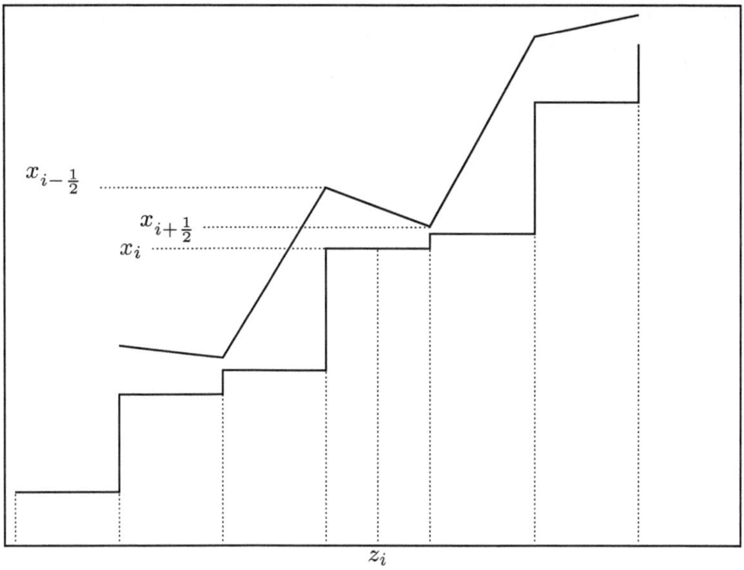

Fig. 5.16 Comparison of first- and second-order approximations

$$r_i = \frac{\dfrac{\bar{x}_{i+1} - \bar{x}_i}{z_{i+1} - z_i}}{\dfrac{\bar{x}_i - \bar{x}_{i-1}}{z_i - z_{i-1}}} = \frac{\bar{x}_{i+1} - \bar{x}_i}{\bar{x}_i - \bar{x}_{i-1}}$$

and will allow to use a low-order approximation in spatial regions where there is a high solution activity and a higher order approximation in regions where the solution is smoother. More specifically, the slope limiter has properties imposed by the requirement that the resulting numerical scheme has to be TVD:

- $\phi(r) \geqslant 0$ if $r \geqslant 0$. Indeed $r \geqslant 0$ expresses the solution monotonicity which could be destroyed by $\phi(r) < 0$.
- $\phi(r) = 0$ if $r < 0$. Indeed $r < 0$ is the signal that there is an abrupt change in the sign of the solution slope, and in this case the approximation is limited to first-order.

More specifically, a TVD requirement can be expressed by (see [21, 22], for more details)

$$0 \leqslant \phi(r) \leqslant \min(2r, 2)$$

which is represented in Fig. 5.17. A smaller portion of this TVD region corresponds to second-order schemes

- $r \leqslant \phi(r) \leqslant 2r$ for $0 \leqslant r < 1$
- $\phi(1) = 1$ for $r = 1$
- $1 \leqslant \phi(r) \leqslant r$ for $0 \leqslant r \leqslant 1$
- $1 \leqslant \phi(r) \leqslant 2$ for $r > 2$

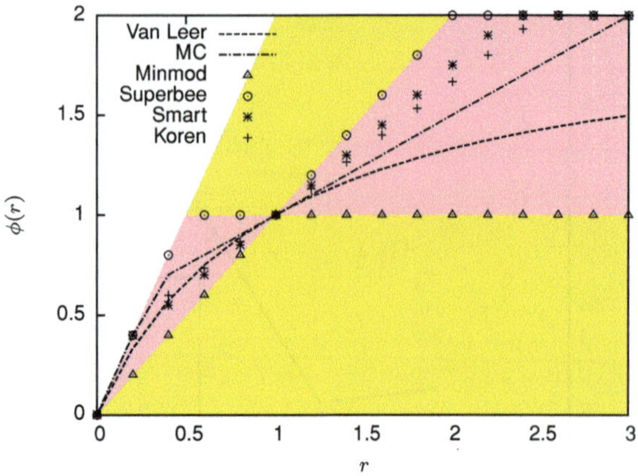

Fig. 5.17 Colored TVD regions—*purple* corresponds to second-order schemes—and representation of 6 popular slope limiters

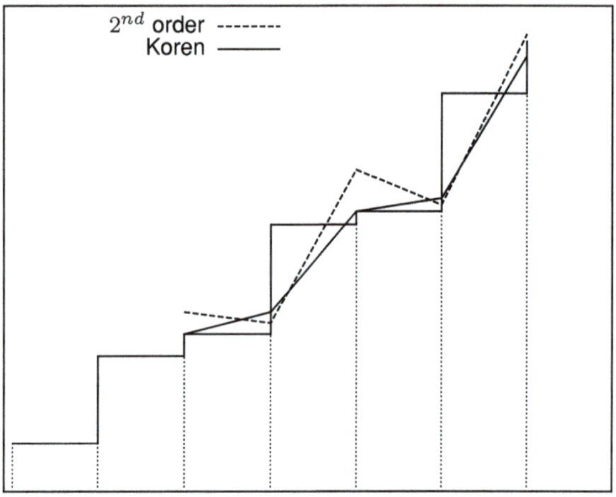

Fig. 5.18 Comparison of first-order, plain second-order, and Koren-limited approximations

A list of a few popular limiters (there are many more, see for instance the nice report by Zijlema and Wesselin [23]) is given in the following:

- Van Leer [24]: $\phi(r) = \frac{r+|r|}{1+|r|}$
- Monotonized central (MC) [25]: $\phi(r) = \max\left(0, \min\left(2r, \frac{1+r}{2}, 2\right)\right)$
- Minmod [26]: $\phi(r) = \max(0, \min(1, r))$
- Superbee [26]: $\phi(r) = \max(0, \min(2r, 1), \min(r, 2))$
- Smart [27]: $\phi(r) = \max\left(0, \min\left(4r, \frac{1+3r}{4}, 2\right)\right)$
- Koren [28]: $\phi(r) = \max\left(0, \min\left(2r, \frac{1+2r}{3}, 2\right)\right)$

In analogy with Fig. 5.16, the effect of the Koren limiter on the solution profile is shown in Fig. 5.18. As expected, the slope limiter avoids the appearance of oscillations, while allowing increased accuracy.

As an example, the code of the Koren flux limiter is given in `koren_slope_limiter_fz.m`, in the generalized case of a possibly nonuniform grid. This code is used to solve Burgers equation with a small viscosity coefficient $\mu = 0.000001$.

```
function [fz] = koren_slope_limiter_fz(n, z, t, x, flux,dflux_dx)

delta      = 1.0e-05;
dz         = zeros(n,1);
dz(1)      = (z(2)-z(1))/2;
dz(2:n-1)  = (z(3:n)-z(1:n-2))/2;
dz(n)      = (z(n)-z(n-1))/2;
valdfdx    = feval(dflux_dx, t, x);

% computation of the fz derivative at the first left
% boundary point
fz(1) = (feval(flux,t,x(2)) -...
         feval(flux,t,x(1)))/(z(2)-z(1));

% computation of the fz derivative at the second left
% boundary point: fz(2) depends on the sign of valdfdx(2)
if valdfdx (2) >= 0
    fz (2) = (feval(flux,t,x(2)) -...
              feval(flux,t,x(1)))/(z(2)-z(1));
else
    % computation of r(2) and phi(2)
    if abs(x(2)-x(3)) < delta
        phi(2) = 0;
    else
        r(2) = ((x(1)-x(2))/(z(1)-z(2)))/...
               ((x(2)-x(3))/(z(2)-z(3)));
        phi(2) = max(0, min([2*r(2) (1+2*r(2))/3  2]));
    end
    % computation of r(3) and phi(3)
    if abs(x(3)-x(4)) < delta
        phi(3) = 0;
    else
        r(3)   = ((x(2)-x(3))/(z(2)-z(3)))/...
                 ((x(3)-x(4))/(z(3)-z(4)));
        phi(3) = max(0, min([2*r(3) (1+2*r(3))/3  2]));
    end
    % computation of xL and xR for the second left boundary point
    xL     = x(2)-dz(2)*(x(3)-x(2))*phi(2)/(dz(2)+dz(3));
    xR     = x(3)-dz(3)*(x(4)-x(3))*phi(3)/(dz(3)+dz(4));
    fz(2) = (feval(flux,t,xR)-feval(flux,t,xL))/dz(2);
end
% computation of the fz derivative at the interior points 3,
% ..., n-2 : fz(i) depends on the sign of valdfdx(i)
for i=3:n-2
    if valdfdx(i) >= 0
        % computation of r(i) and phi(i)
        if abs(x(i)-x(i-1)) < delta
            phi(i) = 0;
        else
            r(i)    = ((x(i+1)-x(i))/(z(i+1)-z(i)))/...
                      ((x(i)-x(i-1))/(z(i)-z(i-1)));
            phi(i) = max(0, min([2*r(i) (1+2*r(i))/3  2]));
        end
        % computation of r(i-1) and phi(i-1)
        if abs(x(i-1)-x(i-2)) < delta
            phi(i-1) = 0;
        else
            r(i-1)    = ((x(i)-x(i-1))/(z(i)-z(i-1)))/...
```

```
                       ((x(i-1)-x(i-2))/(z(i-1)-z(i-2)));
            phi(i-1) = max(0, min([2*r(i-1)...
                          (1+2*r(i-1))/3  2]));
        end
        % computation of xL and xR for the interior points
        xL = x(i-1) + dz(i-1)*(x(i-1)-x(i-2))*phi(i-1)/...
             (dz(i-1)+dz(i-2));
        xR = x(i)+dz(i)*(x(i)-x(i-1))*phi(i)/...
             (dz(i)+dz(i-1));
        fz(i) = (feval(flux,t,xR)-feval(flux,t,xL))/dz(i);
    else
        % computation of r(i) and phi(i)
        if abs(x(i)-x(i+1)) < delta
            phi(i) = 0;
        else
            r(i)   = ((x(i-1)-x(i))/(z(i-1)-z(i)))/...
                     ((x(i)-x(i+1))/(z(i)-z(i+1)));
            phi(i) = max(0, min([2*r(i) (1+2*r(i))/3  2]));
        end
        % computation of r(i+1) and phi(i+1)
        if abs(x(i+1)-x(i+2)) < delta
            phi(i+1) = 0;
        else
            r(i+1)   = ((x(i)-x(i+1))/(z(i)-z(i+1)))/...
                       ((x(i+1)-x(i+2))/(z(i+1)-z(i+2)));
            phi(i+1) = max(0, min([2*r(i+1)...
                       (1+2*r(i+1))/3  2]));
        end
        % computation of xL and xR for the interior points
        xL     = x(i)-dz(i)*(x(i+1)-x(i))*phi(i)/...
                 (dz(i)+dz(i+1));
        xR     = x(i+1)-dz(i+1)*(x(i+2)-x(i+1))*...
                 phi(i+1)/(dz(i+1)+dz(i+2));
        fz(i) = (feval(flux,t,xR)-feval(flux,t,xL))/dz(i);
    end
end
% computation of the fz derivative at the penultimate
% point: fz(n-1) depends on the sign of valdfdx(n-1)
if valdfdx(n-1) < 0
    fz(n-1) = (feval(flux,t,x(n)) -...
               feval(flux,t,x(n-1)))/(z(n)-z(n-1));
else
    % computation of r(n-1) and phi(n-1)
    if abs(x(n-1)-x(n-2)) < delta
        phi(n-1) = 0;
    else
        r(n-1)   = ((x(n)-x(n-1))/(z(n)-z(n-1)))/...
                   ((x(n-1)-x(n-2))/(z(n-1)-z(n-2)));
        phi(n-1) = max(0, min([2*r(n-1)...
                   (1+2*r(n-1))/3  2]));
    end
    % computation of r(n-2) and phi(n-2)
    if abs(x(n-2)-x(n-3)) < delta
        phi(n-2)=0;
    else
        r(n-2) = ((x(n-1)-x(n-2))/(z(n-1)-z(n-2)))/...
                 ((x(n-2)-x(n-3))/(z(n-2)-z(n-3)));
        phi(n-2) = max(0, min([2*r(n-2)...
                   (1+2*r(n-2))/3  2]));
    end
    % computation of xL and xR for the penultimate point
    xL       = x(n-2)+dz(n-2)*(x(n-2)-x(n-3))*phi(n-2)/...
               (dz(n-2)+dz(n-3));
    xR       = x(n-1)+dz(n-1)*(x(n-1)-x(n-2))*phi(n-1)/...
               (dz(n-1)+dz(n-2));
    fz(n-1) = (feval(flux,t,xR)-feval(flux,t,xL))/dz(n-1);
```

```
end
% computation of the fz derivative at the right boundary
% point
fz(n) = (feval(flux,t,x(n))-feval(flux,t,x(n-1)))/...
        (z(n)-z(n-1));
fz    = fz';
```

Function koren_slope_limiter_fz.m Implementation of the ODE system

A satisfactory solution can be obtained with 300 finite volumes (cells), which is much less than the number of points required by a finite difference scheme (see for instance Fig 5.14 which uses 701 grid points). The solution is also well behaved, capturing the sharp fronts without introducing too much artificial dissipation, nor spurious oscillations. The price to pay is an increased computational load, as the discretization of the spatial operators can no longer be formulated with simple matrix multiplications as in FD schemes. Figures 5.19 and 5.20 shows this numerical solution in a 2D and 3D plot, respectively. Although upwind schemes are very efficient, they require the determination of the direction of flow. This determination is easy in scalar problems, such as the one studied in this section, but more intricate for systems of equations. Whereas for linear systems, it is possible to compute the eigenvalues (this corresponds to a diagonalization of the system), nonlinear problems require the determination of the characteristic structure through the solution of *Riemann problems*. We will not enter into details here, but just mention that Riemann problems correspond to problems with piecewise constant initial conditions, as the first-order approximation shown in Fig. 5.16. The Russian mathematician Godunov proposed in 1959 [29] an efficient first-order upwind scheme based on a finite volume discretization, and the solution of Riemann problems corresponding to averaged values in each cells. The method consists of determining the kind of waves emanating from each cell interface, and in reconstructing a global solution by piecing together the set of local solutions. The solution of the local Riemann problems have to be computed on a limited time interval, such that the characteristic lines do not intersect (this corresponds to the CFL condition of Courant, Friedrichs and Lewy). When applied to the linear advection equation, the method of characteristics shows that Godunov method is equivalent to a 2-point upwind finite difference scheme. For Burgers equation, it is also possible to find exact solutions using more advanced concepts including the Rankine-Hugoniot condition and the entropy condition (see [30], for an analysis of Burgers equation).

If we consider the inviscid Burgers equation

$$x_t = -x x_z$$

with initial conditions (defining a Riemann problem)

$$x(z,0) = \begin{cases} x^L & z < 0 \\ x^R & z > 0 \end{cases}$$

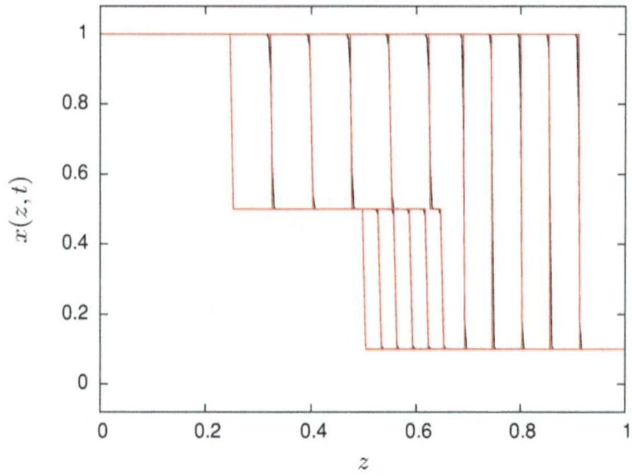

Fig. 5.19 Numerical solution of Burgers equation in conservation form ($\mu = 0.000001$) at times $t = 0, 0.1, \ldots, 1$ with $N = 300$ finite volume and a Koren slope limiter. *Red curves* exact solution. *Black curves* numerical solution

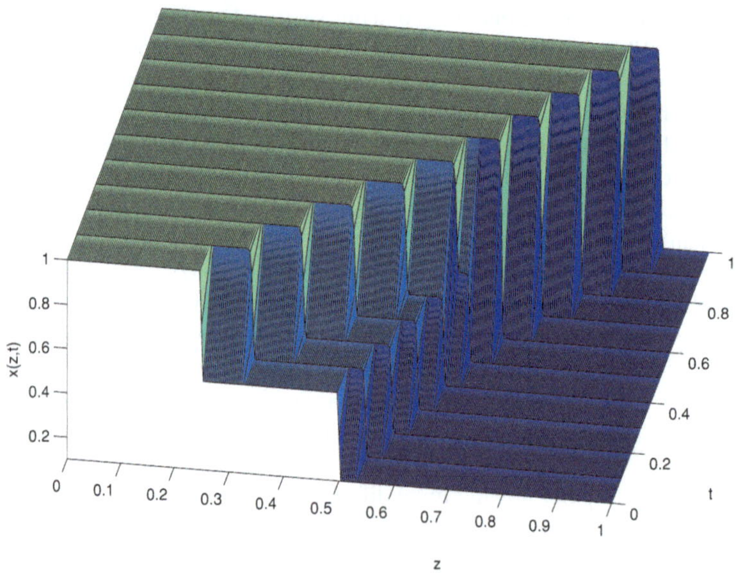

Fig. 5.20 3D representation of the numerical solution of Burgers equation in conservation form ($\mu = 0.000001$) with $N = 300$ using the finite volume method and a Koren slope limiter

there are two possible cases:

(a) $x_L > x_R$. The solution is unique and is given by

$$x(z,t) = \begin{cases} x_L & z < st \\ x_R & z > st \end{cases} \quad \text{with} \quad s = \frac{x_L + x_R}{2}$$

(b) $x_L < x_R$. The solution is not unique, but the vanishing viscosity solution is given by

$$x(z,t) = \begin{cases} x_L & z < x_L t \\ z/t & x_L t \leqslant z \leqslant x_R t \\ x_R & z > x_R t \end{cases}$$

This leads to a simple numerical scheme in conservation form (given in a fully discrete form, i.e., both time and space are discretized)

$$\frac{x_i^{k+1} - x_i^k}{\Delta t} = -\frac{F(x_i^k, x_{i+1}^k) - F(x_{i-1}^k, x_i^k)}{\Delta z}$$

where the flux function $F(x_L, x_R) = F(\tilde{x})$ with

(a) $x_L > x_R$

$$\tilde{x} = \begin{cases} x_L & s > 0 \\ x_R & s < 0 \end{cases}$$

(b) $x_L < x_R$

$$\tilde{x} = \begin{cases} x_L & x_L > 0 \\ 0 & x_L \leqslant 0 \leqslant x_R \\ x_R & x_R < 0 \end{cases}$$

A numerical solution (inspired from a MATLAB code by Landajuela [31]) for a Riemann problem with initial condition $x(z,0) = 0.5$ for $z < 0.5$ and $x(z,0) = 1$ for $z > 0.5$ is represented in Fig. 5.21, for $N = 200$ and $\Delta t = 0.01$. For other hyperbolic conservation laws and higher order schemes, Riemann solvers are required (see the books by LeVeque and Toro [4, 32, 33]) as well as a reconstruction of the solution using piecewise polynomial approximation, that we will not discuss in this introductory-level text.

To avoid the solution of Riemann problems, *nonoscillatory central schemes* have been proposed, which are simpler, yet very efficient. We will briefly discuss a family of methods [34] that can be viewed as an extension of the first-order Lax-Friedrich scheme [35].

$$x_i^{k+1} = \frac{x_{i+1}^k + x_{i-1}^k}{2} - \frac{\Delta t}{2\Delta z}\left(f\left(x_{i+1}^k\right) - f\left(x_{i-1}^k\right)\right)$$

and their further extension by Kurganov and Tadmor [36] to a semi-discrete (i.e., in the MOL framework) second-order scheme. This latter scheme can be used as a convenient general purpose MOL solver for nonlinear convection-diffusion PDE systems, as it is not tied to the eigenstructure of the problem. The only information that is required is the local propagation speed of the wave at the cell interfaces $z_i \pm \frac{\Delta z}{2}$.

$$a_{i\pm\frac{1}{2}}^k = \max\left\{\rho\left[\frac{\partial f}{\partial x}\left(x_{i\pm\frac{1}{2}}^-\right)\right], \rho\left[\frac{\partial f}{\partial x}\left(x_{i\pm\frac{1}{2}}^+\right)\right]\right\}$$

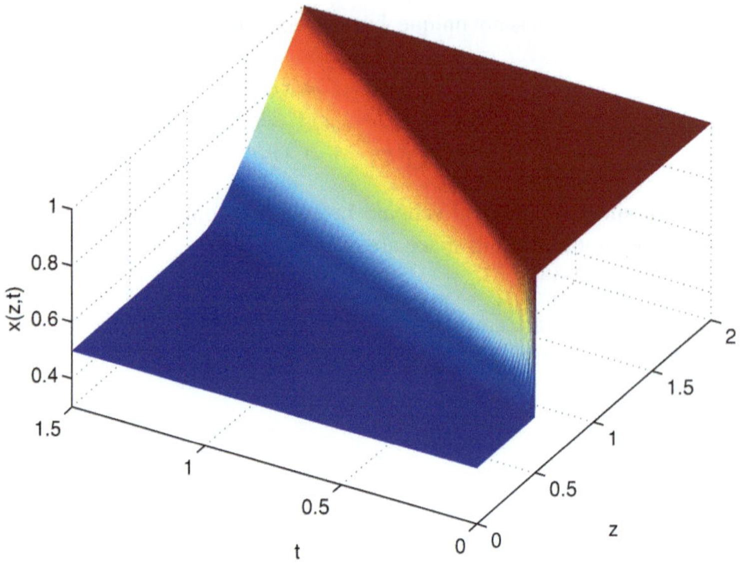

Fig. 5.21 Numerical solution of inviscid Burgers equation with $N = 200$ finite volume and Godunov method

where ρ denotes the spectral radius , i.e., the largest modulus of the eigenvalues of the Jacobian matrix evaluated in

$$x^{+}_{i+\frac{1}{2}} = x^{k}_{i+1} - \frac{\Delta z}{2}\,(x_z)^{k}_{i+1}$$

$$x^{-}_{i+\frac{1}{2}} = x^{k}_{i} + \frac{\Delta z}{2}\,(x_z)^{k}_{i}$$

The scheme is given by (see [36] for the derivation):

$$\begin{aligned}\frac{\mathrm{d}x_i}{\mathrm{d}t} = -\frac{1}{2\Delta z}\Bigg\{&\left[f\left(x^{+}_{i+\frac{1}{2}}\right) + f\left(x^{-}_{i+\frac{1}{2}}\right)\right] - \left[f\left(x^{+}_{i-\frac{1}{2}}\right) + f\left(x^{-}_{i-\frac{1}{2}}\right)\right]\\ &- a_{i+\frac{1}{2}}\left[x^{+}_{i+\frac{1}{2}} - x^{-}_{i+\frac{1}{2}}\right] + a_{i-\frac{1}{2}}\left[x^{+}_{i-\frac{1}{2}} - x^{-}_{i-\frac{1}{2}}\right]\Bigg\}\end{aligned}$$

with

$$x^{+}_{i+\frac{1}{2}} = x_{i+1} - \frac{\Delta z}{2}\,(x_z)_{i+1} \qquad x^{-}_{i+\frac{1}{2}} = x_{i} + \frac{\Delta z}{2}\,(x_z)_{i}$$

$$x^{+}_{i-\frac{1}{2}} = x_{i} - \frac{\Delta z}{2}\,(x_z)_{i} \qquad x^{-}_{i-\frac{1}{2}} = x_{i-1} + \frac{\Delta z}{2}\,(x_z)_{i-1}$$

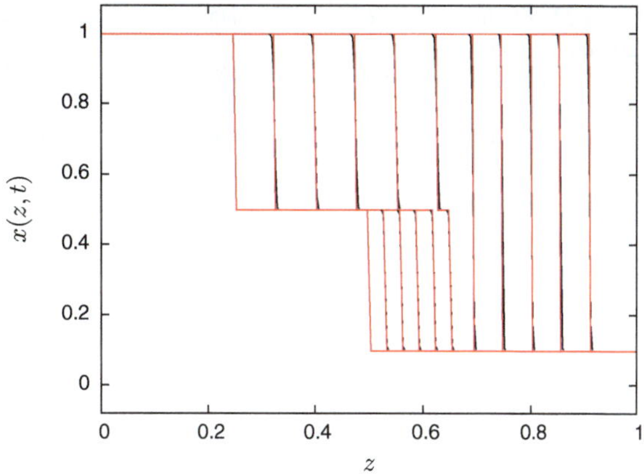

Fig. 5.22 Numerical solution of Burgers equation in conservation form $\mu = 0.000001$ at times $t = 0, 0.1, \ldots, 1$ with $N = 200$ finite volume and Kurganov-Tadmor centered scheme. *Red curves* exact solution. *Black curves* numerical solution

This scheme can be proved to be TVD in the 1D scalar case, using a minmod limiter

$$(x_z)_i = \min\text{mod}\left(\theta\frac{x_{i+1}-x_i}{z_{i+1}-z_i}, \frac{x_{i+1}-x_{i-1}}{z_{i+1}-z_{i-1}}, \theta\frac{x_i-x_{i-1}}{x_i-z_{i-1}}\right) \qquad 1 \leqslant \theta \leqslant 2$$

$$\min\text{mod}(\alpha_1, \alpha_2, \ldots) = \begin{cases} \min\limits_j (\alpha_j) & \text{if } \alpha_j > 0 \; \forall j \\ \max\limits_j (\alpha_j) & \text{if } \alpha_j < 0 \; \forall j \\ 0 & \text{otherwise} \end{cases}$$

Figure 5.22 shows the solution to Burgers equation with a small viscosity coefficient $\mu = 0.000001$. A quite accurate solution can already be obtained with 200 finite volumes (cells).

We now consider a few additional PDE applications, in the form of systems of 2 or 3 PDEs, in order to illustrate some additional features of the methods introduced in this chapter. We start with a jacketed tubular reactor in which an irreversible, exothermic, first-order reaction takes place. The jacket with water temperature $T_w(z,t)$ includes a heating and a cooling section [37].

$$\frac{\partial T}{\partial t} = -v\frac{\partial T}{\partial z} + D_1\frac{\partial^2 T}{\partial z^2} - \frac{\Delta H}{\rho c_p}k_0 c e^{-\frac{E}{RT}} + \frac{4h}{\rho c_p d}(T_w - T)$$

$$\frac{\partial c}{\partial t} = -v\frac{\partial c}{\partial z} + D_2\frac{\partial^2 c}{\partial z^2} - k_0 c e^{-\frac{E}{RT}}$$

In these energy- and mass-balance equations, c $[mol/L]$ and T $[K]$ denote respectively the reactant concentration and the temperature, D_1 $[m^2/s]$ and D_2 $[m^2/s]$ are the thermal diffusion and mass dispersion coefficients, v $[m/s]$ is the fluid velocity, $-\Delta H$ $[J/kmol]$ is the heat of reaction ($\Delta H < 0$ for an exothermic reaction)

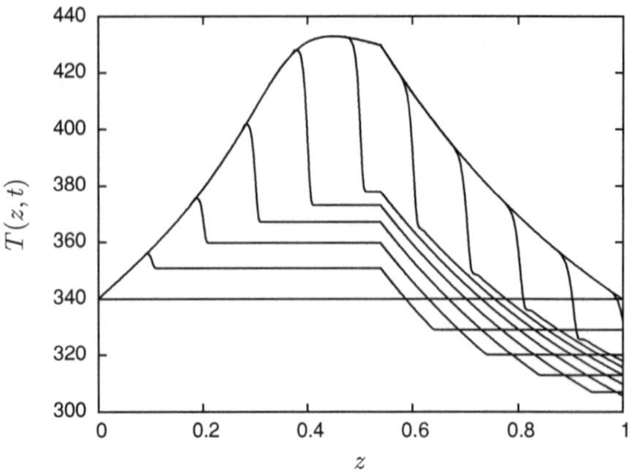

Fig. 5.23 Evolution of the temperature profile in the jacketed tubular reactor with $P_e = 10^6$ at $t = 0, 1, \ldots, 50$ computed using an upwind scheme and Van Leer slope limiter with 600 finite volumes

and ρ $[kg/m3]$, c_p $[J/(kg\,K)]$, k_0 $[1/s]$, E $[J/mol]$, R $[J/(mol\,K)]$, h $[W/(m^2K)]$ and d $[m]$ are the fluid density, the specific heat, the kinetic constant, the activation energy, the ideal gas constant, the heat transfer coefficient, and the reactor diameter, respectively. The tubular reactor has a length $L = 1$ $[m]$. The jacket fluid temperature T_w $[K]$ is a piecewise constant function with a change from a maximum value $T_{w,\max} = 400$ $[K]$ to a minimum $T_{w,\min} = 280$ $[K]$ at a transition location $z_{sw} = 0.54$ $[m]$. The code and full set of parameters can be found in the companion software.

These PDEs are supplemented by boundary and initial conditions [38]

$$\begin{aligned} D_1 \frac{\partial T(0,t)}{\partial z} &= v(T(0,t) - T_{in}) \qquad \frac{\partial T(L,t)}{\partial z} = 0 \\ D_1 \frac{\partial c(0,t)}{\partial z} &= v(c(0,t) - c_{in}) \qquad \frac{\partial c(L,t)}{\partial z} = 0 \end{aligned}$$

with $T_{in} = 340$ $[K]$ and $c_{in} = 0.02$ $[mol/l]$.

$$\begin{aligned} T(z,0) &= T_0(z) = T_{in} \\ C(z,0) &= c_0(z) = 0 \end{aligned}$$

The PDEs are nonlinear and coupled because of the source terms which include the reaction kinetics, but the transportation terms are linear and uncoupled. We are therefore in a situation—quite frequent in chemical engineering applications—where the convective terms can be approximated using any of the flux limiters that we have discussed. We consider a high Peclet number corresponding to a convection-dominated problem, i.e., $P_e = 10^6$, and first present results obtained with 600 finite volumes and an upwind scheme with a Van Leer limiter (Figs. 5.23 and 5.24) followed by results obtained with the same number of finite volumes and a centered limiter of Kurganov-Tadmor (Fig. 5.25). The results obtained with the latter are quite similar but requires much less computational expense (about half the computation time). Time integration is achieved with the explicit solver `ode45`.

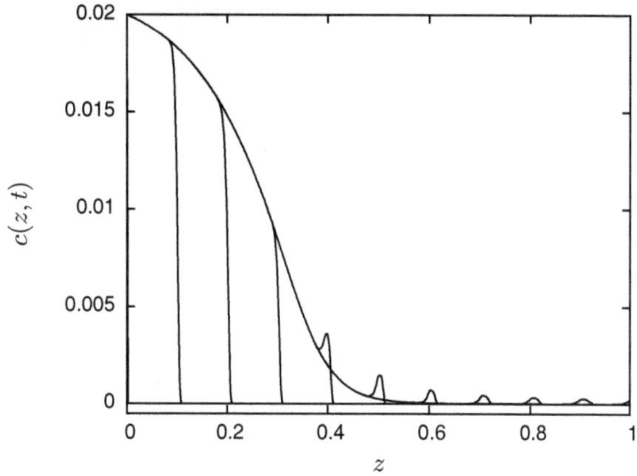

Fig. 5.24 Evolution of the concentration profile in the jacketed tubular reactor with $P_e = 10^6$ at $t = 0, 1, \ldots, 50$ computed using an upwind scheme and Van Leer slope limiter with 600 finite volumes

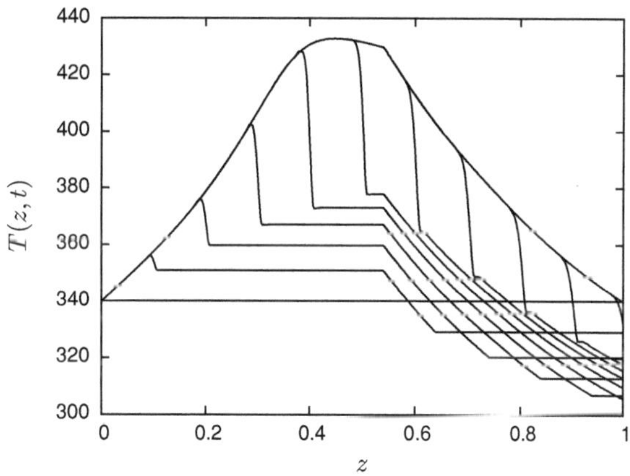

Fig. 5.25 Evolution of the temperature profile in the jacketed tubular reactor with $P_e = 10^6$ at $t = 0, 1, \ldots, 50$ computed using an upwind scheme and Kurganov-Tadmor centered scheme with 600 finite volumes

We now consider a nonlinear system of conservation laws, for which the use of our upwind slope limiters is no longer adequate (our MATLAB functions are appropriate only for scalar problems, or linear systems, as in the previous test examples). This system of PDEs corresponds to the so-called *Sod shock tube* problem, named after Gary Sod [39], and which is a popular test for the accuracy of computational fluid codes. Shock tubes are experimental devices used to study gas flow (compressible flow phenomena and blast waves) under various conditions of temperature and pressure. The underlying model is given by *Euler equations* which represent the conservation of mass, momentum, and energy, and corresponds to the Navier-Stokes equations for an inviscid flow. In one spatial dimension (as all our PDE examples so far), these PDEs are given by

$$\frac{\partial \rho}{\partial t} + \frac{\partial m}{\partial z} = 0$$

$$\frac{\partial m}{\partial t} + \frac{\partial}{\partial z}\left[\frac{m^2}{\rho} + p\right] = 0$$

$$\frac{\partial e}{\partial t} + \frac{\partial}{\partial z}\left[\frac{m}{\rho}(e + p)\right] = 0$$

where $\rho(z,t)$ denotes gas density, $m(z,t) = \rho(z,t)u(z,t)$ represents momentum ($u(z,t)$ is the gas velocity), $p(z,t) = (\gamma - 1)(e - (m^2/2\rho))$ is pressure and $e(z,t)$ is energy per unit length.

Introducing the expression of the pressure in the PDEs as well as small linear diffusion terms to regularize the problem and ease the numerical solution (the diffusion terms avoid the formation of true shocks which would be difficult to resolve), we can write the PDEs in the following form:

$$\frac{\partial \rho}{\partial t} = -\frac{\partial m}{\partial z} + \varepsilon \frac{\partial^2 u}{\partial z^2}$$

$$\frac{\partial m}{\partial t} = -\frac{\partial}{\partial z}\left[(\gamma - 1)e - (\gamma - 3)\frac{m^2}{2\rho}\right] + \varepsilon \frac{\partial^2 m}{\partial z^2}$$

$$\frac{\partial e}{\partial t} = -\frac{\partial}{\partial z}\left[\frac{m}{\rho}\left(\gamma e - (\gamma - 1)\frac{m^2}{2\rho}\right)\right] + \varepsilon \frac{\partial^2 e}{\partial z^2}$$

The parameter values are $\gamma = 1.4$ (ideal gas) and $\varepsilon = 0.001$. The initial conditions considered by Sod [39] define a Riemann problem

$$\rho(z,0) = \begin{cases} 1 & z \le 0.5 \\ 0.125 & z > 0.5 \end{cases}$$

$$m(z,0) = 0$$

$$e(z,0) = \begin{cases} 2.5 & z \le 0.5 \\ 0.25 & z > 0.5 \end{cases}$$

This corresponds to the situation where the tube is divided in two equal parts by a diaphragm. The gas is initially at rest (i.e., $m(z,0) = 0$), and has higher density and pressure in the left part of the tube (the pressure to the left is $p(z,0) = 1$, while $p(z,0) = 0.1$ to the right, giving the above-mentioned values for $e(z,0)$). At time $t = 0$, the diaphragm is broken and waves propagate through the tube. The representation of a closed tube is achieved thanks to reflection boundary conditions

$$\begin{array}{ll} \rho_z(0,t) = 0 & \rho_z(1,t) = 0 \\ m(0,t) = 0 & m(1,t) = 0 \\ e_z(0,t) = 0 & e_z(1,t) = 0 \end{array}$$

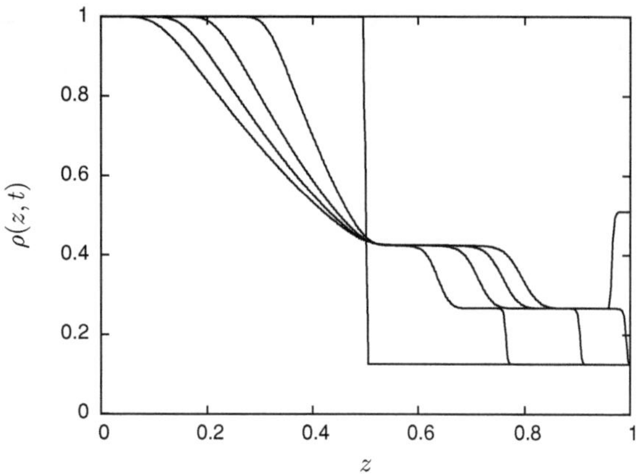

Fig. 5.26 Evolution of the density profile in the tube at $t = 0, 0.15, 0.23, 0.28, 0.32$, computed using Kurganov-Tadmor centered scheme with 800 finite volumes

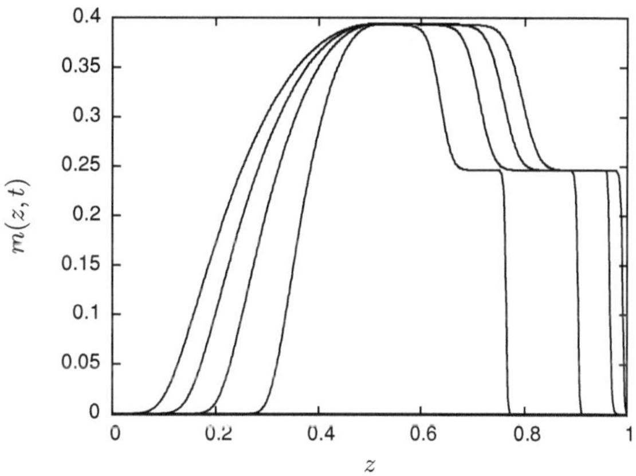

Fig. 5.27 Evolution of the momentum profile in the tube at $t = 0, 0.15, 0.23, 0.28, 0.32$, computed using Kurganov-Tadmor centered scheme with 800 finite volumes

Waves will therefore propagate while interacting and "bounce" on the boundaries. Figs. 5.26, 5.27 and 5.28 show the numerical solution obtained with a centered scheme of Kurganov-Tadmor and a discretization into 800 finite volumes.

5.7 Grid Refinement

Another approach to deal with steep moving fronts is to adapt the spatial grid in such a way that the grid points are concentrated in regions of high solution activity and are not "wasted" in smoother parts. There are a variety of mechanisms for grid adaptation, leading to what could be called *Adaptive Method of Lines* (as reviewed in the book edited by the first two authors [40]). In this section, we will introduce a grid refinement method based on the *equidistribution principle*.

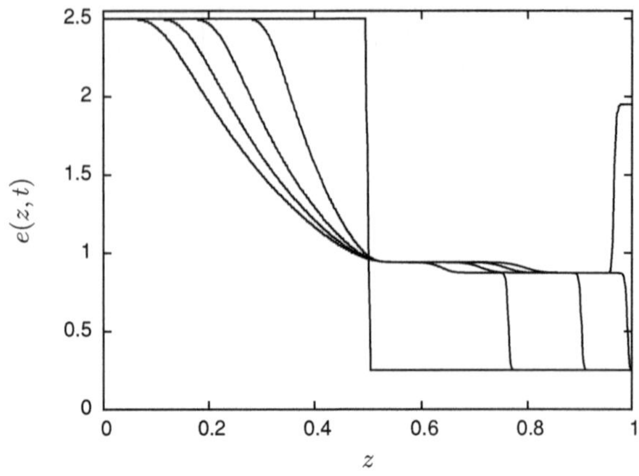

Fig. 5.28 Evolution of the energy profile in the tube at $t = 0, 0.15, 0.23, 0.28, 0.32$, computed using Kurganov-Tadmor centered scheme with 800 finite volumes

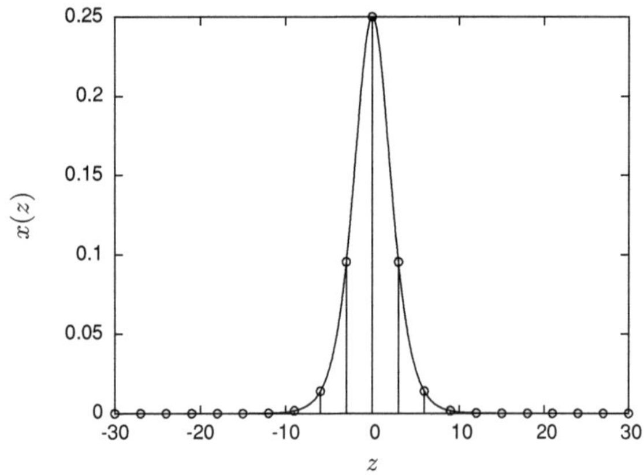

Fig. 5.29 Representation of a spatial profile using a uniform grid

The underlying idea is to rearrange the grid point location at a time t in such a way that a specified quantify is equidistributed. For instance, if we consider a wave that we would like to discretize using a grid with regular spacing as in Fig. 5.29, we see that the resolution in the steep fronts is much coarser than in the flat parts of the solution. To increase the resolution in the steep fronts, and get an equivalent level of discretization everywhere, one could relocate the grid points so as to equidistribute the arc-length of the solution profile, i.e.,

$$m(x) = \sqrt{(\alpha + \|x_z\|_2^2}$$

where $\alpha > 0$ ensures that the function $m(x)$ is strictly positive and acts as a regularization parameter which forces the existence of at least a few nodes in flat parts of the solution.

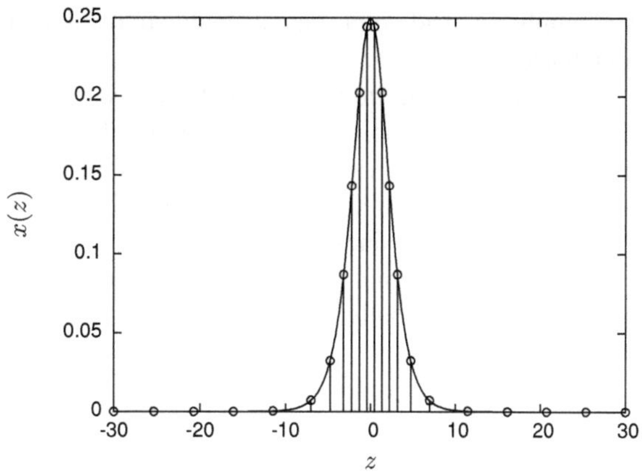

Fig. 5.30 Representation of a spatial profile using a nonuniform *grid* concentrated in the wave fronts

This leads to the grid distribution shown in Fig. 5.30, where each interval on the z axis corresponds to an interval on the solution graph which has a (more or less—due to regularization of the grid) equal length.

The equidistribution principle consists in enforcing that on each spatial grid interval

$$\int_{z_{i-1}}^{z_i} m(x)\mathrm{d}z = \int_{z_i}^{z_{i+1}} m(x)\mathrm{d}z = c, \quad 1 \le i \le N-1$$

where c is some constant.

The function $m(x)$ is called a *monitor function*. It can involve the first-derivative as in the above example, but also second-order derivative terms (taking account of the solution curvature).

To solve an IBVP problem, the procedure is as follows:

- Definition of an initial spatial grid
- Approximation of the spatial derivatives on the spatial grid;
- Time integration of the resulting semi-discrete ODEs on a time interval;
- Adaptation of the spatial grid at the end of the time interval;
- Interpolation of the solution on the new grid to produce new initial conditions for the next time interval.

The adaptation of the grid points is therefore not continuous in time, but requires time integration to be halted periodically to update the grid.

This procedure has several advantages. Especially, the PDE solution and grid adaptation mechanisms are uncoupled and can be programmed in separate MATLAB functions. Furthermore, grid adaptation can involve a change in the location of the grid points but also a change in their numbers (grid refinement). *Grid refinement* allows spatial accuracy to be controlled as grid points are added or deleted depending on the evolution of spatial activity of the solution. For instance, the birth of new moving fronts can be easily accommodated, whereas a dynamic moving grid with a fixed number of grid points can be unable to resolve unforeseen spatial activity with enough accuracy.

However, the procedure also has drawbacks. As the grid movement is not continuous in time, grid points are suboptimally located, i.e., they can for instance lag behind the front if the grid is not updated sufficiently frequently. Moreover, periodic solver restarts (each time the grid is adapted/refined) involve some computational overhead (and usually one-step solvers, such as implicit Runge-Kutta or Rosenbrock algorithms, which do not require the past solution history, will perform better than multistep solvers). Finally, interpolation to produce new initial conditions is an additional source of errors.

All in one, the procedure is simple to implement and will work successfully on problems of moderate complexity. In the following, we use an algorithm proposed in [40–42], which allows to refine the spatial grid based on the equidistribution of a monitor function, and a regularization mechanism proposed in [43]. Regularization ensures that the grid is locally bounded with respect to a constraint,

$$\frac{1}{K} \leq \frac{\Delta z_i}{\Delta z_{i-1}} \leq K, \quad 1 \leq i \leq N, \qquad K \geq 1$$

To ensure this property, the monitor function is modified, keeping its maximal values, but increasing its minimum values, in a procedure called *padding*. The resulting padded monitor function is then equidistributed yielding a grid whose ratios of consecutive grid steps are bounded as required. In practice, the padding is chosen (there is, in principle, an infinity of possibilities to achieve padding) so that the equidistributing grid has adjacent steps with constant ratios equal to the maximum allowed.

The MATLAB implementation involves the following issues:

- The Rosenbrock solver ode23s is usually the preferred algorithm as it does not require the past solution history;
- The solver is periodically halted, after a fixed number of time steps, using an `Events` function, which monitors the evolution of the number of steps;
- The solution is interpolated using cubic splines as implemented in function `spline` in order to generate initial conditions on the new grid.

As an example, we consider the computation of the solution to the classical Korteweg-de-Vries equation, which was originally introduced by Korteweg and deVries [44] to describe the behavior of small amplitude shallow-water waves in one space dimension

$$x_t = -6xx_z - x_{zzz}, \qquad -\infty \leqslant z \leqslant \infty, \quad t \geqslant 0, \quad x(z,0) = x_0(z)$$

which combines the effect of nonlinearity and dispersion. In the following, the propagation of a single soliton

$$x(z,t) = 0.5\ s\ \text{sech}^2\left[0.5\sqrt{s}(z - st)\right]$$

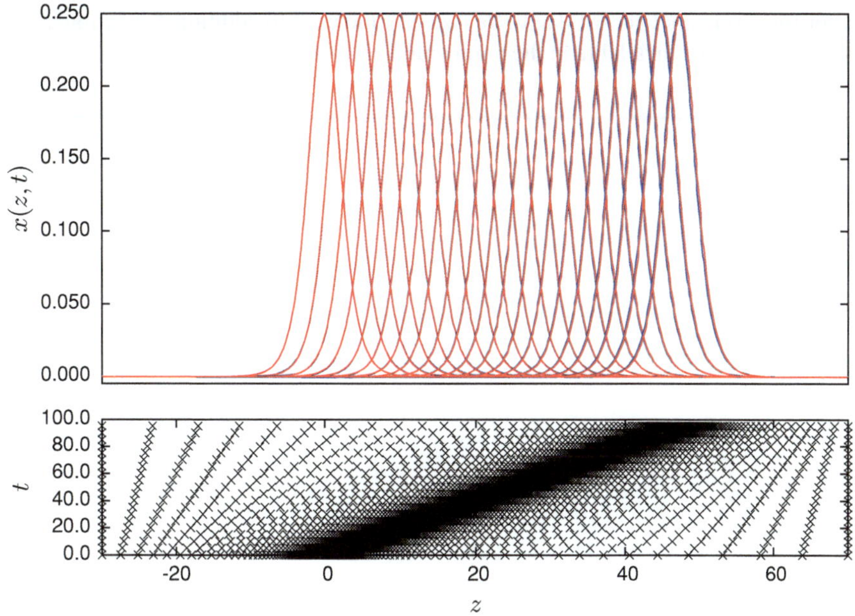

Fig. 5.31 Propagation of a single soliton solution of the Korteweg-de Vries equation, graphed at $t = 0, 5, \ldots, 100$. *Red curves* exact solution. *Blue curves* numerical solution. Grid adaptation is shown in the lower subplot

is studied. The first-order derivative term is computed using with a three-point centered differentiation matrix D_1, whereas the third-order derivative term is computed using stagewise differentiation, i.e., $x_{zzz} = D_1(D_1(D_1 x)))$.

For grid refinement, a monitor in the form

$$m(x) = \sqrt{(\alpha + \|x_z\|_2^2}$$

is used, where the parameter α can be used to avoid that too few grid points are allocated to flat parts of the solution.

Figure 5.31 shows the propagation of a single soliton with $s = 0.5$ on the time interval [0, 100].

Time integration is performed using `ode23s` with `AbsTol=10^{-3}` and `RelTol=10^{-3}`. Grid refinement occurs every 10 integration steps and the following parameter values are selected for regularization: $\alpha = 0.0$, $c = 0.005$ and $K = 1.1$. The average number of grid points for this simulation is 155. The grid refinement evolution is shown in the lower part of Fig. 5.31. The grid points are concentrated in the soliton and follow its progression. Script `main_korteweg_devries.m` shows the main program, which calls the grid refinement function `agereg`. This latter function is not shown due to its size and the calls to additional functions. The library of codes is available in the companion software.

```
close all
clear all

% Start a stopwatch timer
tic

% Set global variables
global s
global z0 zL nz D1
global nsteps maxsteps tprint tflag tout

% Spatial grid
z0     = -30.0;
zL     = 70.0;
nz     = 201;
nzout = nz;
dz     = (zL-z0)/(nz-1);
z      = [z0:dz:zL]';

% Initial conditions
s = 0.5;
x = kdv3_exact(z,0);

% parameters of the adaptive grid
npdes = 1;
nzmax = 1001;
alpha = 0;
beta  = 100;
tolz  = 0.005;
bound = 1.1;
imesh = 0;
ilim  = 0;

% refine the initial grid
[z_new, nz_new, ier, tolz_new] = agereg(z,x,npdes,...
                 nzmax,alpha,beta,tolz,bound,imesh,ilim);

% interpolate the dependent variables
x_new = spline(z,x,z_new);
x      = x_new';
z      = z_new';
nz     = nz_new;
tolz  = tolz_new;

% differentiation matrix
D1 = three_point_centered_D1(z);

% call to ODE solver
t0     = 0;
tf     = 100;
dt     = 5;
yout  = x;
zout  = z;
```

```
nzout = [nzout ; nz];
tout  = t0;

% solver to stop after this many steps:
maxsteps = 10;

% initial situation
figure(1)
subplot('position',[0.1 0.3 0.8 0.6])
plot(z,x);
ylabel('x(z,t)');
axis([-30 70 0 0.3])
hold on
subplot('position',[0.1 0.08 0.8 0.17])
plot(z,t0*ones(nz,1),'.b')
ylabel('t');
xlabel('z');
axis([-30 70 0 tf])
hold on

%
tk = t0;
tspan = [t0 tf];
tprint = dt;
while tk <= tf-1.e-5
    % initialize step counter
    nsteps = 0;

    % do the integration for maxsteps steps in a loop
    % until t becomes larger than tf
    options = odeset('RelTol',1e-3,'AbsTol',1e-6);
    options = odeset(options,'Events',@count_steps);
    options = odeset(options,'JPattern',jpattern(nz));
    [t,y,te,ye,ie] = ode23s(@kdv3_adaptive_pde,tspan,...
                                 x,options);
    %
    tk    = t(end);
    tspan = [tk tf];
    x     = [];
    x     = y(end,:);

    % refine the grid
    [z_new,nz_new,ier,tolz_new]=agereg(z,x,npdes,nzmax,...
                       alpha,beta,tolz,bound,imesh,ilim);

    % interpolate the dependent variables
    x_new = spline(z,x,z_new);
    x     = x_new';
    yout  = [yout ; x];

    %
    z     = z_new';
    nz    = nz_new;
    tolz  = tolz_new;
    zout  = [zout ; z];
    nzout = [nzout ; nz];
    tout  = [tout ; tk];
```

```
    % plot intermediate results
    if tflag >= 0
        figure(1)
        subplot('position',[0.1 0.3 0.8 0.6])
        plot(z,x);
        %
        yexact=kdv3_exact(z,tk);
        plot(z,yexact(1:length(z)),'r')
        %
        subplot('position',[0.1 0.08 0.8 0.17])
        plot(z,tk*ones(nz,1),'.b')
        tprint = tprint + dt;
    end

    % compute a new differentiation matrix
    D1 = three_point_centered_D1(z);

end

% read the stopwatch timer
tcpu=toc
nav = sum(nzout)/length(nzout)
```

Script main_korteweg_devries.m Main program to solve Korteweg–de Vries equation with a grid refinement method.

The code in `count_steps.m` shows the function Events that allows the solver to be halted periodically, after a given number of integration steps, or at a plot instant.

```
function [value,isterminal,direction] = count_steps(t,x)
%
global maxsteps tprint tflag nsteps
%
% update step counter
nsteps = nsteps + 1;

if nsteps < maxsteps || t < tprint
    value = 1;
else
    value = 0;
    tflag = t - tprint;
end
isterminal = 1;
direction  = 0;
```

Function count_steps.m "Events" function to halt periodically the time integrator.

We now consider another example, which consists of a system of two coupled PDEs modeling flame propagation, as originally discussed by Dwyer and Sanders [45].

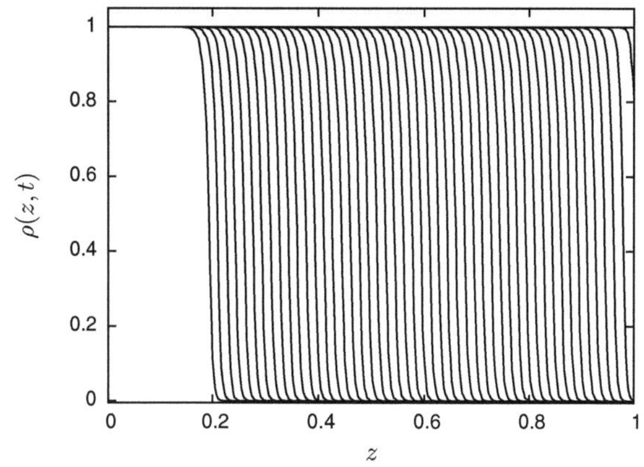

Fig. 5.32 Propagation of a density wave (from *left* to *right*) graphed at $t = 0, \ldots, 0.006$ (with time intervals 0.006/50), computed on a uniform grid with 401 *grid* points

$$\begin{array}{l} \rho_t = \rho_{zz} - N_{DA}\rho \\ T_t = T_{zz} + N_{DA}\rho \end{array}, \quad 0 < z < 1, \quad t > 0$$

where $\rho(z, t)$ is the gas density, $T(z, t)$ the temperature and $N_{DA} - 3.52 \times 10^6 e^{-4/T}$.

In this problem, a heat source located at one end of the spatial domain generates a moving flame front. The initial conditions are given by

$$\begin{array}{l} \rho(z, 0) = 1 \\ T(z, 0) = 0.2 \end{array}, \quad 0 \leqslant z \leqslant 1$$

whereas the boundary conditions include Neumann and Dirichlet conditions

$$\begin{array}{ll} \rho_z(0, t) = 0 & \rho_z(1, t) = 0 \\ T_z(0, t) = 0 & T(1, t) = f(t) \end{array}, \quad t \geqslant 0$$

The function $f(t)$ represents the heat source located on the left end

$$f(t) = \begin{cases} 0.2 + \dfrac{t}{2 \times 10^{-4}} & \text{if } t \leqslant 2 \times 10^{-4} \\ 1.2 & \text{if } t > 2 \times 10^{-4} \end{cases}$$

This heat source generates a flame front that propagates from left to right at an almost constant speed. This problem is not very difficult to solve but would require a relatively fine grid in order to resolve the flame front, as shown in Figs. 5.32 and 5.33 where 401 grid points are used together with a 3-point centered finite difference approximation.

Grid refinement can be used to track the flame front. This is shown in Figs. 5.34 and 5.35 where the equidistribution of a monitor function related to the solution

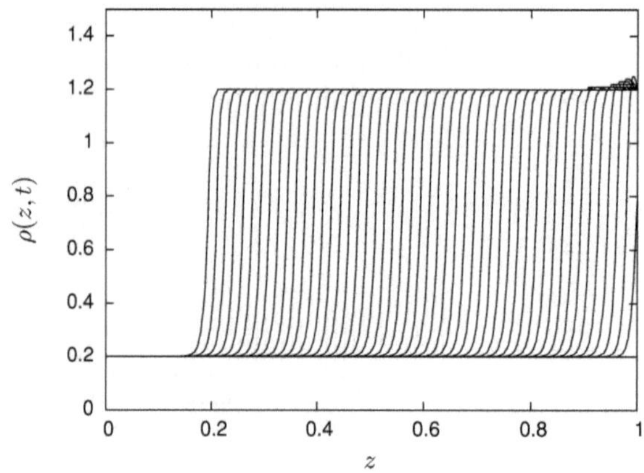

Fig. 5.33 Propagation of a temperature wave (from *left* to *right*) graphed at $t = 0, \ldots, 0.006$ (with time intervals 0.006/50), computed on a uniform *grid* with 401 *grid* points

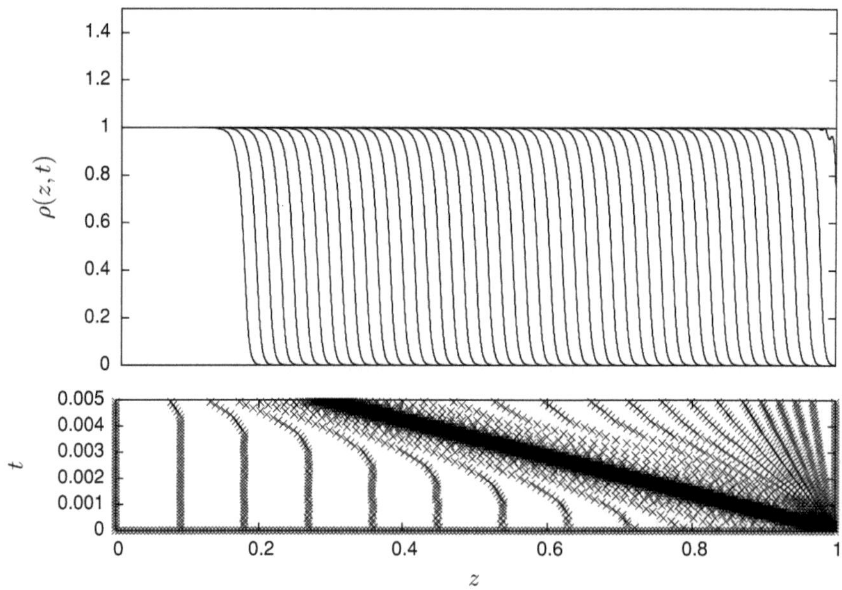

Fig. 5.34 Propagation of a density wave (from *left* to *right*) graphed at $t = 0, \ldots, 0.006$ (with time intervals 0.006/50), computed with a grid refinement algorithm. *Grid* adaptation is shown in the *lower* subplot

curvature is used

$$m(x) = \sqrt{\alpha + \|x_{zz}\|_\infty}$$

where $\|x_{zz}\|_\infty$ represents the maximum value taken by the second-order derivative in the spatial domain (it represents the infinity-norm). The lower subplots show that the grid points nicely concentrate in the front region. An average number of nodes of 152

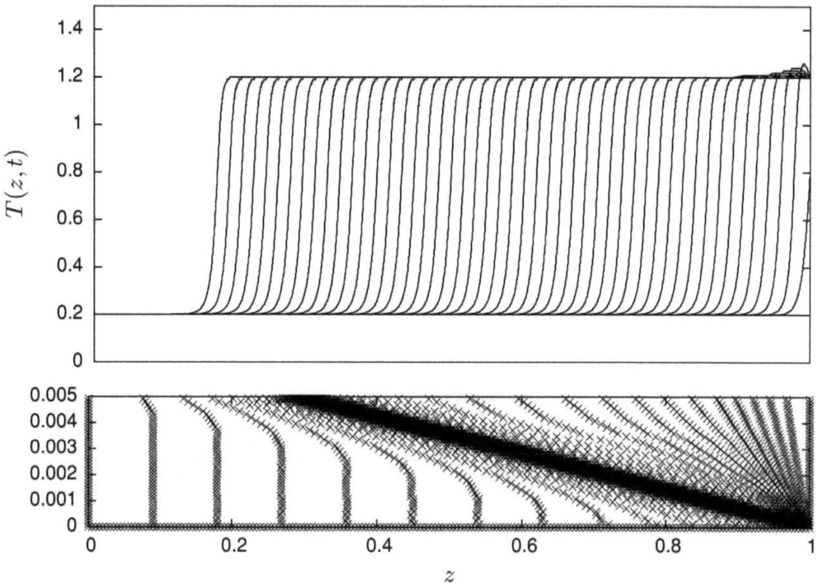

Fig. 5.35 Propagation of a temperature wave (from *left* to *right*) graphed at $t = 0, \ldots, 0.006$ (with time intervals 0.006/50), computed with a *grid* refinement algorithm. *Grid* adaptation is shown in the *lower* subplot

is used. As in the previous example, time integration is performed using `ode23s` with `AbsTol = 10^-3` and `RelTol=10^-3`. The following parameter values are selected for regularization: $\alpha = 0.05$, $c = 0.02$, and $K = 1.4$. A major difference with respect to the previous example is that grid adaptation occurs at regular time intervals ($t_f/50$) instead of an adaptation after a specified number of integration steps. The grid update interval has to be selected short enough so that the grid adapts well and does not lag too much behind of the front movement. Figure 5.36 shows what would happen if the grid adaptation interval is too long, e.g., $t_f/10$.
The algorithm is still able to compute a solution, but the spatial profiles suffer from important distortion.

5.8 An Additional PDE Application

We conclude this chapter with an application related to 2-phase flow in porous medium as described by Buckley-Leverett equation [46]

$$\frac{\partial x}{\partial t} + f(x) = 0$$

with the flux function $f(x) = \frac{x^2}{x^2+M(1-x)^2}$.

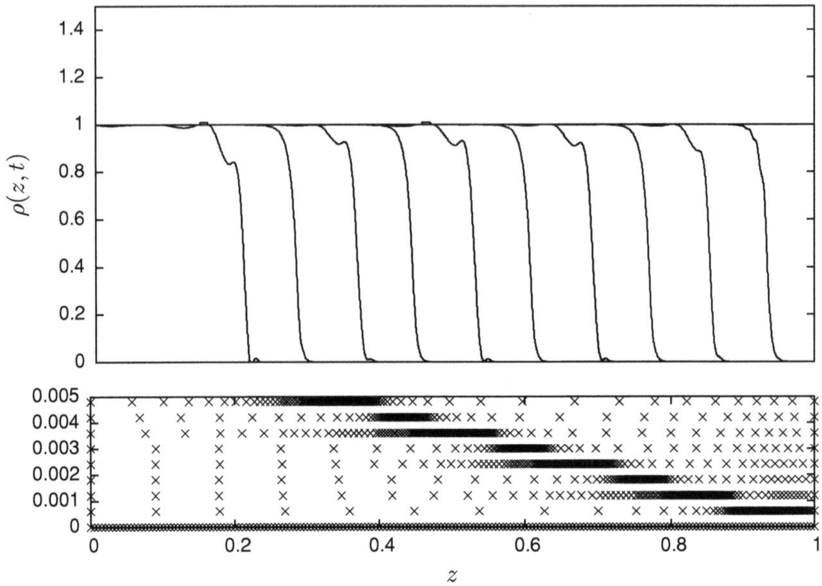

Fig. 5.36 Propagation of a density wave (from *left* to *right*) graphed at $t = 0, \ldots, 0.006$ (with time intervals 0.006/10), computed with a *grid* refinement algorithm. *Grid* adaptation is shown in the *lower* subplot

This mass balance equation finds applications in the petroleum industry, as a model of secondary recovery by water drive in oil reservoir. In this context, $x(z,t)$ represents water saturation ($x = 1$ corresponds to pure water and $x = 0$ corresponds to oil only). The flux function f is the water fractional flow function and $M > 0$ is the water over oil viscosity ratio. When a capillarity diffusion term is taken into account, the mass balance becomes

$$\frac{\partial x}{\partial t} + f(x) = \varepsilon \frac{\partial}{\partial z}\left[\nu(x)\frac{\partial}{\partial z}x\right]$$

with a diffusion coefficient, vanishing in $x = 0$ and $x = 1$, $\nu(x) = 4x(1-x)$ and a small parameter $\varepsilon = 0.01$.

If now gravitational effects are also taken into account, the flux function becomes (with $M = 1$)

$$f(x) = \frac{x^2}{x^2 + (1-x)^2}(1 - 5(1-x)^2)$$

We first solve the problem with the s-shaped flux function $f(x) = \frac{x^2}{x^2+M(1-x)^2}$, and with an initial condition

$$x(z,0) = \begin{cases} 1 - 3z & \text{if } \ 0 \leqslant z \leqslant 1/3 \\ 0 & \text{if } \ 1/3 < z \leqslant 1 \end{cases}$$

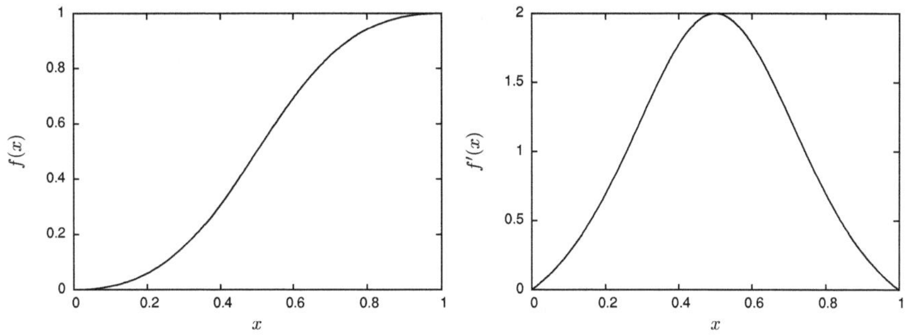

Fig. 5.37 Flux function $f(x) = \frac{x^2}{x^2+M(1-x)^2}$ with $M = 1$ and its derivative

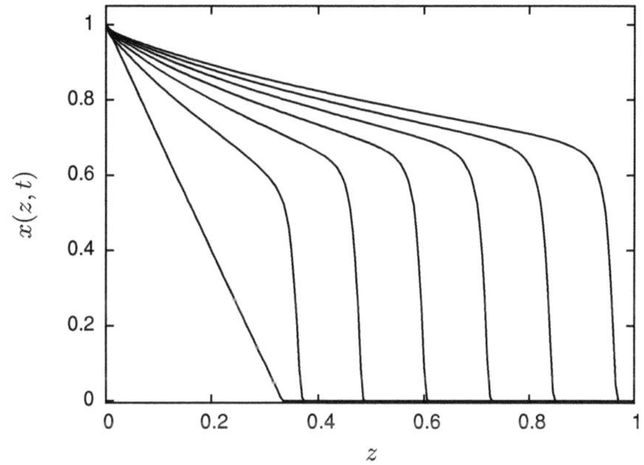

Fig. 5.38 Solution of Buckley-Leverett using an upwind Superbee limiter with 200 finite volumes graphed at $t = 0, 0.1, \ldots, 0.6$

The flux function has a derivative equal to $f'(x) = 2Mx(1-x)/\left(x^2 + M(1-x)^2\right)^2$ which is always positive (see Fig. 5.37).

This information is exploited in the numerical solution using a slope limiter. Figure 5.38 shows the solution obtained with an upwind Superbee slope limiter and 200 finite volumes.

The flux function $f(x) = \frac{x^2}{x^2+(1-x)^2}(1-5(1-x)^2)$ is more delicate to handle as the derivative $f'(x) = \frac{2x(1-x)(((10x-15)x+15)x-4)}{\left(x^2+M(1-x)^2\right)^2}$ changes sign, as shown in Fig. 5.39. It is therefore convenient to use a centered limiter of Kurganov-Tadmor, as shown in Fig. 5.40. Of course, an upwind scheme does also the job very well, but requires the information on the "direction of the wind" (which is given by the sign of the derivative of the flux function, and is easy to compute in this scalar example).

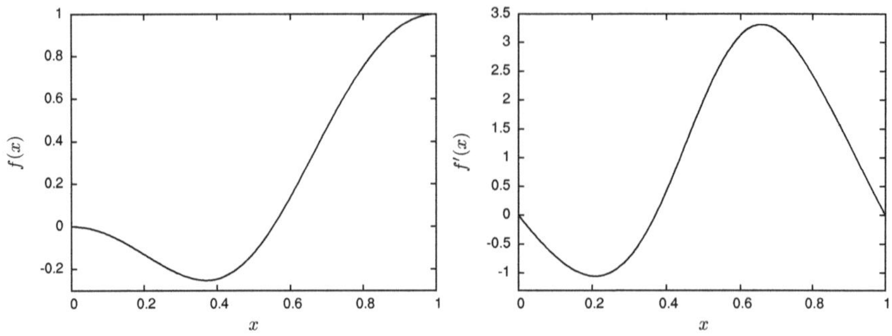

Fig. 5.39 Flux function $f(x) = x^2/x^2 + (1-x)^2(1-5(1-x)^2)$ and its derivative

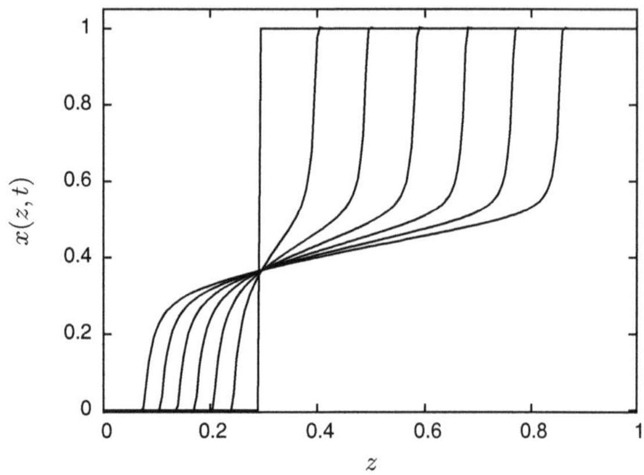

Fig. 5.40 Solution of Buckley-Leverett using a centered Kurganov-Tadmor with 200 finite volumes graphed at $t = 0, 0.04, \ldots, 0.28$

5.9 Summary

In this chapter, dedicated to problems with traveling wave solutions, we have first reviewed a few important theoretical concepts including conservation laws, the method of characteristics, the method of vanishing viscosity, and transformation-based methods. Then, we have focused attention on several numerical schemes:

- The classical upwind finite difference schemes;
- An operator splitting approach exploiting in a sequential way the previous concepts;
- Slope limiters which ensure an oscillation-free solution (TVD schemes);
- Grid refinement which allows to concentrate grid points in spatial regions where they are needed.

Although upwinding is central to the solution of convection-dominated problems, there also exists a class of centered schemes, reminiscent to Lax-Wendroff scheme, which has a very broad applicability and avoids the use of Riemann solvers for general nonlinear hyperbolic conservation laws.

These methods are tested with various application examples, including Burgers equation, a tubular bioreactor, a jacketed chemical reactor with an exothermic reaction, Sod shock tube problem, Korteweg-de Vries equation, a flame propagation problem, and Buckley-Leverett equation.

References

1. Sarra S (2003) The method of characteristics with applications to conservation laws. J Online Math Appl 3:8
2. Oleinik OA (1957) Discontinuous solutions of non-linear differential equations. AMS Transl Ser 2(26):95–172
3. Kruzkov S (1970) First-order quasilinear equations with several space variables. Math USSR Sb 10:217–273
4. LeVeque RJ (1990) Numerical methods for conservation laws. Lectures in Mathematics. ETH-Zurich. Birkhauser, Basel
5. Hopf E (1950) The partial differential equation $u_t + uu_x = \mu u_{xx}$. Commun Pure Appl Math 3:201–230
6. Sachdev PL (1978) A generalised cole-hopf transformation for nonlinear parabolic and hyperbolic equations. Z für Angew Math Phys 29(6):963–970
7. Salas AH, Gómez CA (2010) Application of the cole-hopf transformation for finding exact solutions to several forms of the seventh-order KdV equation. Math Prob Eng. doi:10.1155/2010/194329
8. Sakai K, Kimura I (2005) A numerical scheme based on a solution of nonlinear advection-diffusion equations. J Comput Appl Math 173:39–55
9. Malfliet W, Hereman W (1996) The tanh method: i. exact solutions of nonlinear evolution and wave equations. Phys Scr 54:563–568
10. He JH, Wu XH (2006) Exp-function method for nonlinear wave equations. Chaos Solitons Fractals 30:700–708
11. Misirli Y, Gurefe E (2010) The exp-function method to solve the generalized burgers-fisher equation. Nonlinear Sci Lett A Math Phys Mech 1:323–328
12. Gurefe Y, Misirli E (2011) Exp-function method for solving nonlinear evolution equations with higher order nonlinearity. Comput Math Appl 61(8):2025–2030
13. Hundsdorfer W, Verwer J (2003) Numerical solution of time-dependent advection-diffusion-reaction equations. Springer series in computational mathematics, vol 33. Springer, Berlin
14. Simpson M, Landman K, Clement T (2005) Assessment of a non-traditional operator split algorithm for simulation of reactive transport. Math. Comput Simul 70:44–60
15. Lanser D, Verwer JG (1999) Analysis of operator splitting for advection-diffusion-reaction problems from air pollution modelling. J Comput Appl Math 111:201–216
16. Renou S, Perrier M, Dochain D, Gendron S (2003) Solution of the convection-dispersion-reaction equation by a sequencing method. Comput Chem Eng 27:615–629
17. Laabissi M, Winkin JJ, Dochain D, Achhab ME (2005) Dynamical analysis of a tubular biochemical reactor infinite-dimensional nonlinear model. In: 44th IEEE conference on decision and control and European control conference, pp 5965–5970
18. Logist F, Saucez P, Van Impe J, Vande-Wouwer A (2009) Simulation of (bio)chemical processes with distributed parameters using matlab. Chem Eng J 155:603–616
19. Lax PD (1973) Hyperbolic systems of conservation laws and the mathematical theory of shock waves. Society for Industrial and Applied Mathematics, Philadelphia
20. Harten A (1983) High-resolution schemes for hyperbolic conservation laws. J Comput Phys 49:357–385
21. Sweby PK (1984) High resolution schemes using flux-limiters for hyperbolic conservation laws. SIAM J Num Anal 21(5):995–1011

22. Tannehil JC, Anderson DA, Pletcher RH (1997) Computational fluid mechanics and heat transfer, 2nd edn. Taylor and Francis, Wasington DC
23. Zijlema M, Wesselin P (1995) Higher order flux-limiting methods for steady-state, multidimensional, convection-dominated flow. Technical Report 95–131, TU Delft
24. van Leer B (1974) Towards the ultimate conservative difference scheme II: monotonicity and conservation combined in a second order scheme. J Comput Phys 14(4):361–370
25. Van Leer B (1997) Towards the ultimate conservative difference scheme III: upstream-centered finite-difference schemes for ideal compressible flow. J Comput Phys 23(3):263–275
26. Roe PL (1986) Characteristic-based schemes for the Euler equations. Ann Rev Fluid Mech 18:337–365
27. Gaskell PH, Lau AKC (1988) Curvature-compensated convective transport: smart, a new boundedness-preserving transport algorithm. Int J Num Meth Fluids 8(6):617–641
28. Koren B (1993) A robust upwind discretisation method for advection, diffusion and source terms. In: Vreugdenhil CB, Koren B (eds) Numerical methods for advection-diffusion problems. Vieweg, Braunschweig, pp 117–138
29. Godunov SK (1959) A difference scheme for numerical solution of discontinuous solution of hydrodynamic equations. Math Sb 47:271–306
30. Obertscheider C (2001) Burgers equation. Projektarbeit, Universität Wien Mathematik und Physik
31. Landajuela M (2011) Burgers equation. Technical report, BCAM Internship report
32. Toro EF (1999) Riemann solvers and numerical methods for fluid dynamics. Springer, Berlin
33. LeVeque RJ (2002) Finite volume methods for hyperbolic problems. Cambridge University Press, Cambridge
34. Nessyahu H, Tadmor E (1990) Non-oscillatory central differencing for hyperbolic conservation laws. J Comput Phys 87(2):408–463
35. Lax PD (1954) Weak solutions of nonlinear hyperbolic equations and their numerical computation. Comm Pure Appl Math 7:159
36. Kurganov A, Tadmor E (2000) New high-resolution central schemes for nonlinear conservation laws and convection-diffusion equations. J Comput Phys 160:241–282
37. Logist F, Smets I, Van Impe J (2005) Optimal control of dispersive tubular chemical reactors: Part i. In: 16th IFAC World Congress, Prague
38. Danckwerts P (1953) Continuous flow systems. Chem Eng Sci 2:1–13
39. Sod GA (1978) A survey of several finite difference methods for systems of nonlinear hyperbolic conservation laws. J Comput Phys 27:1–31
40. Vande Wouwer A, Saucez P, Schiesser WE (2001) Adaptive method of lines. Chapman Hall/CRC, Boca Raton
41. Steinebach G (1995) Die Linienmethode und row-verfahren zur abfluß- und prozeßsimulation in fließfließgewässern am beispiel von rhein und mosel. Ph.D. thesis, Darmstadt Technical University, Germany
42. Saucez P, Vande Wouwer A, Schiesser WE (1996) Some observations on a static spatial remeshing method based on equidistribution principles. J Comput Phys 128:274–288
43. Kautsky J, Nichols NK (1980) Equidistributing meshes with constraints. SIAM J Sci Stat Comput 1:499–511
44. Korteweg DJ, deVries G (1895) On the change of form of long waves advancing in a rectangular canal, and on a new type of long stationary wave. Philos Mag 39:442–443
45. Dwyer HA, Sanders BR (1978) Numerical modeling of unsteady flame propagation. Technical report, Report SAND77-8275, Sandia National Lab, Livermore
46. Buckley SE, Leverett MC (1942) Mechanism of fluid displacement in sands. Trans AIME 146:107–116

Further Readings

47. LeVeque RJ (1990) Numerical methods for conservation laws. Lectures in mathematics. ETH-Zurich, Birkhauser, Basel
48. LeVeque RJ (2002) Finite volume methods for hyperbolic problems. Cambridge University Press, Cambridge
49. Tannehil JC, Anderson DA, Pletcher RH (1997) Computational fluid mechanics and heat transfer, 2nd edn. Taylor and Francis, Wasington DC
50. Toro EF (1999) Riemann solvers and numerical methods for fluid dynamics. Springer, Berlin
51. Vande Wouwer A, Saucez P, Schiesser WE (2001) Adaptive method of lines. Chapman Hall/CRC Press, Boca Raton
52. Vande Wouwer A, Saucez P, Schiesser WE, Thompson S (2005) A MATLAB implementation of upwind finite differences and adaptive grids in the method of lines. J Comput Appl Math 183:245–258
53. Whitham GB (1999) Linear and Nonlinear Waves. Wiley, Hoboken

Chapter 6
Two Dimensional and Time Varying Spatial Domains

So far in this book, we have focused attention on initial-boundary value problems defined on time-invariant spatial domains in one spatial dimension. These problems are common in practice and are easier to handle than problems in more spatial dimensions. They are also best suited for introducing various numerical techniques. In addition, a 1D formulation can be a good approximation of a more complex problem in many instances, which can be advantageous when using the simulation program in contexts such as optimization or control, where a small computational load is of paramount importance.

In this chapter, we will depart from this simple definition of the spatial domain, and first consider problems in two spatial dimensions, and their treatment with finite difference and finite element approximations. FDs are easily extended on spatial domains with simple geometries such as squares, rectangles, cylinders, etc., but are not appropriate for spatial domains with more complex shapes, where finite element methods are the first choice. We also consider through examples the use of the slope limiters and of the proper orthogonal decomposition technique introduced earlier in this book.

We then briefly address problems defined on time-varying spatial domains. This situation occurs in many practical situations, for instance, a melting snowman whose volume is decreasing due to positive temperatures, or a biofilm on the wall of a tubular reactor whose boundary layer is moving as bacterial population develops and colonizes the available space. In this chapter, this problematic is introduced based on a freeze-drying application example.

6.1 Solution of Partial Differential Equations in More than 1D Using Finite Differences

In this section we would like to introduce a few simple finite difference methods that can be used to solve problems in two spatial dimensions on domains with simple geometries. What we mean by a simple geometry is a convex domain in Cartesian or polar coordinates, such as a square, a rectangle, a quadrilateral, a disk. Finite

A. Vande Wouwer et al., *Simulation of ODE/PDE Models with MATLAB®, OCTAVE and SCILAB*, DOI: 10.1007/978-3-319-06790-2_6,

difference schemes are easy to understand and to apply in 1D but unfortunately they cannot be extended efficiently on spatial domains with complex shapes.

We will depart from the notation used in the previous chapters where x is the dependent variable and z the spatial coordinate (which is a notation inspired from systems theory where the state variable is usually denoted x) to avoid using a subscript notation for the several spatial coordinates $z_1, z_2, \ldots$. In agreement with the notational habits in the numerical analysis community, we will use u to denote the dependent variable, and x, y, z for the Cartesian coordinates. This way, the general problem under consideration in this chapter reads

$$\mathbf{u}_t = \mathbf{f}\left(\mathbf{u}, \mathbf{u}_x, \mathbf{u}_y, \mathbf{u}_{xx}, \mathbf{u}_{yy}, \ldots, x, y, t\right) \tag{6.1}$$

where $\mathbf{u}(x, y, t)$ can be scalar or vector. The spatial domain $\Omega(x, y)$ is assumed convex and time invariant, i.e, the spatial boundary Γ does not move or change shape along time.

PDE (6.1) is complemented by an initial condition

$$\mathbf{u}(x, y, 0) = \mathbf{u}^0(x, y) \tag{6.2}$$

and by conditions along the boundary Γ. These conditions can basically be of three types:

- Dirichlet

$$\mathbf{u}(x_\Gamma, y_\Gamma, t) = \mathbf{g}_D(t) \tag{6.3}$$

- Neumann

$$\nabla\mathbf{u}(x_\Gamma, y_\Gamma, t) = \mathbf{g}_N(t) \tag{6.4}$$

- Robin

$$k_0\nabla\mathbf{u}(x_\Gamma, y_\Gamma, t) + k_1\mathbf{u}(x_\Gamma, y_\Gamma, t) = \mathbf{g}_R(t) \tag{6.5}$$

where (x_Γ, y_Γ, t) is an arbitrary point along Γ and ∇ is the gradient operator (i.e. in 1D, $\nabla u = \frac{\partial u}{\partial z}$). A mix of these conditions can be used on different portions of Γ.

We will now study a few examples of increasing complexity and use finite difference methods to elaborate a solution.

6.1.1 The Heat Equation on a Rectangle

Consider the heat equation on a rectangular spatial domain

$$T_t = k(T_{xx} + T_{yy}) \tag{6.6}$$

with $k = 10$, defined on

$$\Omega = \{(x, y) : 0 \geqslant x \geqslant 1; \quad 0 \geqslant y \geqslant 2\} \tag{6.7}$$

for $0 \leqslant t \leqslant 0.01$, and with the following initial conditions:

$$T(x, y, 0) = 0 \tag{6.8}$$

and boundary conditions

$$T(x, 0, t) = T(0, y, t) = 0 \tag{6.9}$$

$$T(x, 2, t) = T(1, y, t) = T_M = 100 \tag{6.10}$$

We assume that in the points (1, 0) and (0, 2), BC (6.10) prevails.

The coding of the solution to this problem is analog to the codes previously detailed: a main program defines the model parameters, the spatial domain, the finite difference matrices, calls an ODE solver, then displays the results, whereas an ODE function computes the time derivatives of the dependent variables based on the model equations, and the finite difference operators. The codes are listed in the script `heat_equation_main` and `heat_equation_pde`.

```
close all
clear all

% start a stopwatch timer
tic

% set global variables
global k TM
global nx ny D2x D2y

% model parameters
TM = 100;
k = 10;

% spatial grid
x  = [0:0.01:1];
y  = [0:0.02:2];
nx = length(x);
ny = length(y);
nv = nx*ny;

% initial conditions
T0(1:nv,1) = zeros(nv,1);

% finite difference (FD) approximation of the spatial derivatives
D2x = five_point_centered_D2(x);
D2y = five_point_centered_D2(y);

% call to ODE solver
t = [0:0.01/3:0.01];
[tout,Tout] = ode45(@heat_equation_pde,t,T0);

% impose BCs
for k = 1:length(tout)
```

```
    Tout(k,1:nx) = 0;
    Tout(k,1:nx:nx*(ny-1)+1) = 0;
    Tout(k,nx:nx:nx*ny) = TM;
    Tout(k,nx*(ny-1)+1:nx*ny)= TM;
end

% plot results
for j = 1:length(tout)
    figure
    u(ny:-1:1,:) = reshape(Tout(j,1:nv),nx,[])';
    surf(x,y(ny:-1:1),u,'EdgeColor','none');
    axis([0 1 0 2 0 100])
    xlabel('x','FontSize',14);
    ylabel('y','FontSize',14);
    zlabel('T(x,y)','FontSize',14)
    set(gca,'FontSize',14)
end

% isotherms
for j = 1:length(t)
    figure
    u(ny:-1:1,:) = reshape(Tout(j,1:nv),nx,[])';
    [C,h] = contour(x,y(ny:-1:1),u);
    set(h,'ShowText','on','TextStep',get(h,'LevelStep')*2)
    axis([-.1 1.1 -.1 2.1])
    xlabel('x','FontSize',14); ylabel('y','FontSize',14)
    set(gca,'FontSize',14)
end

% read the stopwatch timer
toc
```

Script heat_equation_main Main program for the solution of the heat equation on a rectangular domain.

```
function Tt = heat_equation_pde(t,T)

% set global variables
global k TM
global nx ny D2x D2y

% boundary conditions
T(1:nx,1)              = 0;
T(1:nx:nx*(ny-1)+1,1)  = 0;
T(nx:nx:nx*ny,1)       = TM;
T(nx*(ny-1)+1:nx*ny,1) = TM;

% spatial derivatives
[Txx Tyy] = second_order_derivatives_2D(T,nx,ny,D2x,D2y);

% temporal derivatives
Tt = k*(Txx+Tyy);
```

Function heat_equation_pde PDE function for the solution of the heat equation on a rectangular domain.

We can note the following points, which are specific to the 2D nature of the problem:

1. The rectangular spatial domain is easily defined by

```
% spatial grid
x = [0:0.01:1];
y = [0:0.02:2];
nx = length(x);
ny = length(y);
nv = nx*ny;
```

2. The discretized solution is a matrix with `ny` lines and `nx` columns, which is stored as a column vector `u` with `nv` elements. The storage proceeds line by line, the first line corresponding to the bottom limit of the spatial domain, and the last one (the `ny`th line) to the upper limit.

   This column-vector storage is initiated by the definition of the initial condition:

```
% initial conditions
T0(1:nv,1) = zeros(nv,1);
```

   and leads to some index manipulation when handling the boundary conditions in the PDE function:

```
% boundary conditions
T(1:nx,1)               = 0;
T(1:nx:nx*(ny-1)+1,1)   = 0;
T(nx:nx:nx*ny,1)        = TM;
T(nx*(ny-1)+1:nx*ny,1)  = TM;
```

   It also necessitates reshaping the output of the solver for further graphing:

```
u(ny:-1:1,:) = reshape(Tout(j,1:nv),nx,[])';
surf(x,y(ny:-1:1),u,'EdgeColor','none');
```

3. The construction of the finite difference approximation is essentially based on the 1D finite differences library that has been developed in Chap. 3

```
D2x = five_point_centered_D2(x);
D2y = five_point_centered_D2(y);
```

   There is flexibility in the selection of the discretization scheme in each direction, and for instance an alternative to the previous choice would be:

```
D2x = five_point_centered_D2(x);
D2y = three_point_centered_D2(y);
```

   The differentiation matrices `D2x` and `D2y` are used by the function `second_order_derivatives_2D`, which uses an intermediate matrix `v`, and provides column vectors `uxx` and `uyy` with the appropriate storage order.

```
function [uxx uyy] = second_order_derivatives_2D(u,nx,ny,D2x,D2y)

% use finite differences in 1D along x and along y
v   = reshape(u,nx,[]);
uxx = reshape(D2x*v,nx*ny,1);
uyy = reshape(v*D2y',nx*ny,1);
```

Function second_order_derivatives_2D Function to compute the second-order derivatives on the 2D spatial domain.

Figure 6.1 shows the 3D plot of the solution at time $t = 0.01$ s while Fig. 6.2 is a 2D graph displaying isotherms of the solution at the same time.

6.1.2 Graetz Problem with Constant Wall Temperature

The first example has shown the possibility to use 1D finite difference approximations to compute derivatives on rectangular domains. This possibility of course also exists for other domains defined by orthogonal coordinates, such as the following problem defined in a cylinder. This problem initially considered by Graetz [1] describes the energy balance of an incompressible Newtonian fluid entering a tube with a wall at constant temperature (which can be larger or lower than the initial fluid temperature). It is assumed that the flow is laminar. In normalized variables, the problem can be written as

$$T_t = -P_e v(r) T_z + T_{zz} + \frac{1}{r} T_r + T_{rr} \tag{6.11}$$

where $T(r, z, t)$ denotes the temperature, $v(r) = 2(1 - (r/R)^2)$ is the parabolic velocity profile, and $P_e = Rv_0/\alpha$ is Peclet number, where R is the tube radius, v_0 is the velocity on the center line (the maximum fluid velocity) and $\alpha = \lambda/(\rho c_p)$ is the thermal diffusivity. As a particular example, we will consider $R = 1$, $L = 30$, $P_e = 60$.

Figure 6.3 represents the spatial domain for this problem.

The previous energy balance PDE is supplemented with the following initial conditions:

$$T(r, z, 0) = 0 \tag{6.12}$$

and boundary conditions

$$T_r(0, z, t) = 0 \quad \text{(radial symmetry)} \tag{6.13}$$

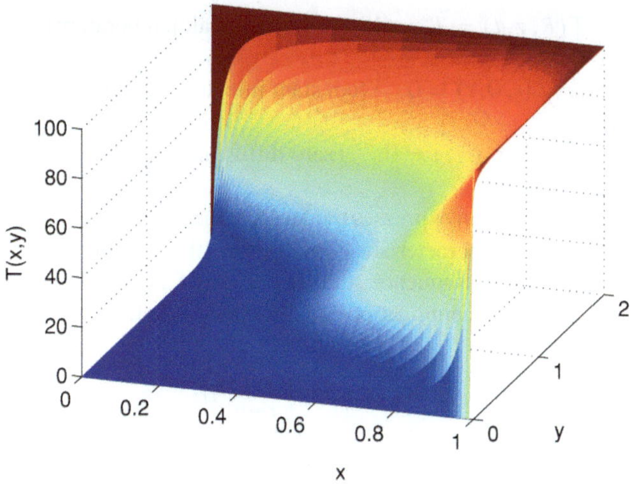

Fig. 6.1 Numerical solution of the 2D heat equation at time $t = 0.01$—3D representation

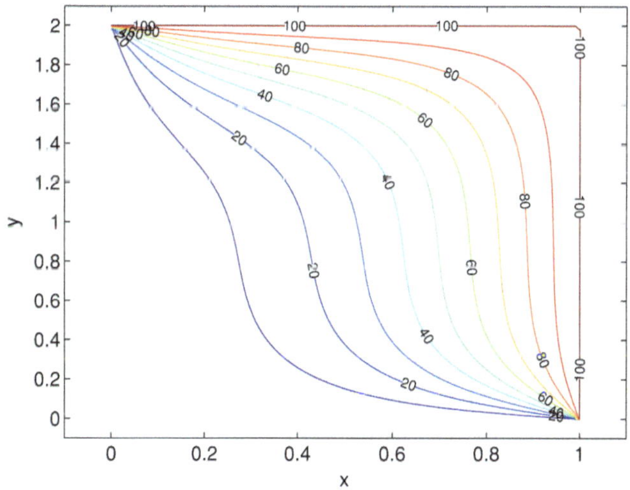

Fig. 6.2 Numerical solution of the 2D heat equation at time $t = 0.01$—isotherm representation

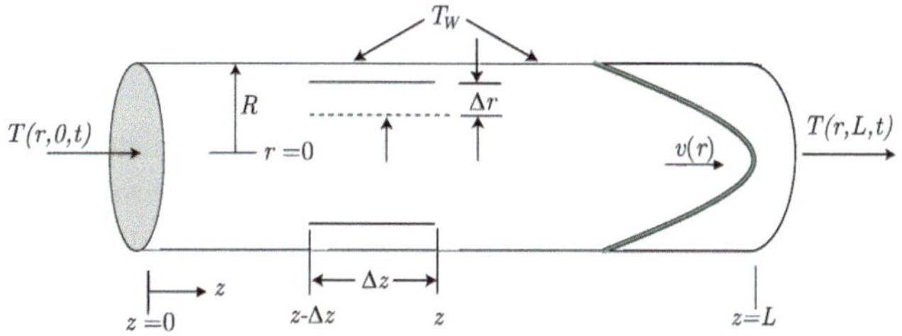

Fig. 6.3 Spatial domain for Graetz problem

$$T(R, z, t) = T_w = 1 \qquad \text{(constant wall temperature)} \tag{6.14}$$

$$T(r, 0, t) = 0 \qquad \text{(constant inlet temperature)} \tag{6.15}$$

$$T_z(r, L, t) = 0 \qquad \text{(zero diffusion at outlet)} \tag{6.16}$$

The first boundary condition expresses the fact that the temperature profile is radially symmetric, an important characteristic, which allows the consideration of two spatial coordinates only. As a consequence, the term T_r/r of Eq. (6.11) is undeterminate (0/0) on the tube axis. The application of L'Hôpital's rule leads to

$$\lim_{r \to 0} = \frac{1}{r} T_r = \lim_{r \to 0} T_{rr} \tag{6.17}$$

so that PDE (6.11) can be expressed as

$$T_t = -P_e v(r) T_z + T_{zz} + 2T_{rr} \tag{6.18}$$

along the center line.

The program structure is very similar to the previous example. The code is available in the companion library. Here, we just stress a few points:

- The spatial grid is defined by

```
% spatial grid
z0 = 0; zL = 30;
nz = 101; dz = (zL-z0)/(nz-1); z = [z0:dz:zL];
%
r0 = 0; rR = 1;
nr = 101; dr = (rR-r0)/(nr-1); r = [r0:dr:rR];
```

- The finite difference schemes are selected according to the nature (convective or diffusive) of the material transportation phenomena

```
% finite difference (FD) approximation of the spatial
derivatives
D1z = four_point_biased_upwind_D1(z,1);
D1r = five_point_centered_D1(r);
D2z = five_point_centered_D2(z);
D2r = five_point_centered_D2(r);
```

- The first-order derivatives are computed with the function listed in `first_order_derivatives_2D` (which is almost identical to the function `second_order_derivatives_2D`, but it is used for clarity and convenience).

```
function [ux uy] = first_order_derivatives_2D(u,nx,ny,...
                                              D1x,D1y)

% Use finite differences in 1D along x and along y
v  = reshape(u,nx,[]);
ux = reshape(D1x*v,nx*ny,1);
uy = reshape(v*D1y',nx*ny,1);
```

Function first_order_derivatives_2D Function to compute the first-order derivatives on the 2D spatial domain.

- Neumann BCs (6.13) and (6.16) can be expressed—in an approximate way—as Dirichlet BCs using two-point finite difference schemes (a zero slope means that the values of the variable in the two neighboring grid points are the same). With BCs (6.14–6.15) they form the following set of Dirichlet BCs:

```
% boundary conditions
T(1:nz,1)  = T(nz+1:2*nz,1);
T(nv-nz+1:nv,1)  = 1;
T(1:nz:nv-nz+1,1)  = 0;
T(nz:nz:nv,1)  = T(nz-1:nz:nv-1,1);
```

- The PDE is coded so as to take account of the initial undetermination along the central line, and the use of l'Hôpital rule (the vector `r_inv` contains the values of $1/r$)

```
% temporal derivatives
Tt                  = -Pe*v.*Tz + Tzz + Trr;
Tt(nz+1:nz*nr,1)  = Tt(nz+1:nz*nr,1) +...
                      r_inv(nz+1:nz*nr,1).*Tr(nz+1:nz*nr,1);
Tt(1:nz,1)          = Tt(1:nz,1) + Trr(1:nz,1);
```

- The problem symmetry leads us to express the solution on half of the actual spatial domain (i.e. for $0 \leqslant r \leqslant R$), and at the time of plotting the results a mirror image around the r-axis has to be created

```
[Z R]  = meshgrid(z,r(nr:-1:1));
[Z1 R1]  = meshgrid(z,-r);
```

and in a loop

```
u(nr:-1:1,:)  = reshape(Tout(j,1:nv),nz,[])';
%
surf(Z,R,u,'EdgeColor','none')%[.95 .95 .95]
hold
surf(Z1,R1,u(nr:-1:1,:),'EdgeColor','none')
%[.95 .95 .95]
```

The solution at time $t = 0.05$, $t = 0.15$ and $t = 0.35$ is displayed in Figs. 6.4, 6.5 and 6.6, respectively.

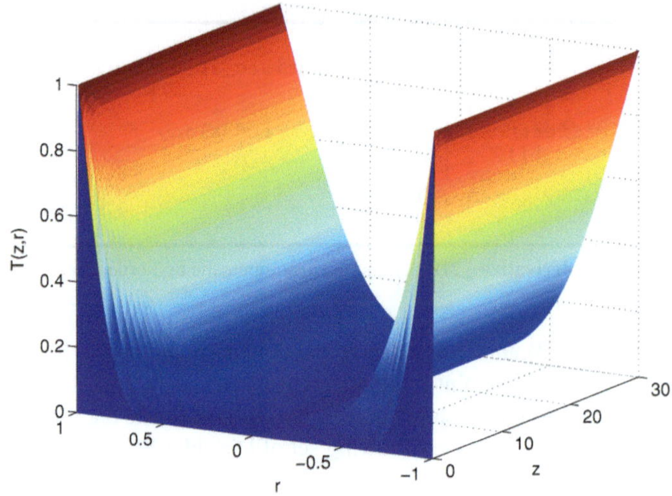

Fig. 6.4 Numerical solution of the Graetz problem at $t = 0.05$

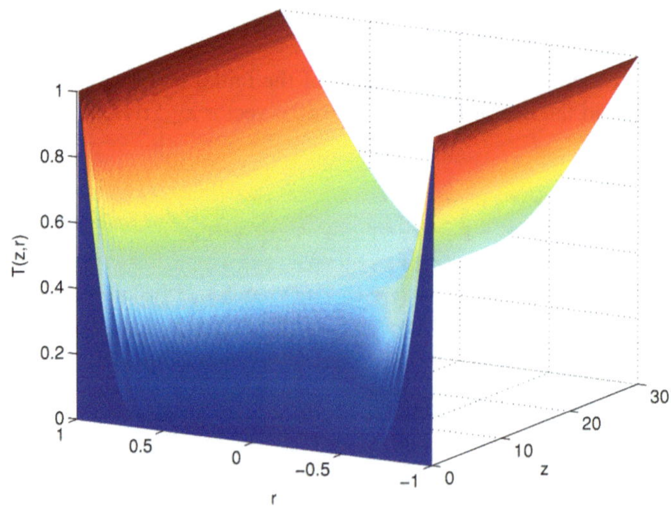

Fig. 6.5 Numerical solution of the Graetz problem at $t = 0.15$

6.1.3 Tubular Chemical Reactor

Based on the results obtained with Graetz problem, we can now consider a more challenging IBVP related to a tubular chemical reactor, represented in Fig. 6.7, and for which we will study the numerical simulation results both in 1D and 2D. In 2D, the reactor equations consist of one mass-balance PDE, including convection,

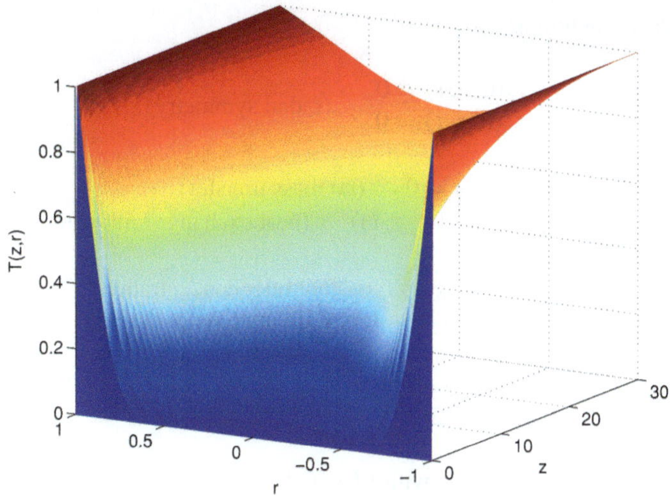

Fig. 6.6 Numerical solution of the Graetz problem at $t = 0.35$

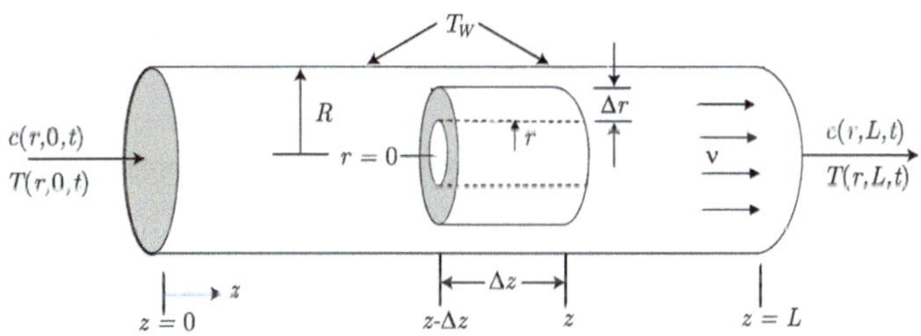

Fig. 6.7 Tubular reactor spatial domain

diffusion and chemical reaction, and one energy balance PDE with the counterpart terms.

$$
\begin{aligned}
c_t &= -v(r)c_z + D\left(c_{zz} + c_{rr} + \frac{1}{r}c_r\right) - r(c, T) \\
T_t &= -v(r)T_z + \frac{\lambda}{\rho c_p}\left(T_{zz} + T_{rr} + \frac{1}{r}T_r\right) + \frac{-\Delta H}{\rho c_p} r(c, T)
\end{aligned}
\tag{6.19}
$$

where $r(c, T) = k0 \exp\left(-\frac{E}{RT}\right)c^2$ and with a fluid flow velocity profile, which can be parabolic as in Graetz problem, $v(r) = v_{\max}\left(1 - (r/R)^2\right)$ or constant, $v(r) = v_{\max}/2$. These equations are supplemented by initial conditions

$$
\begin{aligned}
c(r, z, 0) &= c_0(r, z) \\
T(r, z, 0) &= T_0(r, z)
\end{aligned}
\tag{6.20}
$$

and boundary conditions

$$\begin{aligned} c_r(0, z, t) &= 0 \\ T_r(0, z, t) &= 0 \end{aligned} \quad \text{(radial symmetry)} \tag{6.21}$$

$$\begin{aligned} c_r(R, z, t) &= 0 \quad \text{(no mass transfer)} \\ T_r(R, z, t) &= \tfrac{h}{\lambda}(T_w - T(R, z, t)) \quad \text{(heat exchange with wall)} \end{aligned} \tag{6.22}$$

$$\begin{aligned} c(r, 0, t) &= c_{in} \quad \text{(constant inlet concentration)} \\ T(r, 0, t) &= T_{in} \quad \text{(constant inlet temperature)} \end{aligned} \tag{6.23}$$

$$\begin{aligned} c_z(r, L, t) &= 0 \\ T_z(r, L, t) &= 0 \end{aligned} \quad \text{(zero diffusion at outlet)} \tag{6.24}$$

Again the use of l'Hôpital's rule with BCs (6.21) allows expressing the PDEs along the center line as

$$\begin{aligned} c_t &= -vc_z + D\,(c_{zz} + 2c_{rr}) - r(c, T) \\ T_t &= -vT_z + \frac{\lambda}{\rho c_p}\,(T_{zz} + 2T_{rr}) - \frac{\Delta H}{\rho c_p} r(c, T) \end{aligned} \quad \text{in } r = 0 \tag{6.25}$$

There is obviously no mass transfer through the wall, as expressed by BC (6.22), so that the mass-balance PDE at the wall ($r = R$) reduces to:

$$c_t = -v(r)c_z + D(c_{zz} + c_{rr}) - r(c, T)$$

This way, some of the BCs are naturally accounted for. As for BC (6.22) describing the heat exchange at the wall, different coding could be used as introduced in Chap. 3. In our implementation, we have used a transformation of the BCs into an ODE with fast dynamics (adjustable via the parameter `gamma`).

```
T_t(nr,j) = gamma * ((h/lambda) * (Tw-T(nr,j))-Tr(nr,j));
```

In 1D, the reactor equations can be written as follows

$$\begin{aligned} c_t &= -vc_z + Dc_{zz} - r(c, T) \\ T_t &= -vT_z + \frac{\lambda}{\rho c_p} T_{zz} - \frac{\Delta H}{\rho c_p} r(c, T) + \frac{1}{\rho c_p}\frac{2h}{R}(T_w - T) \end{aligned} \tag{6.26}$$

with initial and boundary conditions

$$\begin{aligned} c(z, 0) &= c_0(z) \\ T(z, 0) &= T_0(z) \end{aligned} \tag{6.27}$$

$$\begin{aligned} c(0, t) &= c_{in} \\ T(0, t) &= T_{in} \end{aligned} \quad \text{(constant inlet temperature and concentration)} \tag{6.28}$$

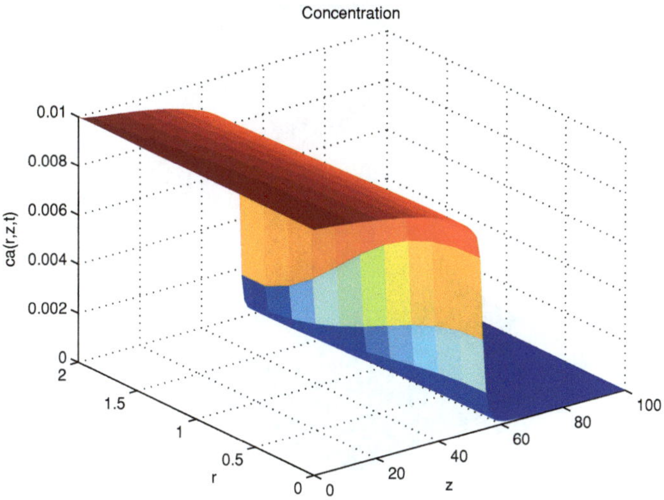

Fig. 6.8 Steady-state concentration profiles in the tubular reactor with a parabolic velocity profile

$$\begin{aligned} c_z(L,t) &= 0 \\ T_z(L,t) &= 0 \end{aligned} \quad \text{(zero diffusion at outlet)} \tag{6.29}$$

We see that the heat exchange phenomenon is no longer formulated as a boundary condition, but as a source term in the energy-balance PDE. The multiplicative coefficient $2/R$ expresses the ratio of an elementary surface area (i.e. heat flows through an element of the wall surface $2\pi R\Delta z$) to the corresponding volume of the tube section (i.e. the PDE is obtained by expressing balances around an element of volume $\pi R^2 \Delta z$).

Figures 6.8 and 6.9 show the steady state-concentration and temperature profiles corresponding to a parabolic velocity profile, whose influence is clearly recognizable in the shape of the profiles, whereas Figs. 6.10 and 6.11 shows the same information in the case of a constant velocity. These results are obtained with 101 grid points in the axial direction and 11-points in the radial direction.

On the other hand, Fig. 6.12 presents the time evolution of the temperature profile inside the reactor, modeled as a 1D system. It is apparent that this evolution is close to the temperature on the reactor center line, as predicted by the 2D model. As already mentioned, a 1D representation of a distributed parameter system can be advantageous in terms of simplicity and computational efficiency. In the context of the tubular reactor problem, this approximation will be accurate if the reactor length is large as compared to its radius (in our example this ratio is 50), and if the thermal conductivity is high.

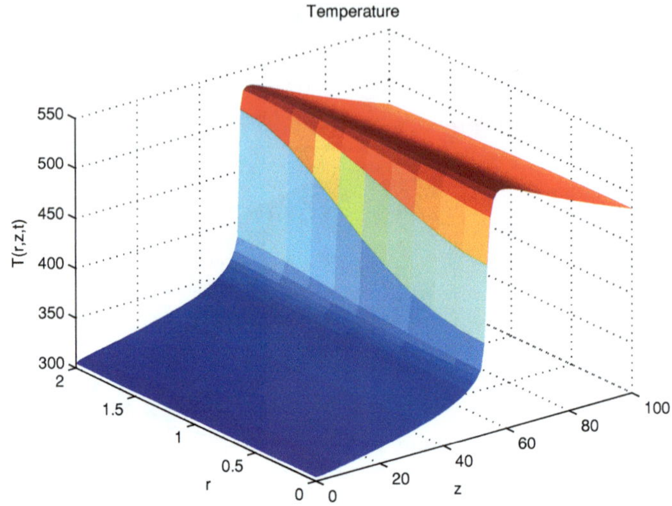

Fig. 6.9 Steady-state temperature profiles in the tubular reactor with a parabolic velocity profile

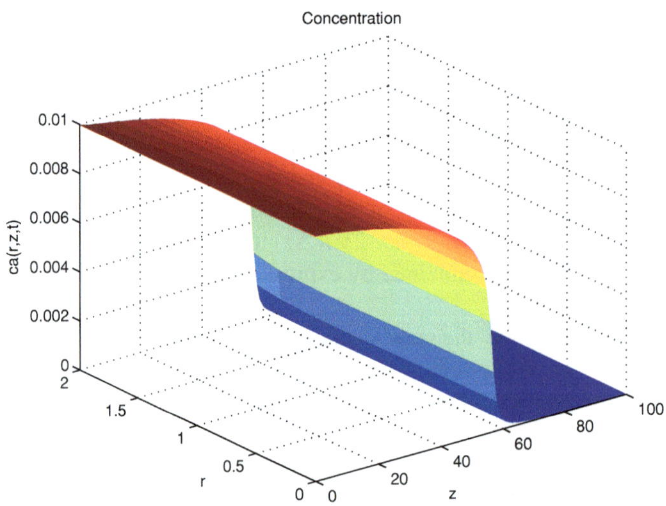

Fig. 6.10 Steady-state concentration profiles in the tubular reactor reactor with a constant velocity profile

6.1.4 Heat Equation on a Convex Quadrilateral

We now consider a more challenging problem: the heat equation previously presented (see Eq. 6.6)

$$T_t = k(T_{xx} + T_{yy})$$

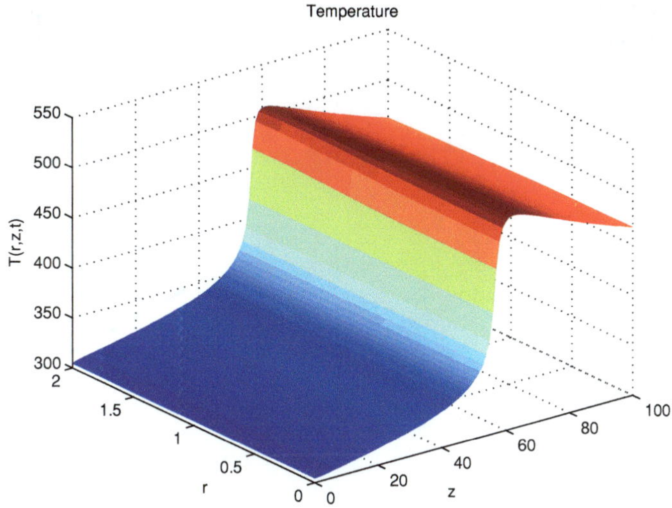

Fig. 6.11 Steady-state temperature profiles in the tubular reactor with a constant velocity profile

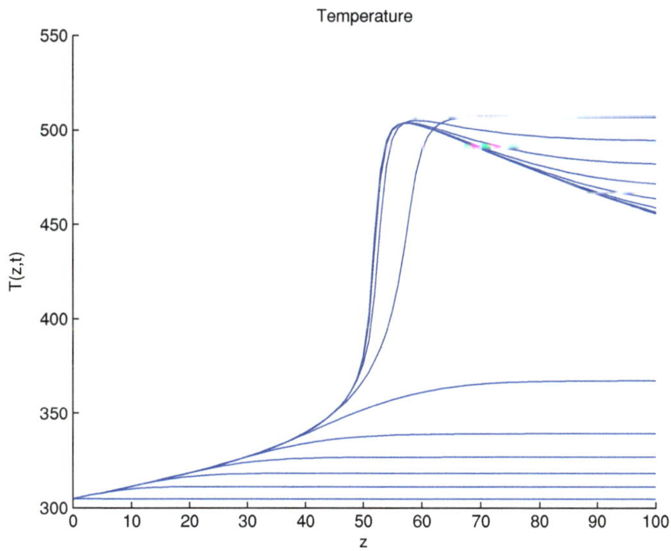

Fig. 6.12 Evolution of the temperature profile at $t = 0, 20, \ldots, 300$ in the tubular reactor, modeled as a 1D system

with $k = 10$, defined on a convex quadrilateral Ω with vertices $(-3, -2)$, $(2, -5)$, $(1, 4)$, $(-2, 1)$ (see Fig. 6.13) The following initial conditions are imposed:

$$T(x, y, 0) = 0 \qquad \forall (x, y) \in \Omega \tag{6.30}$$

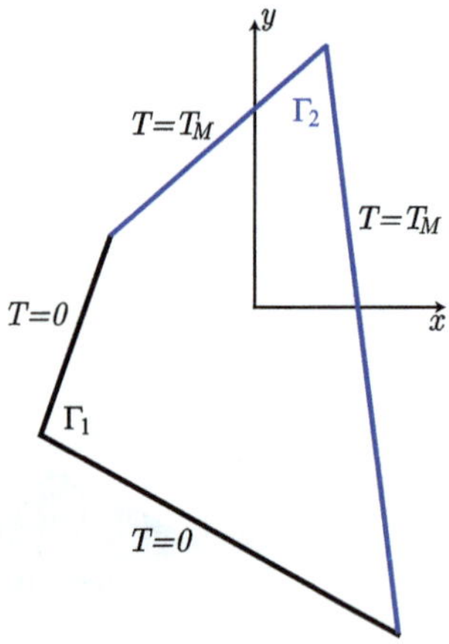

Fig. 6.13 Spatial domain and boundary conditions for the heat equation. Boundaries Γ_1 and Γ_2 are marked in *black* and *blue*, respectively

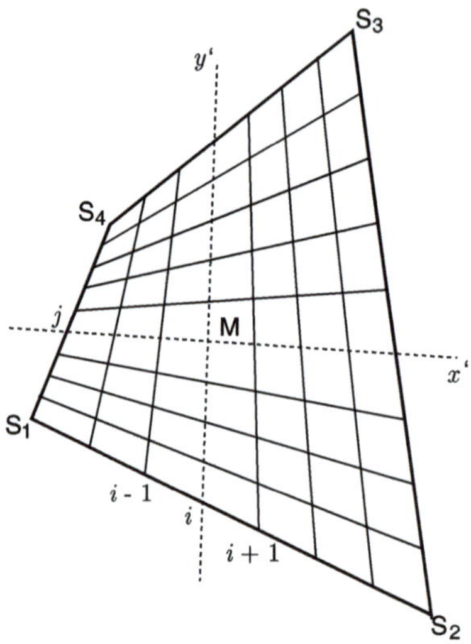

Fig. 6.14 Spatial domain and the spatial grid in the coordinate system (x', y')

Also, the boundary conditions considered are of the form:

$$T(\Gamma_1, t) = 0 \tag{6.31}$$

$$T(\Gamma_2, t) = T_M \tag{6.32}$$

The first step in the numerical solution is the definition of a spatial grid. Unfortunately, a standard grid (in Cartesian coordinates (x, y)) will not match the boundary Γ of the spatial domain Ω. The idea is therefore to build a grid in a new coordinate system (x', y'), as shown in Fig. 6.14, so as to fit well the shape of the spatial domain Ω and its boundary Γ.

With such a grid, it will be easy to evaluate the spatial derivatives, for instance

$$T_{x',x'} = \frac{T_{i-1,j} - 2T_{i,j} + T_{i+1,j}}{\Delta x'^2}$$

However, we still need to find a systematic way to build the grid, and to compute the original derivatives.

This computation will be made possible by considering a reference spatial domain Ω_{ref} in a normalized coordinate system ξ and η:

$$\Omega_{\text{ref}} = \{(\xi, \eta) : 0 \leqslant \xi \leqslant 1, 0 \leqslant \eta \leqslant 1\}$$

Each point $(x, y) \in \Omega$ will be associated to a point $(\xi, \eta) \in \Omega_{\text{ref}}$ through an injective transformation, so that the following relations exist:

$$\begin{aligned} x &= x(\xi, \eta); \quad y = y(\xi, \eta) \\ \xi &= \xi(x, y); \quad \eta = \eta(x, y) \end{aligned} \tag{6.33}$$

These relations are obtained by matching the vertices of Ω and Ω_{ref}, starting with S_1 (see Fig. 6.14) which is the left-most vertice (if there are two vertices with the same x-coordinate, then the one with the smaller y-coordinate is selected) and is associated to the vertice (0, 0) of Ω_{ref}. The other vertices are associated in the anti-clockwise direction ($S_1 \leftrightarrow (0, 0)$, $S_2 \leftrightarrow (1, 0)$, $S_3 \leftrightarrow (1, 1)$, $S_4 \leftrightarrow (0, 1)$,)

If (x_i, y_i) are the coordinates of S_i with $i = 1, \ldots, 4$, the change of variables can be expressed by

$$\begin{aligned} x &= a_0 + a_1\xi + a_2\eta + a_3\xi\eta \\ y &= b_0 + b_1\xi + b_2\eta + b_3\xi\eta \end{aligned} \tag{6.34}$$

so that

$$\begin{aligned} a_0 &= x_1; \quad a_1 = x_2 - x_1; \quad a_2 = x_4 - x_1; \quad a_3 = (x_3 + x_1) - (x_4 + x_2) \\ b_0 &= y_1; \quad b_1 = y_2 - y_1; \quad b_2 = y_4 - y_1; \quad b_3 = (y_3 + y_1) - (y_4 + y_2) \end{aligned} \tag{6.35}$$

The change of variables (6.34) allows us to express the dependent variable $u(x, y)$ as $u(\xi, \eta) = u(\xi(x, y), \eta(x, y))$, and the derivatives

$$\begin{aligned} u_x &= u_\xi \xi_x + u_\eta \eta_x \\ u_y &= u_\xi \xi_y + u_\eta \eta_y \end{aligned} \tag{6.36}$$

$$\begin{aligned} u_{xx} &= u_{\xi\xi}(\xi_x)^2 + u_{\eta\eta}(\eta_x)^2 + u_{\xi\eta}(2\xi_x\eta_x) + u_\xi(\xi_{xx})u_\eta(\eta_{xx}) \\ u_{yy} &= u_{\xi\xi}(\xi_y)^2 + u_{\eta\eta}(\eta_y)^2 + u_{\xi\eta}(2\xi_y\eta_y) + u_\xi(\xi_{yy})u_\eta(\eta_{yy}) \end{aligned} \tag{6.37}$$

To evaluate these expressions, we need ξ_x, ξ_y, ξ_{xx}, ξ_{yy} and η_x, η_y, η_{xx}, η_{yy} which are given by Pozrikidis [2]

$$\xi_x = \frac{y_\eta}{d(\xi,\eta)}; \quad \xi_y = -\frac{x_\eta}{d(\xi,\eta)}; \quad \eta_x = -\frac{y_\xi}{d(\xi,\eta)}; \quad \eta_y = \frac{x_\xi}{d(\xi,\eta)} \tag{6.38}$$

with

$$d(\xi,\eta) = x_\xi y_\eta - x_\eta y_\xi \tag{6.39}$$

Using (6.34), we can find

$$\begin{aligned} d(\xi,\eta) &= (a_1b_2 - a_2b_1) + (a_1b_3 - a_3b_1)\xi + (a_3b_2 - a_2b_3)\eta \\ y_\eta &= b_2 + b_3\xi \\ x_\eta &= a_2 + a_3\xi \\ y_\xi &= b_1 + b_3\eta \\ x_\xi &= a_1 + a_3\eta \end{aligned} \tag{6.40}$$

The evaluation of the second-order derivatives, such as $\xi_{xx} = \frac{\partial}{\partial x}\left(\frac{y_\eta}{d(\xi,\eta)}\right)$, is a bit laborious, but leads to

$$\begin{aligned} \xi_{xx} &= \tfrac{2y_\eta y_\xi}{d^3}(a_3b_2 - a_2b_3) \\ \xi_{yy} &= \tfrac{2x_\eta x_\xi}{d^3}(a_3b_2 - a_2b_3) \\ \eta_{xx} &= \tfrac{2y_\eta y_\xi}{d^3}(a_1b_3 - a_3b_1) \\ \eta_{yy} &= \tfrac{2x_\eta x_\xi}{d^3}(a_1b_3 - a_3b_1) \end{aligned} \tag{6.41}$$

We now have all the tools required to tackle the solution of Eq. (6.6) on Ω:

- Based on the location of the vertices of Ω, and the selection of a grid of points on the unit square (ξ,η), use Eqs. (6.35) and (6.38–6.41) to evaluate ξ_x, ξ_y, ξ_{xx}, ξ_{yy} and η_x, η_y, η_{xx}, η_{yy}
- Compute T_{xx} and T_{yy} using (6.37), and replacing $T_{\xi\xi}$, $T_{\eta\eta}$, T_ξ, T_η, $T_{\xi\eta}$ with classical finite difference formulas.

The full code is given in the companion software. Here we just list in `heat_equation_quadrilateral_domain_pde` the function used to compute the time derivatives based on Eqs. (6.36, 6.37), as well as a function (`mixed_second_order_derivatives_2D`) used to evaluate the mixed derivatives appearing in Eq. (6.37), i.e. $u_{\xi\eta}$.

```
function Tt = heat_equation_quadrilateral_domain_pde(t,T)

% set global variables
global k TM
global nksi neta D1_ksi D1_eta D2_ksi D2_eta
global ksi_x ksi_y eta_x eta_y ksi_xx ksi_yy eta_xx eta_yy

% boundary conditions
```

```
T(1:nksi,1) = 0;
T(1:nksi:nksi*(neta-1)+1,1) = 0;
T(nksi:nksi:nksi*neta,1) = TM;
T(nksi*(neta-1)+1:nksi*neta,1) = TM;

% spatial derivatives
[Tksi Teta] = first_order_derivatives_2D(T,nksi,neta,D1_ksi,...
                                          D1_eta);
[Tksiksi Tetaeta] = second_order_derivatives_2D(T,nksi,neta,...
                                                 D2_ksi,D2_eta);
Tksieta = mixed_second_order_derivatives_2D(T,nksi,neta,D1_ksi,...
                                             D1_eta);
%
Txx = Tksiksi.*(ksi_x.^2) + 2*Tksieta.*ksi_x.*eta_x +...
      Tetaeta.*(eta_x.^2) + Tksi.*ksi_xx + Teta.*eta_xx;
Tyy = Tksiksi.*(ksi_y.^2) + 2*Tksieta.*ksi_y.*eta_y +...
      Tetaeta.*(eta_y.^2) + Tksi.*ksi_yy + Teta.*eta_yy;

% temporal derivatives
Tt = k*(Txx+Tyy);
```

Function heat_equation_quadrilateral_domain_pde Function to evaluate the right-hand side of the heat equation on a quadrilateral domain.

```
function uxy = mixed_second_order_derivatives_2D(u,nx,ny,D1x,D1y)

% use finite differences in 1D along x and along y
v   = reshape(u,nx,[]);
uxy = reshape((D1x*v)*D1y',nx*ny,1);
```

Function mixed_second_order_derivatives_2D Function to compute mixed derivatives on the 2D spatial domain.

6.1.5 A Convection-Diffusion Equation on a Square

So far, it was relatively easy to select an appropriate finite difference scheme depending on the convective or diffusive nature of the phenomena. We will now consider a more complex situation, where caution has to be exercised in this selection. The problem under consideration is taken from [3].

$$u_t = -b_1 u_x - b_2 u_y + \nu(u_{xx} + u_{yy}) + f \tag{6.42}$$

on the spatial domain $\Omega = \{(x, y) : 0 \leqslant x \leqslant 200,\ 0 \leqslant y \leqslant 200\}$ and the time span $0 \leqslant t \leqslant 1$. The initial and boundary conditions are of the form:

$$u(x, y, 0) = 0 \qquad \forall (x, y) \in \Omega \tag{6.43}$$

$$u(\Gamma, t) = 0 \tag{6.44}$$

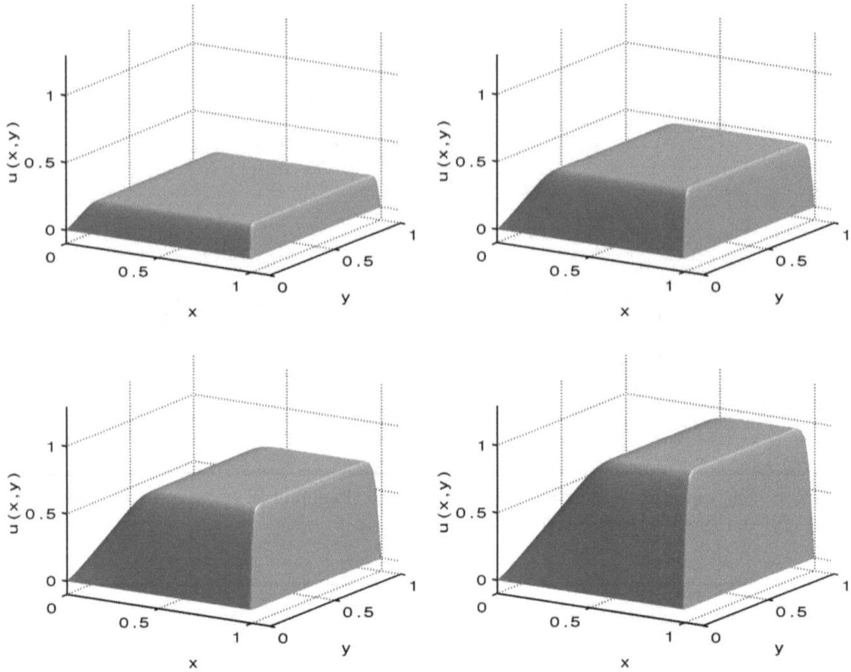

Fig. 6.15 Numerical convection-diffusion problem with $\theta = 0$ at $t = 0.25, 0.5, 0.75, 1$

where Γ is the boundary of Ω. The model parameters are $\nu = 0.001, b_1 = 0.5\cos(\theta)$, $b_2 = 0.5\sin(\theta)$ with $-\pi/2 < \theta < \pi/2$, and $f = 1$. An interesting feature of this problem is the possibility of changing the angle θ, and in turn the direction of the convective terms. This will allow us to discuss the selection of different finite difference schemes, i.e. centered or non centered schemes, in relation with the direction of these terms. We first consider $\theta = 0$, so that the equation reduces to

$$u_t = -u_x + \nu\left(u_{xx} + u_{yy}\right) + f$$

i.e. convection is in the x-direction only. The first order-derivative is computed using a 2-point upwind scheme and the second-order derivatives are computed using 5-point centered schemes. Some results at $t = 0.25, 0.5, 0.75, 1$ are shown in Fig. 6.15.

We next consider $\theta = \pi/2$, so that the equation reduces to

$$u_t = -u_y + \nu\left(u_{xx} + u_{yy}\right) + f$$

i.e. convection is in the y-direction only. Again, the first order-derivative is computed using a 2-point upwind (a higher-order scheme would increase the computational load) scheme and the second-order derivatives are computed using 5-point centered schemes. Some results at $t = 0.25, 0.5, 0.75, 1$ are shown in Fig. 6.16.

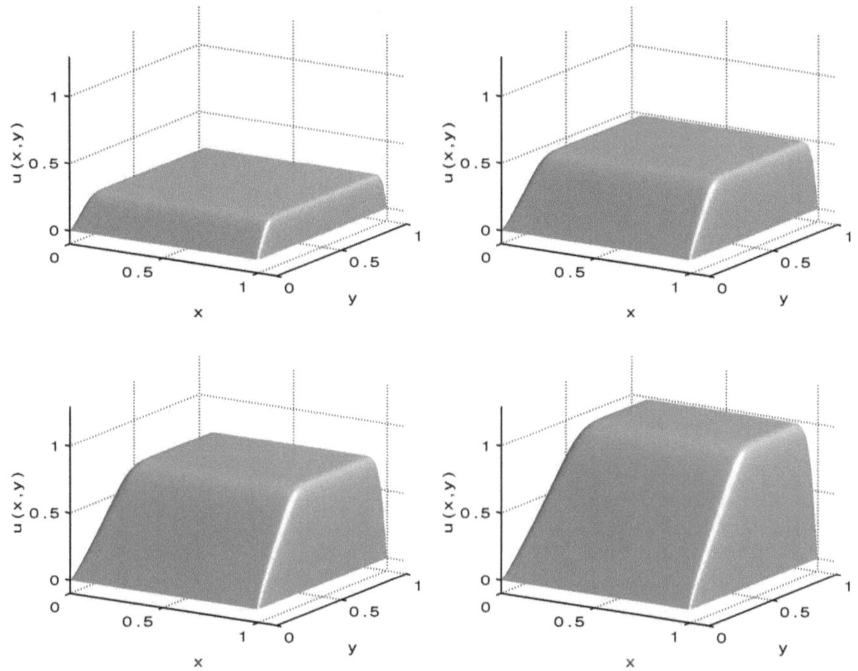

Fig. 6.16 Numerical convection-diffusion problem with $\theta = \pi/2$ at $t = 0.25, 0.5, 0.75, 1$

We now consider $\theta = \pi/4$, so that both convection terms are present in the equation, i.e. convection is in the x- and y-directions. These derivatives can be computed using a 2-point upwind scheme. If $\theta = -\pi/4$, the derivatives have to be computed with 2-point downwind schemes. In fact, the idea is to program the selection of the schemes using the coefficients b_1, b_2 to indicate the correct direction of the fluid velocity.

```
D1x_up = two_point_upwind_D1(x,b1)
D1y_up = two_point_upwind_D1(y,b2)

[ux uy] = first_order_derivatives_2D(u,nx,ny,D1x_up,
          D1y_up);
```

Some results at $t = 0.25, 0.5, 0.75, 1$ are shown in Fig. 6.17 for $\theta = \pi/4$ and Fig. 6.18 for $\theta = -\pi/4$.

What happens if we make the wrong choice for the upwind-downwind schemes? Spurious oscillations will occur as demonstrated in our last example where we pick $\theta = -\pi/3.25 \approx -55°$. Figure 6.19 first shows the results obtained when the choice is well done, i.e. upwind in x and downwind in y (as implemented automatically in the coding lines above). On the other hand, Fig. 6.20 shows the dramatical consequences of reversing the assumed direction in y (i.e. we wrongly assume an upwind scheme for u_y by changing the sign of b_2 in the coding line `D1y_up = two_point_upwind_D1(y,-b2)`).

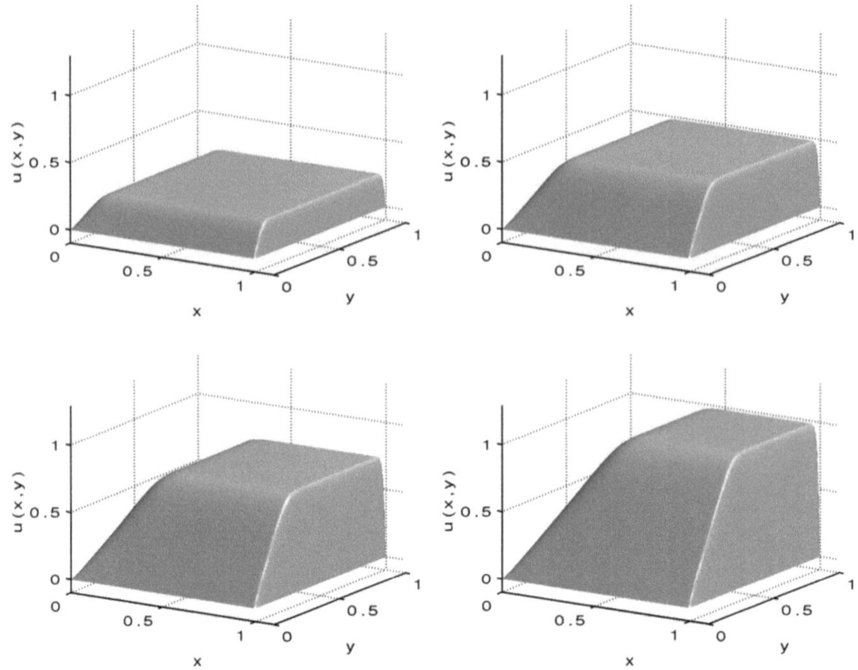

Fig. 6.17 Numerical convection-diffusion problem with $\theta = \pi/4$ at $t = 0.25, 0.5, 0.75, 1$

However, in nonlinear problems, the direction of the movement will depend on the eigenvalues of the Jacobian of the flux function, which can vary with time, so that it will be impossible to choose the correct direction once and for all as we have done in this linear problem. The following example illustrates further this situation.

6.1.6 Burgers Equation on a Square

The following example is a generalized version of Burgers equation [4]:

$$u_t = -\left(u^2\right)_x - \left(u^2\right)_y + \varepsilon\left(v(u)u_x\right)_x + \varepsilon\left(v(u)u_y\right)_y \tag{6.45}$$

We consider an IBVP defined on a spatial domain $\Omega = \{(x, y) : -1.5 \leqslant x \leqslant 1.5, -1.5 \leqslant y \leqslant 1.5\}$ and the time span $0 \leqslant t \leqslant 0.6$, with the model parameters $\varepsilon = 0.5$ and

$$v(u) = \begin{cases} 1 & \text{if } |u| \leqslant 0.5 \\ 0 & \text{otherwise} \end{cases}$$

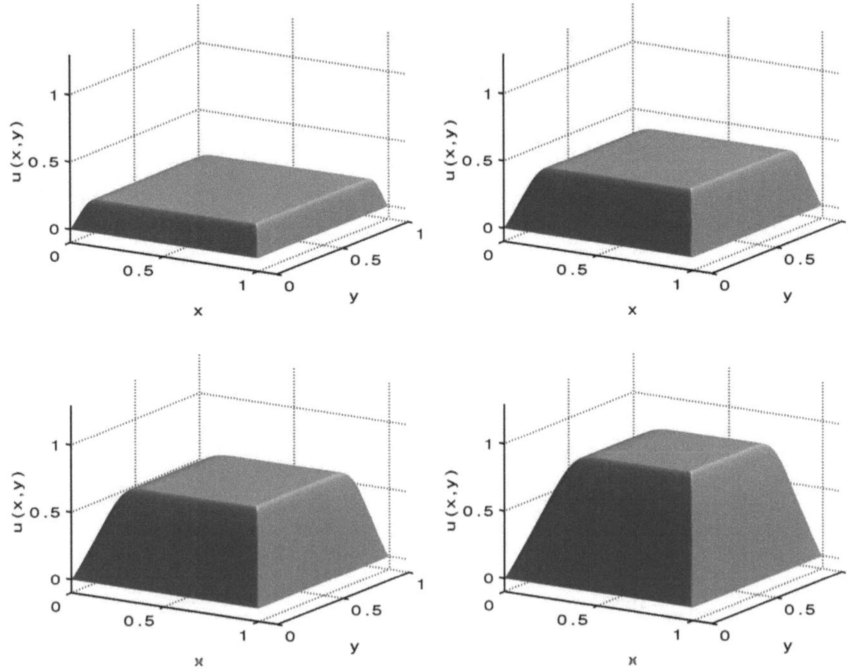

Fig. 6.18 Numerical convection-diffusion problem with $\theta = -\pi/4$ at $t = 0.25, 0.5, 0.75, 1$

The initial condition is shown in Fig. 6.21: $u(x, y, 0) = 0$ except in two disks, one centered in $(-0.5, -0.5)$ with a radius of 0.4, where $u(x, y, 0) = 1$, and the other centered in $(0.5, 0.5)$, also with a radius of 0.4, where $u(x, y, 0) = -1$.

The boundary conditions impose $u(x, y, t) = 0$ along the border Γ.

Figure 6.22 shows the results obtained with finite differences on a square grid of 101×101 grid points, when using 3-point centered finite difference schemes for the first and second-order derivatives.

Spurious oscillations appear from the beginning of the simulation and increase in the course of time. It is therefore appealing to resort to the TVD schemes introduced in Chap. 5 , and in particular the central scheme of Kurganov and Tadmor. Figures 6.23, 6.24 and 6.25 show the nice results obtained at 3 different time instants.

6.2 Solution of 2D PDEs Using Finite Element Techniques

After having discussed several simple 2D PDE problems and their solution with finite difference schemes, we now consider the use of finite elements in 2 spatial dimensions. Finite elements are quite flexible and can accommodate complex geometries. In the following we introduce the basic concepts through a particular example, namely

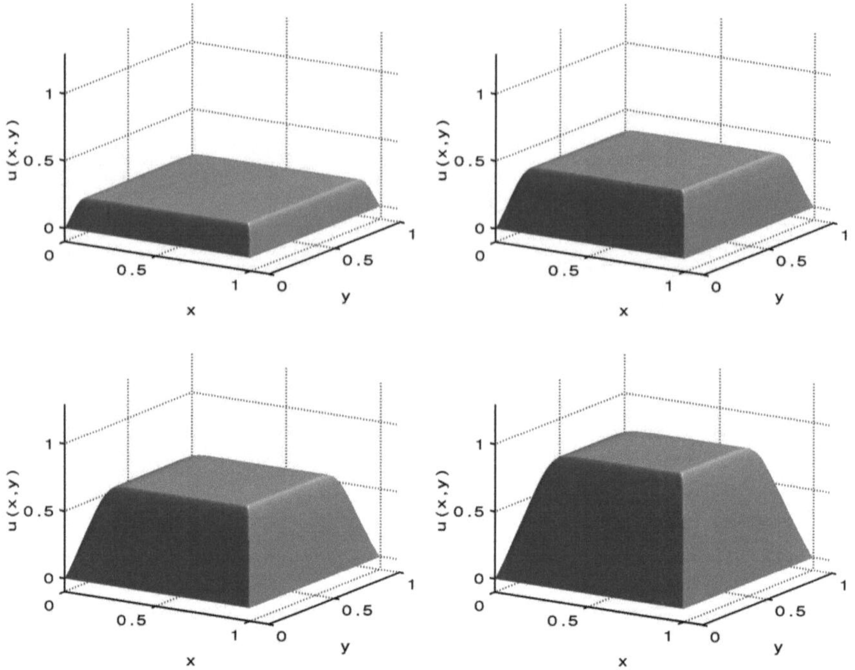

Fig. 6.19 Numerical convection-diffusion problem with $\theta = -\pi/3.25$ at $t = 0.25, 0.5, 0.75, 1$

Fig. 6.20 Numerical convection-diffusion problem with $\theta = -\pi/3.25$ at $t = 1$ when using the wrong scheme for u_y

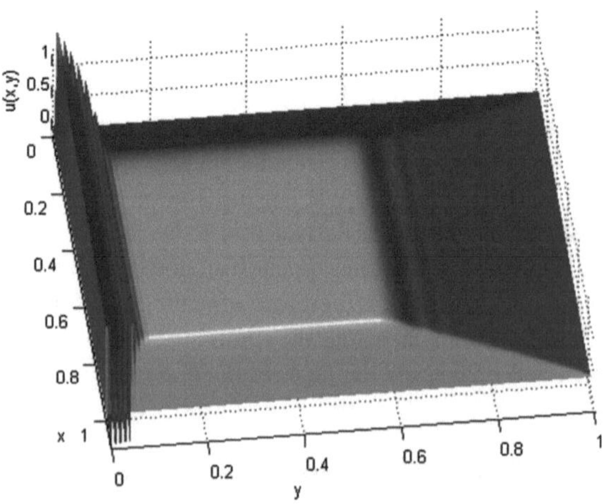

FitzHugh–Nagumo model [5, 6], which is a simplified version of Hodgkin-Huxley model [7] that describes the activation and deactivation dynamics of a spiking neuron. The price to pay for the flexibility of the FEM in 2D is a higher level of complexity of the algorithm, which requires the discretization of the spatial domain into finite

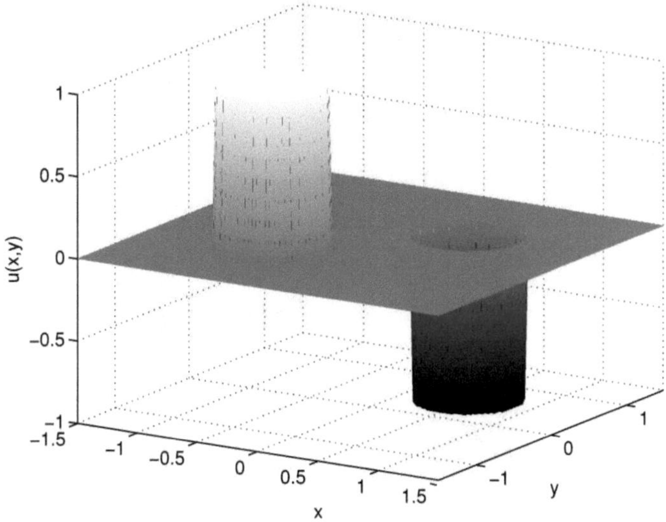

Fig. 6.21 Initial condition of the generalized Burgers equation problem

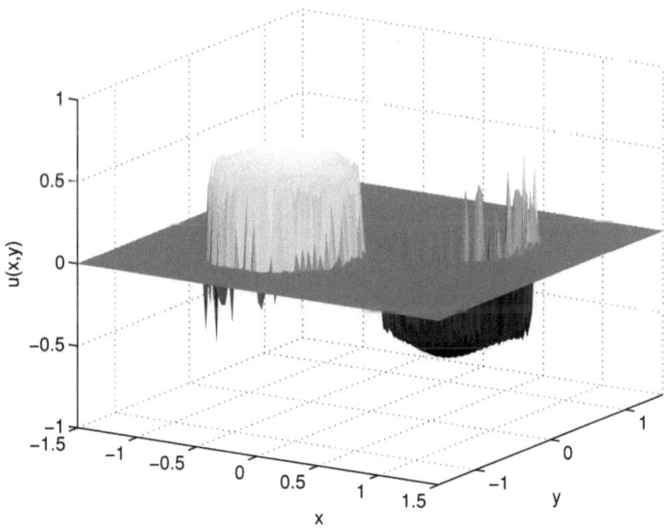

Fig. 6.22 Numerical solution of the generalized Burgers equation problem using centered finite differences at $t = 0.1$

elements (possibly with different geometries), the evaluation of integrals and the construction of several matrices. The ambition of the following example is to explain the general procedure, but not to provide the reader with a library of codes that can easily be adapted to other examples.

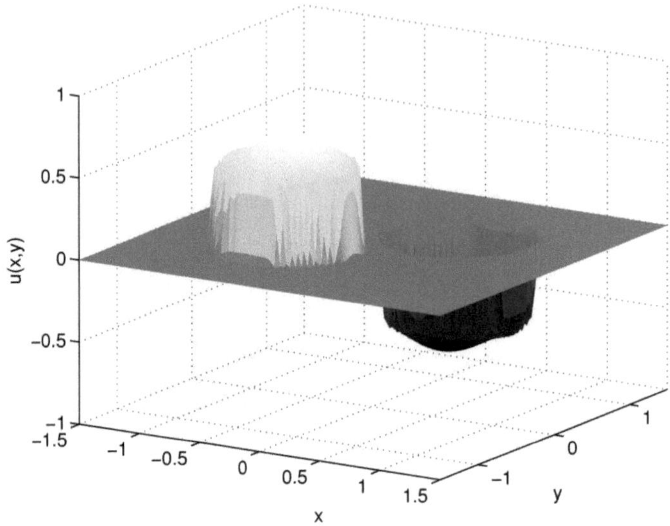

Fig. 6.23 Numerical solution of the generalized Burgers equation problem using centered nonoscillatory schemes at $t = 0.1$

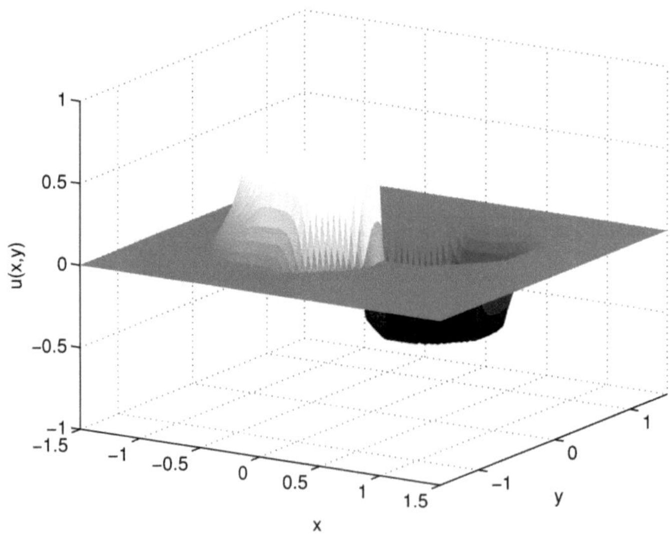

Fig. 6.24 Numerical solution of the generalized Burgers equation problem using centered nonoscillatory schemes at $t = 0.3$

6.2.1 FitzHugh-Nagumo's Model

The spatio-temporal evolution of chemical or electrochemical signals is at the origin of many biological phenomena related with cell growth and distribution as well as with cell communication—see [8]. For example, nervous signals or normal heart

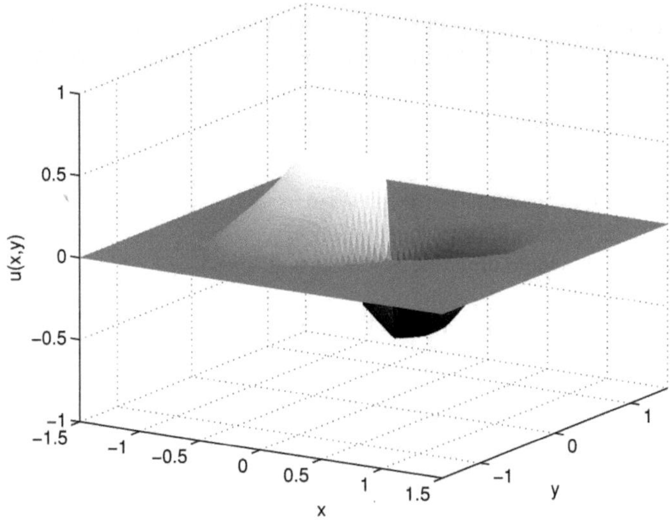

Fig. 6.25 Numerical solution of the generalized Burgers equation problem using centered nonoscillatory schemes at $t = 0.6$

activity are represented in the form of periodic flat pulses (fronts) traveling along neural axons and tissues [7]. On the other hand, heart arrhythmia and other nervous dysfunctions are produced by a regular front being broken into a wandering spiral. In the last instance and after a chain breaking process, the spiral waves would transform into a disorganized set of multiple spirals continually created and destroyed, forming the characteristic pattern of fibrillation [9]. FitzHugh–Nagumo's (FHN) model [5, 6] is able to explain, at least at a qualitative level, these phenomena.

The model equations are of the form:

$$\frac{\partial u}{\partial t} = \kappa \left(\frac{\partial^2 u}{\partial x^2} + \frac{\partial^2 u}{\partial y^2} \right) + g(u) - v, \quad g(u) = u(a-u)(u-1), \tag{6.46}$$

$$\frac{\partial v}{\partial t} = \kappa\rho \left(\frac{\partial^2 v}{\partial x^2} + \frac{\partial^2 v}{\partial y^2} \right) + \varepsilon(\beta u - \gamma v + \delta). \tag{6.47}$$

where u is associated with the membrane potential while v is associated with the contribution to the membrane current of some ions like Na^+, K^+, etc. Parameter κ corresponds to the diffusion coefficient while ρ represents the ratio of the diffusivities of v and u. As pointed out in [10], in biological systems $\rho \approx 0$, thus the diffusion term in Eq. (6.47) can be neglected. In this model, zero-flux boundary conditions are considered:

$$|\mathbf{n} \cdot \nabla u = \mathbf{n} \cdot \nabla v = 0|_\Gamma . \tag{6.48}$$

The spatial domain for this problem is the square surface $\Omega = \{(x, y)/0 < x < 200\, \forall\, y \text{and } 0 < y < 200\, \forall x\}$ with boundary $\Gamma = \{(x, y)/x = \pm 200\, \forall y \text{ and } y = \pm 200\, \forall x\}$.

As explained in Chap. 4, several steps can be distinguished in the construction of the finite element solution, namely:

- Derivation of the variational or weak form
- Element-wise approximation of the solution
- Extension to the whole domain
- Spatial discretization
- Definition of the basis functions and construction of the several matrices.

These steps are also applied when 2D and 3D spatial domains are considered. The main differences are the discretization of the spatial domain and the solution of the spatial integrals involved in this technique. Let us now see how these steps apply to the 2D FitzHugh-Nagumo system.

6.2.1.1 Derivation of the Variational or Weak Form

As explained in Sect. 4.3.4 the FEM makes use of the so-called weak form to reduce the order of the second-order spatial derivatives. The general procedure in 2D is essentially the same as in the 1D case. For the sake of clarity, focus will be on the derivation of the FEM for Eq. (6.46). The same procedure is easily applied to Eq. (6.47). As in the 1D case, the model equations (in this case Eq. (6.46)) are multiplied by an arbitrary test function w, the result is integrated over the spatial domain and Green's first identity is applied to the terms with second-order spatial derivative:

$$\iint_{\Omega} w \frac{\partial u}{\partial t} \mathrm{d}x\mathrm{d}y = \iint_{\Omega} w\kappa \left(\frac{\partial^2 u}{\partial x^2} + \frac{\partial^2 u}{\partial y^2} \right) \mathrm{d}x\mathrm{d}y + \iint_{\Omega} w\,(g(u) - v + p)\,\mathrm{d}x\mathrm{d}y \tag{6.49}$$

By means of Green's first identity, the first term of the RHS of Eq. (6.49) can be expressed as:

$$\iint_{\Omega} w\kappa \left(\frac{\partial^2 u}{\partial x^2} + \frac{\partial^2 u}{\partial y^2} \right) \mathrm{d}x\mathrm{d}y = \int_{\Gamma} w\mathbf{n} \cdot \kappa \nabla u \mathrm{d}\Gamma - \iint_{\Omega} \nabla w \cdot (\kappa \nabla u)\, \mathrm{d}x\mathrm{d}y,$$

Substituting this expression into Eq. (6.49), yields

$$\iint_{\Omega} w \frac{\partial u}{\partial t} \mathrm{d}x\mathrm{d}y + \iint_{\Omega} \nabla w \cdot (\kappa \nabla u)\, \mathrm{d}x\mathrm{d}y = \int_{\Gamma} w\mathbf{n} \cdot \kappa \nabla u \mathrm{d}\Gamma + \iint_{\Omega} w\,(g(u) - v + p)\,\mathrm{d}x\mathrm{d}y,$$

and introducing the zero-flux boundary conditions (6.48), gives:

$$\iint_{\Omega} w\frac{\partial u}{\partial t}\mathrm{d}x\mathrm{d}y + \iint_{\Omega} \nabla w \cdot (\kappa \nabla u)\,\mathrm{d}x\mathrm{d}y = \iint_{\Omega} w\,(g(u) - v + p)\,\mathrm{d}x\mathrm{d}y, \tag{6.50}$$

Note that the second-order spatial derivatives have been eliminated from the formulation.

6.2.1.2 Element-Wise Approximation of the Solution

In order to construct the FEM solution in a systematic way, it is more convenient to consider the weak form over an arbitrary finite element e_i, with volume Ω_{e_i}, rather than the whole domain Ω. Accordingly, the weak form can be expressed as:

$$\iint_{\Omega_{e_i}} w^{e_i}\frac{\partial u^{e_i}}{\partial t}\mathrm{d}x\mathrm{d}y + \iint_{\Omega_{e_i}} \nabla w^{e_i} \cdot \left(\kappa \nabla u^{e_i}\right)\mathrm{d}x\mathrm{d}y = \iint_{\Omega_{e_i}} w^{e_i}\left(g(u)^{e_i} - v^{e_i} + p^{e_i}\right)\mathrm{d}x\mathrm{d}y, \tag{6.51}$$

where w^{e_i} is the test function in the element e_i and the solution u^{e_i} is approximated over each element, so that:

$$u^{e_i}(x, y) = \sum_{j=1}^{n} U_j^{e_i}\phi_j^{e_i}(x, y), \tag{6.52}$$

with $U_j^{e_i}$ being the values of the solution at the nodes of the finite element e_i. $\phi_i^{e_i}$ corresponds to the basis functions over the same element and n is the number of nodes of element e_i which is $n = 3$ in the case of linear Lagrange triangular elements.

Substituting expression (6.52) into (6.51) and choosing the test functions w so as to coincide with the basis functions ϕ^{e_i} (as it is the case in the Galerkin formulation), the following set of n equations is obtained:

$$\iint_{\Omega_{e_i}} \phi_k^{e_i}\frac{\partial \sum_{j=1}^{n} U_j^{e_i}\phi_j^{e_i}}{\partial t}\mathrm{d}x\mathrm{d}y + \iint_{\Omega_{e_i}} \nabla\phi_k^{e_i} \cdot \left(\kappa\nabla\sum_{j=1}^{n} U_j^{e_i}\phi_j^{e_i}\right)\mathrm{d}x\mathrm{d}y$$

$$= \iint_{\Omega_{e_i}} \phi_k^{e_i}\left(g(u)^{e_i} - v^{e_i} + p^{e_i}\right)\mathrm{d}x\mathrm{d}y; \quad k = 1, 2, \ldots, n,$$

or in a more compact (matrix) form:

$$\sum_{j=1}^{n} \mathbf{M}_{kj}^{e_i}\frac{\mathrm{d}U_j^{e_i}}{\mathrm{d}t} + \sum_{j=1}^{n} \kappa\mathbf{D}_{2,kj}^{e_i}U_j^{e_i} = \mathbf{F}_k^{e_i}; \quad k = 1, 2, \ldots, n, \tag{6.53}$$

where the matrices in the former expression are, as in the 1D case:

$$\mathbf{M}_{kj}^{e_i} = \iint\limits_{\Omega_{e_i}} \phi_k^{e_i} \phi_j^{e_i} \mathrm{d}x\mathrm{d}y; \quad \mathbf{D}_{2,kj}^{e_i} = \iint\limits_{\Omega_{e_i}} \nabla\phi_k^{e_i} \cdot \nabla\phi_j^{e_i} \mathrm{d}x\mathrm{d}y;$$

$$\mathscr{F}_k^{e_i} = \iint\limits_{\Omega_{e_i}} \phi_k^{e_i} \left(g(u)^{e_i} - v^{e_i} + p^{e_i}\right) \mathrm{d}x\mathrm{d}y; \qquad k = 1, \ldots, n. \tag{6.54}$$

6.2.1.3 Extension to the Whole Domain

The formulation presented in the previous step was derived for an arbitrary isolated element of the mesh (e_i). In order to obtain the desired solution of Eqs. (6.46)–(6.48), such formulation must be extended to the whole domain. To this purpose, all the isolated elements must be first ordered (numbered) and then assembled. The first part consists of assigning a number ($e^1, e^2, \ldots, e^p$) to each element and to each node. Although the numbers can be arbitrarily assigned to each element, an appropriate order may help to improve the efficiency of the algorithms when solving the final ODE system.

Finally, in order to assemble the elements, one should note that the value of the field in the shared nodes must be the same (continuity of the solution). This step and the subsequent ones (spatial discretization and the shape of the basis functions) will be illustrated in the sequel.

6.2.1.4 Spatial Discretization

Let us first use, for illustrative purposes, an extremely coarse spatial discretization of the spatial domain. Different types of finite elements can be used (triangular, square,...). Triangular elements will be chosen in the following because of their versatility to represent irregular domains. The procedure can be extended to other kind of finite elements.

Figure 6.26 presents the coarse spatial discretization consisting of six finite elements (denoted as e_i with $i = 1, 2, \ldots, 6$) and seven spatial nodes. Of course the numerical solution obtained with this discretization would not be accurate enough and a much larger number of finite elements would be necessary. In this figure, red is used to enumerate the nodes using the global notation (element independent) while blue numbers indicate the node number inside a given finite element (local notation). In general, the nodes inside a finite element are numbered counterclockwise.

Following the notation of previous sections, state variables in the whole spatial domain will be denoted by $u(x, y, t)$ and $v(x, y, t)$. The notation $u^{e_i}(x, y, t)$, $v^{e_i}(x, y, t)$ is used to indicate the state variables inside a given finite element e_i. The value of the fields at a given spatial node (x_j, y_j) of element e_i is denoted by $U_j^{e_i}(t)$, $V_j^{e_i}(t)$.

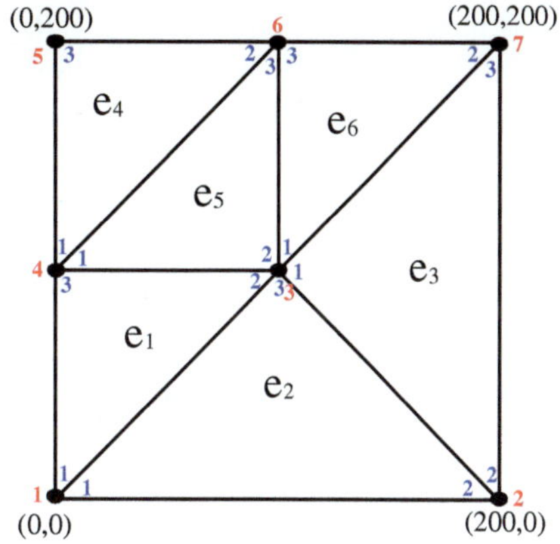

Fig. 6.26 Coarse spatial discretization for the FitzHugh-Nagumo problem used for illustrative purposes

6.2.1.5 Assembly of Elements

Note in Fig. 6.26, that the first global node (marked in red with number (1) coincides with a) the first local node of element e^1 (marked in blue with number 1 inside e^1) and b) the first local node of element e^2 (marked in blue with number 1 inside e^2). Both elements must be taken into account.

Let us now focus on the first node ($k = 1$) of the element e^1. From now on, the arguments of the variables will be omitted for clarity purposes. In this particular case, Eq. (6.53) reads:

$$\mathbf{M}_{11}^{e_1}\frac{dU_1^{e1}}{dt}+\mathbf{M}_{12}^{e_1}\frac{dU_2^{e1}}{dt}+\mathbf{M}_{13}^{e_1}\frac{dU_3^{e1}}{dt}+\kappa\left(\mathbf{D}_{2,11}^{e_1}U_1^{e1}+\mathbf{D}_{2,12}^{e_1}U_2^{e1}+\mathbf{D}_{2,13}^{e_1}U_3^{e1}\right)=\mathscr{F}_1^{e1} \tag{6.55}$$

Similarly, when considering the first local node ($k = 1$) of the element e^2, Eq. (6.53) leads to the following equation:

$$\mathbf{M}_{11}^{e_2}\frac{dU_1^{e2}}{dt}+\mathbf{M}_{12}^{e_2}\frac{dU_2^{e2}}{dt}+\mathbf{M}_{13}^{e_2}\frac{dU_3^{e2}}{dt}+\kappa\left(\mathbf{D}_{2,11}^{e_2}U_1^{e2}+\mathbf{D}_{2,12}^{e_2}U_2^{e2}+\mathbf{D}_{2,13}^{e_2}U_3^{e2}\right)=\mathscr{F}_1^{e2} \tag{6.56}$$

Furthermore, in order to ensure continuity of the solution, the following relations must be satisfied (see Fig. 6.26)

$$U_1 = U_1^{e1} = U_1^{e2}$$

$$U_2 = U_2^{e2} = U_2^{e3}$$

$$U_3 = U_2^{e1} = U_3^{e2} = U_1^{e3} = U_2^{e4} = U_1^{e6}$$

$$U_4 = U_3^{e1} = U_1^{e4} = U_1^{e5} \tag{6.57}$$

$$U_5 = U_3^{e4}$$

$$U_6 = U_2^{e4} = U_3^{e5} = U_3^{e6}$$

$$U_7 = U_3^{e3} = U_2^{e6}$$

Using expressions (6.57), Eqs. (6.55) and (6.56) can be rewritten as:

$$\mathbf{M}_{11}^{e_1}\frac{dU_1}{dt} + \mathbf{M}_{12}^{e_1}\frac{dU_3}{dt} + \mathbf{M}_{13}^{e_1}\frac{dU_4}{dt} + \kappa\left(\mathbf{D}_{2,11}^{e_1}U_1 + \mathbf{D}_{2,12}^{e_1}U_3 + \mathbf{D}_{2,13}^{e_1}U_4\right) = \mathscr{F}_1^{e_1} \tag{6.58}$$

$$\mathbf{M}_{11}^{e_2}\frac{dU_1}{dt} + \mathbf{M}_{12}^{e_2}\frac{dU_2}{dt} + \mathbf{M}_{13}^{e_2}\frac{dU_3}{dt} + \kappa\left(\mathbf{D}_{2,11}^{e_2}U_1 + \mathbf{D}_{2,12}^{e_2}U_2 + \mathbf{D}_{2,13}^{e_2}U_3\right) = \mathscr{F}_1^{e_2} \tag{6.59}$$

Adding Eqs. (6.58) and (6.59) results into:

$$\begin{aligned}&\left(\mathbf{M}_{11}^{e_1} + \mathbf{M}_{11}^{e_2}\right)\frac{dU_1}{dt} + \mathbf{M}_{12}^{e_2}\frac{dU_2}{dt} + \left(\mathbf{M}_{12}^{e_1} + \mathbf{M}_{13}^{e_2}\right)\frac{dU_3}{dt} + \mathbf{M}_{13}^{e_1}\frac{dU_4}{dt}\\ &+ k\Big[\left(\mathbf{D}_{2,11}^{e_1} + \mathbf{D}_{2,11}^{e_2}\right)U_1 + \left(\mathbf{D}_{2,12}^{e_2}\right)U_2\\ &\quad + \left(\mathbf{D}_{2,12}^{e_1} + \mathbf{D}_{2,13}^{e_2}\right)U_3 + \left(\mathbf{D}_{2,13}^{e_1}\right)U_4\Big] = \mathscr{F}_1^{e_1} + \mathscr{F}_1^{e_2}\end{aligned} \tag{6.60}$$

This expression constitutes one of the seven equations required to obtain the solution (the spatial domain discretization consists of seven nodes which gives us seven unknowns, one per node). The second equation is obtained through the second global node. Note that this node is shared by the second local node ($k = 2$) of e_2 and the second local node ($k = 2$) of e_3. In this case, Eq. (6.53) leads to the following set of equations:

$$\mathbf{M}_{21}^{e_2}\frac{dU_1^{e_2}}{dt} + \mathbf{M}_{22}^{e_2}\frac{dU_2^{e_2}}{dt} + \mathbf{M}_{23}^{e_2}\frac{dU_3^{e_2}}{dt} + \kappa\left(\mathbf{D}_{2,21}^{e_1}U_1^{e_2} + \mathbf{D}_{2,22}^{e_2}U_2^{e_2} + \mathbf{D}_{2,23}^{e_2}U_3^{e_2}\right) = \mathscr{F}_2^{e_2} \tag{6.61}$$

$$\mathbf{M}_{21}^{e_3}\frac{dU_1^{e_3}}{dt} + \mathbf{M}_{22}^{e_3}\frac{dU_2^{e_3}}{dt} + \mathbf{M}_{23}^{e_3}\frac{dU_3^{e_3}}{dt} + \kappa\left(\mathbf{D}_{2,21}^{e_3}U_1^{e_3} + \mathbf{D}_{2,22}^{e_3}U_2^{e_3} + \mathbf{D}_{2,23}^{e_3}U_3^{e_3}\right) = \mathscr{F}_2^{e_3} \tag{6.62}$$

Using the global notation and adding Eqs. (6.61) and (6.62) the second of the seven equations is obtained

$$\begin{aligned}&\mathbf{M}_{21}^{e_2}\frac{dU_1}{dt} + \left(\mathbf{M}_{22}^{e_2} + \mathbf{M}_{22}^{e_3}\right)\frac{dU_2}{dt} + \left(\mathbf{M}_{23}^{e_2} + \mathbf{M}_{21}^{e_3}\right)\frac{dU_3}{dt} + \mathbf{M}_{23}^{e_3}\frac{dU_7}{dt} +\\ &k\left[\mathbf{D}_{2,21}^{e_2}U_1 + \left(\mathbf{D}_{2,22}^{e_2} + \mathbf{D}_{2,22}^{e_3}\right)U_2 + \left(\mathbf{D}_{2,23}^{e_2} + \mathbf{D}_{2,21}^{e_3}\right)U_3 + \mathbf{D}_{2,23}^{e_3}U_7\right] = \mathscr{F}_2^{e_2} + \mathscr{F}_2^{e_3}\end{aligned} \tag{6.63}$$

In order to obtain the third equation, the third global node is used. This node is shared by: the second node ($k = 2$) of e_1; the third node ($k = 3$) of e_2; the first node ($k = 1$) of e_3; the second node ($k = 2$) of e_5 and the first node ($k = 1$) of e_6. Applying the same procedure the following equation is derived:

$$
\begin{aligned}
&(\mathbf{M}^{e1}_{21} + \mathbf{M}^{e2}_{31}) \frac{dU_1}{dt} + (\mathbf{M}^{e2}_{32} + \mathbf{M}^{e3}_{12}) \frac{dU_2}{dt} + (\mathbf{M}^{e1}_{22} + \mathbf{M}^{e2}_{33} + \mathbf{M}^{e3}_{11} + \mathbf{M}^{e5}_{22} + \mathbf{M}^{e6}_{11}) \frac{dU_3}{dt} \\
&+ (\mathbf{M}^{e1}_{23} + \mathbf{M}^{e5}_{21}) \frac{dU_4}{dt} + (\mathbf{M}^{e5}_{23} + \mathbf{M}^{e6}_{13}) \frac{dU_6}{dt} + (\mathbf{M}^{e3}_{13} + \mathbf{M}^{e6}_{12}) \frac{dU_7}{dt} \\
&+ k\Big[\left(\mathbf{D}^{e1}_{2,21} + \mathbf{D}^{e2}_{2,31}\right) U_1 + \left(\mathbf{D}^{e2}_{2,32} + \mathbf{D}^{e3}_{2,12}\right) U_2 + \left(\mathbf{D}^{e1}_{2,22} + \mathbf{D}^{e2}_{2,33} + \mathbf{D}^{e3}_{2,11} + \mathbf{D}^{e5}_{2,22} + \mathbf{D}^{e6}_{2,11}\right) U_3 \\
&+ \left(\mathbf{D}^{e1}_{2,23} + \mathbf{D}^{e5}_{2,21}\right) U_4 + \left(\mathbf{D}^{e5}_{2,23} + \mathbf{D}^{e6}_{2,13}\right) U_6 + \left(\mathbf{D}^{e3}_{2,13} + \mathbf{D}^{e6}_{2,12}\right) U_7\Big] \\
&= \mathscr{F}^{e1}_2 + \mathscr{F}^{e2}_3 + \mathscr{F}^{e3}_1 + \mathscr{F}^{e5}_2 + \mathscr{F}^{e6}_1 \qquad (6.64)
\end{aligned}
$$

The remaining equations are derived by applying the same methodology to the rest of global nodes. The final result is:
For the fourth global node

$$
\begin{aligned}
&\mathbf{M}^{e1}_{31} \frac{dU_1}{dt} + (\mathbf{M}^{e1}_{32} + \mathbf{M}^{e5}_{12}) \frac{dU_3}{dt} + (\mathbf{M}^{e1}_{33} + \mathbf{M}^{e5}_{11} + \mathbf{M}^{e4}_{11}) \frac{dU_4}{dt} + \mathbf{M}^{e4}_{13} \frac{dU_5}{dt} + (\mathbf{M}^{e5}_{13} + \mathbf{M}^{e4}_{12}) \frac{dU_6}{dt} \\
&+ k\Big[\mathbf{D}^{e1}_{2,31} U_1 + \left(\mathbf{D}^{e1}_{2,32} + \mathbf{D}^{e5}_{2,12}\right) U_3 + \left(\mathbf{D}^{e1}_{2,33} + \mathbf{D}^{e5}_{2,11} + \mathbf{D}^{e4}_{2,11}\right) U_4 + \mathbf{D}^{e4}_{2,13} U_5 \\
&+ \left(\mathbf{D}^{e5}_{2,13} + \mathbf{D}^{e4}_{2,12}\right) U_6 + \Big] = \mathscr{F}^{e1}_3 + \mathscr{F}^{e4}_1 + \mathscr{F}^{e5}_1 \qquad (6.65)
\end{aligned}
$$

For the fifth global node

$$
\mathbf{M}^{e4}_{31} \frac{dU_4}{dt} + \mathbf{M}^{e4}_{33} \frac{dU_5}{dt} + \mathbf{M}^{e4}_{32} \frac{dU_6}{dt} + k\left[\mathbf{D}^{e4}_{2,31} U_4 + \mathbf{D}^{e4}_{2,33} U_5 + \mathbf{D}^{e4}_{2,32} U_6\right] = \mathscr{F}^{e4}_3 \qquad (6.66)
$$

For the sixth global node

$$
\begin{aligned}
&(\mathbf{M}^{e5}_{32} + \mathbf{M}^{e6}_{31}) \frac{dU_3}{dt} + (\mathbf{M}^{e4}_{21} + \mathbf{M}^{e5}_{31}) \frac{dU_4}{dt} + \mathbf{M}^{e4}_{23} \frac{dU_5}{dt} + (\mathbf{M}^{e4}_{22} + \mathbf{M}^{e5}_{33} + \mathbf{M}^{e6}_{33}) \frac{dU_6}{dt} + \mathbf{M}^{e6}_{32} \frac{dU_7}{dt} \\
&+ k\Big[\left(\mathbf{D}^{e5}_{2,32} + \mathbf{D}^{e6}_{2,31}\right) U_3 + \left(\mathbf{D}^{e4}_{2,21} + \mathbf{D}^{e5}_{2,31}\right) U_4 + \mathbf{D}^{e4}_{2,23} U_5 + \left(\mathbf{D}^{e4}_{2,22} + \mathbf{D}^{e5}_{2,33} + \mathbf{D}^{e6}_{2,33}\right) U_6 \\
&+ \mathbf{D}^{e6}_{2,32} U_7\Big] = \mathscr{F}^{e4}_2 + \mathscr{F}^{e5}_3 + \mathscr{F}^{e6}_3 \qquad (6.67)
\end{aligned}
$$

For the seventh global node

$$
\begin{aligned}
&\mathbf{M}^{e3}_{32} \frac{dU_2}{dt} + (\mathbf{M}^{e6}_{21} + \mathbf{M}^{e3}_{31}) \frac{dU_3}{dt} + \mathbf{M}^{e6}_{23} \frac{dU_6}{dt} + (\mathbf{M}^{e6}_{22} + \mathbf{M}^{e3}_{33}) \frac{dU_7}{dt} + \\
&k\left[\mathbf{D}^{e3}_{2,32} U_2 + \left(\mathbf{D}^{e6}_{2,21} + \mathbf{D}^{e3}_{2,31}\right) U_3 + \mathbf{D}^{e6}_{2,23} U_6 + \left(\mathbf{D}^{e6}_{2,22} + \mathbf{D}^{e3}_{2,33}\right) U_7\right] = \mathscr{F}^{e6}_2 + \mathscr{F}^{e3}_3 \qquad (6.68)
\end{aligned}
$$

Equations (6.60), (6.63) and (6.64)–(6.68) form a system of seven equations with seven unknown quantities which can be solved given that initial conditions are provided. Note also that this system can be rewritten into a more compact form:

$$\mathbf{M}\frac{\mathrm{d}U}{\mathrm{d}t} + k\mathbf{D}_2 U = \mathscr{F} \tag{6.69}$$

where

$$\mathbf{M} = \begin{bmatrix} \left(\mathbf{M}_{11}^{e1}+\mathbf{M}_{11}^{e2}\right) & \mathbf{M}_{12}^{e2} & \left(\mathbf{M}_{12}^{e1}+\mathbf{M}_{13}^{e2}\right) & \mathbf{M}_{13}^{e1} & 0 & 0 & 0 \\ \mathbf{M}_{21}^{e2} & \left(\mathbf{M}_{22}^{e2}+\mathbf{M}_{22}^{e3}\right) & \left(\mathbf{M}_{23}^{e2}+\mathbf{M}_{21}^{e3}\right) & 0 & 0 & 0 & \mathbf{M}_{23}^{e3} \\ \left(\mathbf{M}_{21}^{e1}+\mathbf{M}_{31}^{e2}\right) & \left(\mathbf{M}_{32}^{e2}+\mathbf{M}_{12}^{e3}\right) & \left(\mathbf{M}_{22}^{e1}+\mathbf{M}_{33}^{e2}+ +\mathbf{M}_{21}^{e5}\right) & \left(\mathbf{M}_{23}^{e1}+\mathbf{M}_{21}^{e5}\right) & 0 & \left(\mathbf{M}_{23}^{e5}+\mathbf{M}_{13}^{e6}\right) & \left(\mathbf{M}_{13}^{e3}+\mathbf{M}_{12}^{e6}\right) \\ \mathbf{M}_{31}^{e1} & 0 & \left(\mathbf{M}_{32}^{e1}+\mathbf{M}_{12}^{e5}\right) & \left(\mathbf{M}_{33}^{e1}+\mathbf{M}_{11}^{e5}+\mathbf{M}_{11}^{e4}\right) & \mathbf{M}_{13}^{e4} & \left(\mathbf{M}_{13}^{e5}+\mathbf{M}_{12}^{e4}\right) & 0 \\ 0 & 0 & 0 & \mathbf{M}_{31}^{e4} & \mathbf{M}_{33}^{e4} & \mathbf{M}_{32}^{e4} & 0 \\ 0 & 0 & \left(\mathbf{M}_{32}^{e5}+\mathbf{M}_{31}^{e6}\right) & \left(\mathbf{M}_{21}^{e4}+\mathbf{M}_{31}^{e5}\right) & \mathbf{M}_{23}^{e4} & \left(\mathbf{M}_{22}^{e4}+\mathbf{M}_{33}^{e5}+\mathbf{M}_{33}^{e6}\right) & \mathbf{M}_{32}^{e6} \\ 0 & \mathbf{M}_{32}^{e3} & \left(\mathbf{M}_{21}^{e6}+\mathbf{M}_{31}^{e3}\right) & 0 & 0 & \mathbf{M}_{23}^{e6} & \left(\mathbf{M}_{22}^{e6}+\mathbf{M}_{33}^{e3}\right) \end{bmatrix} \tag{6.70}$$

$$U = \begin{bmatrix} U_1 \\ U_2 \\ U_3 \\ U_4 \\ U_5 \\ U_6 \\ U_7 \end{bmatrix};$$

$$\mathbf{D}_2 = \begin{bmatrix} \left(\mathbf{D}_{2,11}^{e1}+\mathbf{D}_{2,11}^{e2}\right) & \mathbf{D}_{2,12}^{e2} & \left(\mathbf{D}_{2,12}^{e1}+\mathbf{D}_{2,13}^{e2}\right) & \mathbf{D}_{2,13}^{e1} & 0 & 0 & 0 \\ \mathbf{D}_{2,21}^{e2} & \left(\mathbf{D}_{2,22}^{e2}+\mathbf{D}_{2,22}^{e3}\right) & \left(\mathbf{D}_{2,23}^{e2}+\mathbf{D}_{2,21}^{e3}\right) & 0 & 0 & 0 & \mathbf{D}_{2,23}^{e3} \\ \left(\mathbf{D}_{2,21}^{e1}+\mathbf{D}_{2,31}^{e2}\right) & \left(\mathbf{D}_{2,32}^{e2}+\mathbf{D}_{2,12}^{e3}\right) & \left(\mathbf{D}_{2,22}^{e1}+\mathbf{D}_{2,33}^{e2} +\mathbf{D}_{2,11}^{e3}+\mathbf{D}_{2,22}^{e5}\mathbf{D}_{2,11}^{e6}\right) & \left(\mathbf{D}_{2,23}^{e1}+\mathbf{D}_{2,21}^{e5}\right) & 0 & \left(\mathbf{D}_{2,23}^{e5}+\mathbf{D}_{2,13}^{e6}\right) +\mathbf{D}_{2,12}^{e6} & \left(\mathbf{D}_{2,13}^{e3}\right) \\ \mathbf{D}_{2,31}^{e1} & 0 & \left(\mathbf{D}_{2,32}^{e1}+\mathbf{D}_{2,12}^{e5}\right) & \left(\mathbf{D}_{2,33}^{e1}+\mathbf{D}_{2,11}^{e5} +\mathbf{D}_{2,11}^{e4}\right) & \mathbf{D}_{2,13}^{e4} & \left(\mathbf{D}_{2,13}^{e5}+\mathbf{D}_{2,12}^{e4}\right) & 0 \\ 0 & 0 & 0 & \mathbf{D}_{2,31}^{e4} & \mathbf{D}_{2,33}^{e4} & \mathbf{D}_{2,32}^{e4} & 0 \\ 0 & 0 & \left(\mathbf{D}_{2,32}^{e5}+\mathbf{D}_{2,31}^{e6}\right) & \left(\mathbf{D}_{2,21}^{e4}+\mathbf{D}_{2,31}^{e5}\right) & \mathbf{D}_{2,23}^{e4} & \left(\mathbf{D}_{2,22}^{e4}+\mathbf{D}_{2,33}^{e5} +\mathbf{D}_{2,33}^{e6}\right) & \mathbf{D}_{2,32}^{e6} \\ 0 & \mathbf{D}_{2,32}^{e3} & \left(\mathbf{D}_{2,21}^{e6}+\mathbf{D}_{2,31}^{e3}\right) & 0 & 0 & \mathbf{D}_{2,23}^{e6} & \left(\mathbf{D}_{2,22}^{e6}\right) +\mathbf{D}_{2,33}^{e3} \end{bmatrix}$$

$$\mathscr{F} = \begin{bmatrix} \mathscr{F}_1^{e1}+\mathscr{F}_1^{e2} \\ \mathscr{F}_2^{e2}+\mathscr{F}_2^{e3} \\ \mathscr{F}_2^{e1}+\mathscr{F}_3^{e2}+\mathscr{F}_1^{e3}+\mathscr{F}_2^{e5}+\mathscr{F}_1^{e6} \\ \mathscr{F}_3^{e1}+\mathscr{F}_1^{e4}+\mathscr{F}_1^{e5} \\ \mathscr{F}_3^{e4} \\ \mathscr{F}_2^{e4}+\mathscr{F}_3^{e5}+\mathscr{F}_3^{e6} \\ \mathscr{F}_2^{e6}+\mathscr{F}_3^{e3} \end{bmatrix};$$

The same procedure is followed to obtain the equations for the field $v(x, y, t)$:

$$\mathbf{M}\frac{\mathrm{d}V}{\mathrm{d}t} = \mathscr{H} \tag{6.71}$$

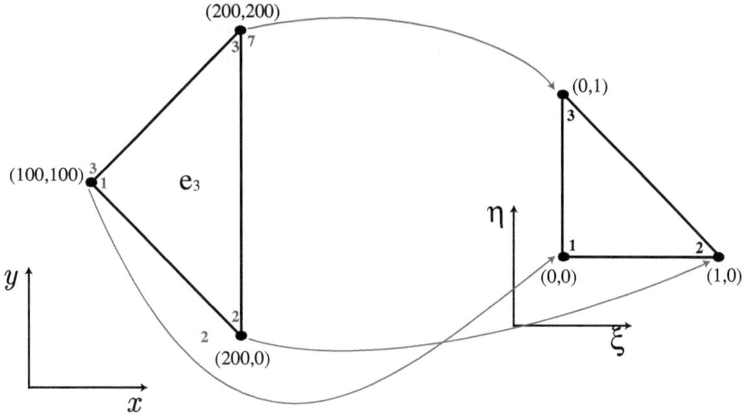

Fig. 6.27 Equivalence between a triangular element in the x, y coordinates and the right-angled triangle in ξ, η

6.2.1.6 Definition of the Basis Functions and Construction of the Several Matrices

In this example, three nodes per element (triangular elements) are defined. Also the edges of the triangle are straight lines. In order to establish a systematic procedure for finding the form of the basis functions and to perform the integrals involved in Eq. (6.54) every triangular element is mapped from the original coordinate system x, y to a right-angled triangle with vertices in the transformed coordinates $(\xi_1, \eta_1) = (0, 0)$, $(\xi_2, \eta_2) = (1, 0)$, $(\xi_3, \eta_3) = (0, 1)$ (see Fig. 6.27 where element e_3 is considered). The mapping is always performed in the same way, i.e, the first node in the original coordinates (x, y) is mapped to (ξ_1, η_1), the second node is mapped to (ξ_2, η_2) and the third node is mapped to (ξ_3, η_3).

The relation between the original coordinates and the transformed ones is computed through the finite element interpolation functions $\phi_i(\xi, \eta)$. In this way, for a given finite element e_i:

$$\begin{aligned} x &= x_1^{e_i}\phi_1(\xi,\eta) + x_2^{e_i}\phi_2(\xi,\eta) + x_3^{e_i}\phi_3(\xi,\eta) \\ y &= y_1^{e_i}\phi_1(\xi,\eta) + y_2^{e_i}\phi_2(\xi,\eta) + y_3^{e_i}\phi_3(\xi,\eta) \end{aligned} \tag{6.72}$$

where the value of the interpolation functions at the element nodes is the Kronecker's delta, i.e, $\phi_i(\xi_j, \eta_j) = \delta_{ij}$.

Considering a single degree of freedom per node, the value of the interpolation functions inside a given element is expressed as a linear combination of the transformed coordinates, i.e,

$$\phi_i(\xi, \eta) = a_i + b_i\xi + c_i\eta \tag{6.73}$$

In order to compute the values of the unknowns a_i, b_i, c_i with $i = 1, 2, 3$ three algebraic linear systems are solved. Such linear systems are obtained from Eq. (6.73), i.e.

$$\begin{bmatrix} \phi_1(\xi_1, \eta_1) \\ \phi_1(\xi_2, \eta_2) \\ \phi_1(\xi_3, \eta_3) \end{bmatrix} = \begin{bmatrix} 1 & \xi_1 & \eta_1 \\ 1 & \xi_2 & \eta_2 \\ 1 & \xi_3 & \eta_3 \end{bmatrix} \begin{bmatrix} a_1 \\ b_1 \\ c_1 \end{bmatrix}$$

which, by substituting the value of the node coordinates and the value of the basis functions at those coordinates results in:

$$\begin{bmatrix} 1 \\ 0 \\ 0 \end{bmatrix} = \begin{bmatrix} 1 & 0 & 0 \\ 1 & 1 & 0 \\ 1 & 0 & 1 \end{bmatrix} \begin{bmatrix} a_1 \\ b_1 \\ c_1 \end{bmatrix} \quad \Longrightarrow \quad \begin{aligned} a_1 &= 1 \\ b_1 &= -1 \\ c_1 &= -1 \end{aligned}$$

Replacing the values of a_1, b_1 and c_1 into Eq. (6.73):

$$\phi_1(\xi, \eta) = 1 - \xi - \eta \tag{6.74}$$

Repeating this procedure with the other two nodes, the following expressions are obtained:

$$\phi_2(\xi, \eta) = \xi \tag{6.75}$$

$$\phi_3(\xi, \eta) = \eta \tag{6.76}$$

The mapping $(x, y) \rightsquigarrow (\xi, \eta)$ is therefore obtained by substituting expressions (6.74)–(6.76) into Eq. (6.72):

$$\begin{aligned} x &= x_1^{e_i} + (x_2^{e_i} - x_1^{e_i})\xi + (x_3^{e_i} - x_1^{e_i})\eta \\ y &= y_1^{e_i} + (y_2^{e_i} - y_1^{e_i})\xi + (y_3^{e_i} - y_1^{e_i})\eta \end{aligned} \tag{6.77}$$

The integrals in Eq. (6.54) can be expressed as integrals in the transformed coordinates (ξ, η) [2]. To this purpose, let us first obtain the relation between $\mathrm{d}x\mathrm{d}y$ and $\mathrm{d}\xi\mathrm{d}\eta$ using the Jacobian of Eq. (6.77):

$$\mathrm{d}x\mathrm{d}y = \det(J)\mathrm{d}\xi\mathrm{d}\eta, \quad \text{with } J = \begin{bmatrix} (x_2^{e_i} - x_1^{e_i}) & (x_3^{e_i} - x_1^{e_i}) \\ (y_2^{e_i} - y_1^{e_i}) & (y_3^{e_i} - y_1^{e_i}) \end{bmatrix}$$

Note that $\det(J) = 2A^{e_i}$ where A^{e_i} is the area of element e_i. In this way, integrals in the mass and diffusion matrices can be computed as:

$$\mathbf{M}_{kj}^{e_i} = \iint_{\Omega_{e_i}} \phi_k^{e_i}\phi_j^{e_i}\mathrm{d}x\mathrm{d}y = 2A^{e_i} \iint \phi_k^{e_i}\phi_j^{e_i}\mathrm{d}\xi\mathrm{d}\eta \tag{6.78}$$

$$\mathbf{D}_{2,kj}^{e_i} = \iint\limits_{\Omega_{e_i}} \nabla\phi_k^{e_i} \cdot \nabla\phi_j^{e_i} \mathrm{d}x\mathrm{d}y = 2A^{e_i} \iint \nabla\phi_k^{e_i} \cdot \nabla\phi_j^{e_i} \mathrm{d}\xi\mathrm{d}\eta \tag{6.79}$$

Let us begin with the contribution of one element e_i to the mass matrix, i.e., Eq. (6.78), where expressions (6.74)–(6.76) will be exploited. Variable χ defined as $\chi = 1 - \xi - \eta$ will be used for convenience. Note that for $k = j = 1$ in (6.78), we obtain $\phi_1\phi_1 = \chi^2$, also for $k = 1,\ j = 2$ we have $\phi_1\phi_2 = \chi\xi$. Going through all $k = 1, 2, 3$ and $j = 1, 2, 3$ the following expression can be derived:

$$\mathbf{M}_{kj} = 2A^{e_i} \iint \begin{bmatrix} \chi^2 & \chi\xi & \chi\eta \\ \xi\chi & \xi^2 & \xi\eta \\ \eta\chi & \eta\xi & \eta^2 \end{bmatrix} \mathrm{d}\xi\mathrm{d}\eta = \frac{A^{e_i}}{12} \begin{bmatrix} 2 & 1 & 1 \\ 1 & 2 & 1 \\ 1 & 1 & 2 \end{bmatrix} \tag{6.80}$$

with the help of the integration formula [2]:

$$\iint \chi^p \xi^q \eta^r d\xi d\eta = \frac{p!q!r!}{(p+q+r+2)!}$$

In order to compute the contribution of one element to the diffusion matrix $\mathbf{D}_2$, the gradients of the basis functions must be computed. For the first basis function the following relations are obtained using the chain rule:

$$\begin{aligned} \frac{\partial\phi_1}{\partial\xi} &= \frac{\partial\phi_1}{\partial x}\frac{\partial x}{\partial\xi} + \frac{\partial\phi_1}{\partial y}\frac{\partial y}{\partial\xi} \\ \frac{\partial\phi_1}{\partial\eta} &= \frac{\partial\phi_1}{\partial x}\frac{\partial x}{\partial\eta} + \frac{\partial\phi_1}{\partial y}\frac{\partial y}{\partial\eta} \end{aligned} \implies \begin{bmatrix} \frac{\partial\phi_1}{\partial\xi} & \frac{\partial\phi_1}{\partial\eta} \end{bmatrix} = \nabla\phi_1 \begin{bmatrix} \frac{\partial x}{\partial\xi} & \frac{\partial x}{\partial\eta} \\ \frac{\partial y}{\partial\xi} & \frac{\partial y}{\partial\eta} \end{bmatrix}$$

which, using the expressions (6.74)–(6.77) results in:

$$\begin{bmatrix} -1 & -1 \end{bmatrix} = \nabla\phi_1 \underbrace{\begin{bmatrix} (x_2^{e_i} - x_1^{e_i}) & (x_3^{e_i} - x_1^{e_i}) \\ (y_2^{e_i} - y_1^{e_i}) & (y_3^{e_i} - y_1^{e_i}) \end{bmatrix}}_{Q} \tag{6.81}$$

The inverse of the 2×2 matrix Q is:

$$Q^{-1} = \frac{1}{|Q|} \begin{bmatrix} (y_3^{e_i} - y_1^{e_i}) & -(y_2^{e_i} - y_1^{e_i}) \\ -(x_3^{e_i} - x_1^{e_i}) & (x_2^{e_i} - x_1^{e_i}) \end{bmatrix} \quad \text{with} \quad |Q| = 2A^{e_i}$$

Multiplying both sides of Eq. (6.81) by Q^{-1} we obtain:

$$\nabla\phi_1 = \frac{1}{2A^{e_i}} \begin{bmatrix} -(y_3^{e_i} - y_2^{e_i}) \\ (x_3^{e_i} - x_2^{e_i}) \end{bmatrix} \tag{6.82}$$

The same logic is applied to find the gradients of ϕ_2 and ϕ_3:

$$\nabla\phi_2 = \frac{1}{2A^{e_i}} \begin{bmatrix} -(y_1^{e_i} - y_3^{e_i}) \\ (x_1^{e_i} - x_3^{e_i}) \end{bmatrix}. \tag{6.83}$$

$$\nabla\phi_3 = \frac{1}{2A^{e_i}} \begin{bmatrix} -(y_2^{e_i} - y_1^{e_i}) \\ (x_2^{e_i} - x_1^{e_i}) \end{bmatrix} \tag{6.84}$$

Choosing $k = j = 1$ in Eq. (6.79) and using expression (6.82)

$$\begin{aligned}\mathbf{D}_{2,11}^{e_i} &= 2A^{e_i} \iint \nabla\phi_1^{e_i} \cdot \nabla\phi_1^{e_i} \mathrm{d}\xi \mathrm{d}\eta = \frac{1}{2A^{e_i}} \left[(y_3 - y_2)^2 + (x_3 - x_2)^2\right] \iint \mathrm{d}\xi \mathrm{d}\eta \\ &= \frac{1}{4A^{e_i}} \left[(y_3 - y_2)^2 + (x_3 - x_2)^2\right]\end{aligned}$$

Extending this to all $k = 1, 2, 3$ and $j = 1, 2, 3$ we find

$$\mathbf{D}_2^{e_i} = \frac{1}{4A^{e_i}} \begin{bmatrix} q_{11} & q_{12} & q_{13} \\ q_{21} & q_{22} & q_{23} \\ q_{31} & q_{32} & q_{33} \end{bmatrix} \tag{6.85}$$

where $q_{11} = (y_3 - y_2)^2 + (x_3 - x_2)^2$, $q_{22} = (y_1 - y_3)^2 + (x_1 - x_3)^2$, $q_{33} = (y_2 - y_1)^2 + (x_2 - x_1)^2$, $q_{12} = q_{21} = (y_3 - y_2)(y_1 - y_3) + (x_3 - x_2)(x_1 - x_3)$, $q_{13} = q_{31} = (y_3 - y_2)(y_2 - y_1) + (x_3 - x_2)(x_2 - x_1)$, $q_{23} = q_{32} = (y_1 - y_3)(y_2 - y_1) + (x_1 - x_3)(x_2 - x_1)$.

Let us now see how the assembly procedure previously described to obtain the global matrices can be implemented in MATLAB. To this purpose the extremely coarse spatial discretization of Fig. 6.26 (including node and element numeration) will be used.

The code starts by defining the connectivity matrix. In this matrix, each row corresponds to a given finite element (ordered from the first to the last one) and contains the global node numbers, stored in the counterclockwise direction. In this way, as indicated in Fig. 6.26 the first element is defined by the global nodes 1, 3 and 4.

```
% Conectivity matrix
CM = [1, 3, 4;
      1, 2, 3;
```

```
        3, 2, 7;
        4, 6, 5;
        4, 3, 6;
        3, 7, 6];
ne = size(CM,1);   % Number of elements
```

For each element, the node order in fact follows the local node numbering (1, 2, 3 in the counterclockwise direction). For instance, the third row of the connectivity matrix (corresponding to the third element) will be 3 (third global node and first local node of element e_3), 2 (second global node and second local node of element e_3) and 7 (seventh global node and third local node of element e_3). The number of elements in the grid coincides with the number of rows of the connectivity matrix.

The next information required to compute the element matrices and to assemble them is the matrix containing coordinates of the global nodes.

```
% Node coordinate matrix
%       x     y
NC = [ 0,    0;    % First global node
      200,   0;    % Second global node
      100, 100;    % Third global node
       0,  100;    % Fourth global node
       0,  200;    % Fifth global node
      100, 200;    % Sixth global node
      200, 200];   % Seventh global node
x  = NC(:,1);      % x-axis coordinates for the global nodes
y  = NC(:,2);      % y-axis coordinates for the global nodes
nd = length(x);    % Number of discretization points (global nodes)
```

Rows correspond to the global nodes (ordered from the first one to the last one) while the columns are the x-axis and the y-axis coordinates.

In order to compute the element matrices, the area of the finite elements have also to be computed. Different formulas can be used, for instance, Heron's formula:

$$A = \sqrt{(s(s-a)(s-b)(s-c))}$$

where a, b and c are the length of the triangle edges and s is the semi-perimeter of the element $s = (a+b+c)/2$.

```
% Triangle elements area (they are stored in a column vector A)
for ii = 1:ne
    p = CM(ii,:); % global nodes defining finite element ii
    a = sqrt( (x(p(1))-x(p(2)))^2 + (y(p(1))-y(p(2)))^2);
    b = sqrt( (x(p(1))-x(p(3)))^2 + (y(p(1))-y(p(3)))^2);
    c = sqrt( (x(p(2))-x(p(3)))^2 + (y(p(2))-y(p(3)))^2);
    s = (a+b+c)/2;
    A(ii,1) = sqrt( (s*(s-a)*(s-b)*(s-c)) ); % Element area
end
```

The next step is to compute the contribution of each element to the mass matrix using Eq. (6.80). The code is divided in two parts, first the part which is the same for all elements is computed and, then, it is multiplied by each element area. The result is stored into an array of matrices (`Mat`).

```
% Element mass matrices
Mi = 1/12*[2 1 1; 1 2 1; 1 1 2];

% Multiplication by the element area
for ii = 1:ne
Mat(ii).M = A(ii)*Mi;
end
```

The more elaborated part of the code is the assembly of the elementary matrices. The global mass matrix will be stored in variable `M`, which is first initialized (this will accelerate the code, especially when the number of grid nodes is large). We have to cover all nodes and for each node:

- First, we have to find the elements that share a given node ii. The command `find(CM==ii)` will find all the elements equal to ii where $ii = 1, 2, \ldots, 7$, and will give us the rows of `CM` where they are found (corresponding to the global node numbers) and the columns (corresponding with the local node position). The results are stored in the variables `gne` and `lnp`. For instance, the fourth global node is shared by the first, the fourth and the fifth finite elements. The command `find(CM==4)` will give us two vectors: `gne=[4,5,1]` (fourth, fifth and first finite elements) and `lnp=[1,1,3]` (fourth global node corresponds with: (i) first local node in finite element e_4, (ii) first local node in finite element e_5 and (iii) third local node in finite element e_1).
- Recall that the contribution of each node to the global matrix was previously stored in an array of 3×3 matrices `Mat.M`. Now each element of `Mat.M` must be ordered according to (6.70). The ordered elements of `Mat.M` will be stored in matrix `Me` where:
  - the number of rows will be the number of elements that share the global node. For instance for global node 4 the number of rows will be 3
  - the number of columns will be 3 since triangular elements have three nodes

  For instance, for the fourth global node, the first row of `Me` is:

$$\texttt{Me}(1,:) = \begin{bmatrix} 0 \; 0 \; 0 \; 833.3 \; 416.7 \; 416.7 \; 0 \end{bmatrix}$$

  because the global nodes of element e_4 are 4, 6 and 5. The second row of `Me` is:

$$\texttt{Me}(2,:) = \begin{bmatrix} 0 \; 0 \; 416.7 \; 833.3 \; 0 \; 416.7 \; 0 \end{bmatrix}$$

  because the global nodes of element e_5 are 4, 3 and 6. The third row of `Me` is:

$$\texttt{Me}(3,:) = \left[416.7\ 0\ 416.7\ 833.3\ 0\ 0\ 0\right]$$

because the global nodes of element e_1 are 1, 3 and 4.

- Finally, all the contributions are added, i.e., the rows of Me are added. For the fourth global node (fourth row in the global mass matrix), the result is:

$$\texttt{M(4,:)} = \left[416.7\ 0\ 833.3\ 2500\ 416.7\ 833.3\ 0\right]$$

```
% Assembly of element mass matrices
M = zeros(nd,nd);    % Mass matrix initialization
for ii = 1:nd        % Number of nodes
    % Find the indexes in the connectivity matrix for shares nodes
    [gne, lnp]    = find(CM==ii);
    % Contribution of each element to the mass matrix
    Me = zeros(length(gne),nd);
    for jj = 1:length(gne)   % Elements that share this node
        for kk = 1:3           % Number of nodes per element
            Me(jj,CM(gne(jj),kk)) = Mat(gne(jj)).M(lnp(jj),kk);
        end
    end
    M(ii,:) = sum(Me,1);
end
```

The final result for the mass matrix using the coarse discretization of Fig. 6.26 is:

$$M = \begin{bmatrix} 2500 & 833.3 & 1250 & 416.7 & 0 & 0 & 0 \\ 833.3 & 3333.3 & 1666.7 & 0 & 0 & 0 & 833.3 \\ 1250 & 1666.7 & 5833.3 & 833.3 & 0 & 833.3 & 1250 \\ 416.7 & 0 & 833.3 & 2500 & 416.7 & 833.3 & 0 \\ 0 & 0 & 0 & 416.7 & 833.3 & 416.7 & 0 \\ 0 & 0 & 833.3 & 833.3 & 416.7 & 2500 & 416.7 \\ 0 & 833.3 & 1250 & 0 & 0 & 416.7 & 2500 \end{bmatrix}$$

The same logic is followed to compute the diffusion matrix. The main difference is the computation of the contribution of each element following expression (6.85).

```
% Element diffusion matrices
for ii = 1:ne
    % Find the global nodes defining this element
    p = CM(ii,:);
    % Find the coordinates (x,y) of the global nodes
    x1 = NC(p(1),1);  % x-coordinate of node 1
    y1 = NC(p(1),2);  % y-coordinate of node 1
    x2 = NC(p(2),1);  % x-coordinate of node 2
    y2 = NC(p(2),2);  % y-coordinate of node 2
    x3 = NC(p(3),1);  % x-coordinate of node 3
    y3 = NC(p(3),2);  % y-coordinate of node 3
    % Matrix quantities (according to the book formula)
```

```
    q11  = (y3-y2)^2 + (x3-x2)^2;
    q22  = (y1-y3)^2 + (x1-x3)^2;
    q33  = (y2-y1)^2 + (x2-x1)^2;
    q12  = (y3-y2)*(y1-y3) + (x3-x2)*(x1-x3);
    q13  = (y3-y2)*(y2-y1) + (x3-x2)*(x2-x1);
    q23  = (y1-y3)*(y2-y1) + (x1-x3)*(x2-x1);
    Mi        = [q11, q12, q13;
                 q12, q22, q23;
                 q13, q23, q33];
    Mat(ii).D2 = 1/(4*A(ii))*Mi;
end


% Assembly of element diffusion matrices
D2 = zeros(nd,nd);  % Diffusion matrix initialization
for ii = 1:nd       % Number of nodes
    % Find the indexes in the connectivity matrix for shares nodes
    [gne,lnp]   = find(CM==ii);
    % Contribution of each element to the mass matrix
    De = zeros(length(gne),nd);
    for jj = 1:length(gne)  % Elements that share this node
        for kk = 1:3            % Number of nodes per element
            De(jj,CM(gne(jj),kk)) = Mat(gne(jj)).D2(lnp(jj),kk);
        end
    end
    D2(ii,:) = sum(De,1);
end
```

6.2.1.7 Simulation of FitzHugh-Nagumo's Model Using a Fine Grid

As mentioned above, the coarse discretization of Fig. 6.26 will not be enough to obtain accurate results (it is only used for illustration purposes). In this section, a much finer grid containing around 2,000 points (see Fig. 6.28) has been used to numerically solve the PDE system (6.46)–(6.48). Coarser grids result into a front-type solution with low resolution while finer grids result essentially into the same solution.

The value of the parameters of system (6.46)–(6.48) considered in this simulation experiment are: $\kappa = 1, a = 0.1, \varepsilon = 0.01, \beta = 0.5, \gamma = 1, \delta = 0$. Initial conditions are taken as:

$$u_0 = \begin{cases} 1 \;\text{ if }\; 0 \leqslant x \leqslant 10 \\ \\ 0 \;\text{ if }\; 10 \leqslant x \leqslant 200 \end{cases} \tag{6.86}$$

$$v_0 = 0, \quad \forall x, y \tag{6.87}$$

The final time chosen for the simulation is $t = 220$. The solution for the dependent variable u at different times is represented in Fig. 6.29. Such solution has the form of a plane front that moves along the x-coordinate. As mentioned at the beginning of

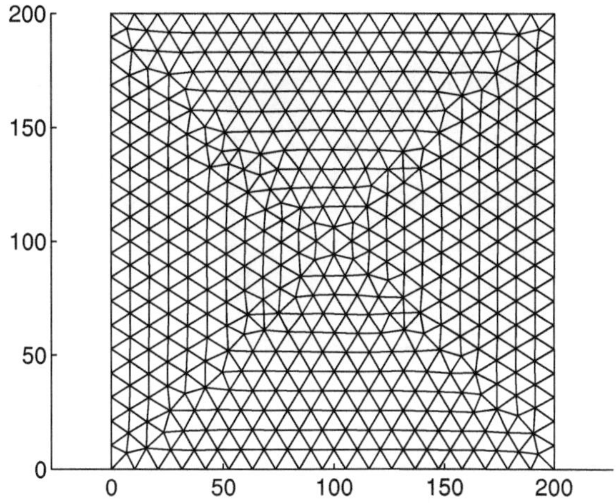

Fig. 6.28 Fine finite element mesh employed to numerically solve the FitzHugh-Nagumo problem

this section, such behavior relates to the normal behavior of the heart. The front, representing an electrical current, moves along the heart tissue producing the contraction of the muscle.

Under some circumstances, for instance if the front encounters an obstacle, the behavior may change. In this application example, such obstacle is simulated by resetting the upper half plane at a given time, i.e, we let the front evolve till $t = 180$ and then we fix $u = 0$ for $y > 100$ and $\forall x$ and continue the simulation. The behavior from that time instant is represented in Fig. 6.30. Figure 6.30a represents the result of resetting the upper half plane creating the initial condition for the new simulation. The other figures are snapshots taken at different times after the perturbation. In this case variable u takes the form of a spiral which is typical of heart problems like arrhythmia.

Practical note: In order to draw 3D figures in MATLAB (using for instance the commands `mesh` or `surf`) a particular grid is required. In this grid, every x spatial coordinate must have the same number of nodes. For instance, if for $x = 10$ we have 20 nodes, the other x-values in the grid must have also 20 nodes. The same holds for the y spatial coordinate.

In order to draw the figure, we first define new spatial coordinates following the previous rule.

```
% Sorted spatial coordinates
xsi  =  0:2:200;
ysi  =  0:2:200;
```

where for $x = 0, 2, 4, \ldots, 200$ we have 100 nodes ($y = 0, 2, 4, \ldots, 200$). Then we use these new coordinates to construct new arrays that can be used with the `mesh` command. Finally, `griddata` is used to interpolate the FEM solution obtained in a given grid into the new structures grid.

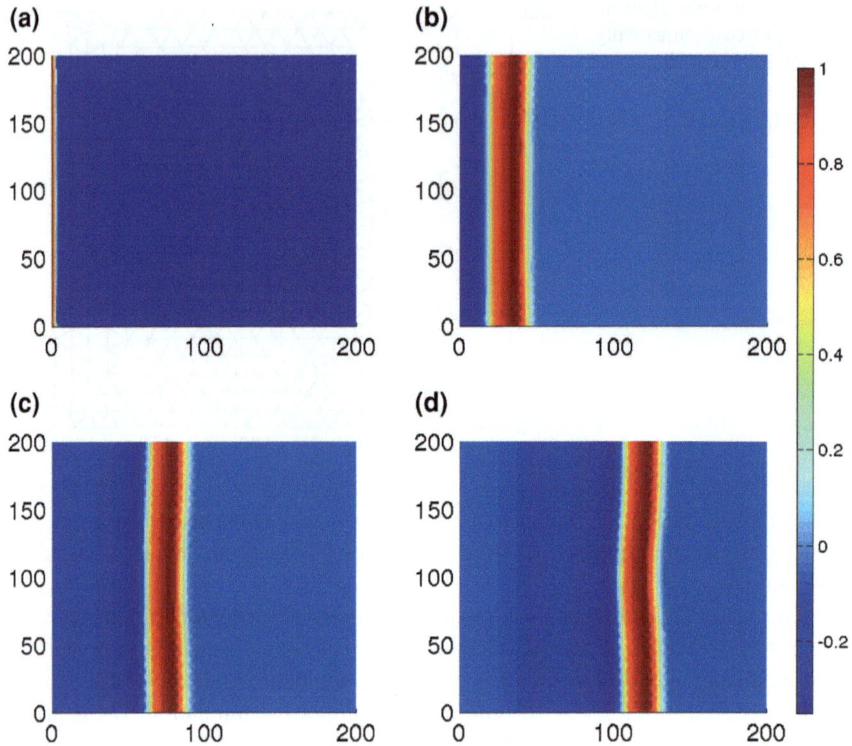

Fig. 6.29 Solution of the FitzHugh-Nagumo problem: spatial distribution of $u(x, y, t)$ at times **a** $t = 0$, **b** $t = 75$, **c** $t = 150$, **d** $t = 220$

```
% New spatial coordinates
[xi,yi] = meshgrid(xsi,ysi);

% Interpolate in the new grid the solution
vvi   = griddata(XX,YY,vv(:,100),xi,yi);
mesh(xi,yi,vvi)
```

Note that, since two dependent variables are considered, such grid implies solving around 4,000 ODEs which requires a large computational effort. In order to overcome such limitation an accurate reduced-order model derived using the POD technique will be developed next.

6.2.2 *Reduced-Order Model for FitzHugh-Nagumo Model*

The POD methodology presented in detail in Chap. 4 begins by approximating the dependent variables by truncated series of the form [11]:

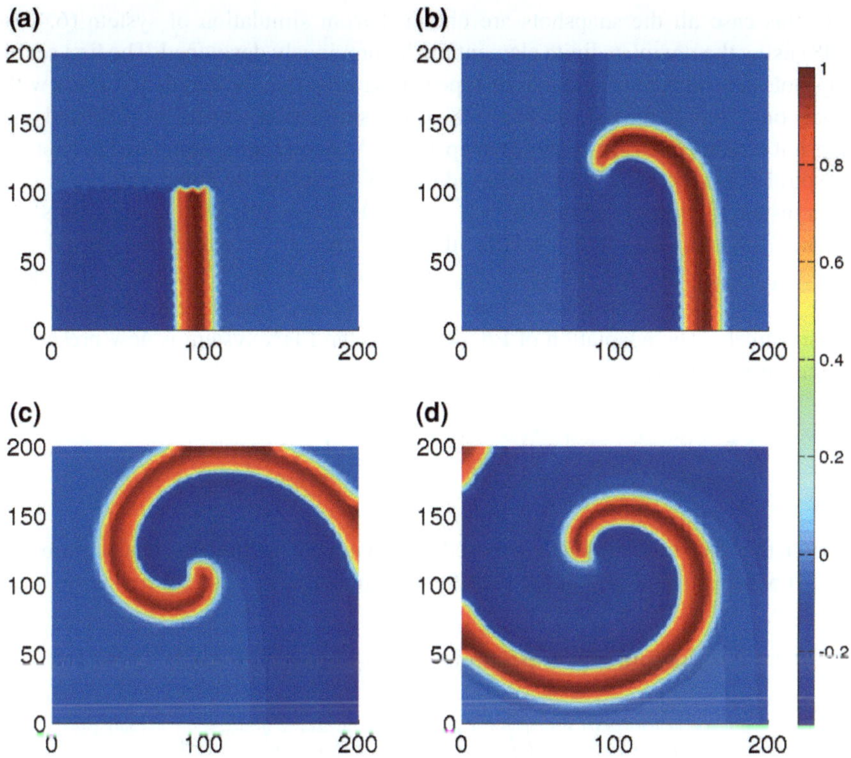

Fig. 6.30 Solution of the FitzHugh-Nagumo problem: spatial distribution of $u(x, y, t)$ at times **a** $t = 0$, **b** $t - 105$, **c** $t = 215$, **d** $t = 320$ after resetting the upper half plane

$$u(t, x, y) \approx \sum_{i=1}^{p_u} m_{u,i}(t)\phi_{u,i}(x, y) \tag{6.88}$$

$$v(t, x, y) \approx \sum_{i=1}^{p_v} m_{v,i}(t)\phi_{v,i}(x, y) \tag{6.89}$$

The basis functions in the POD method ($\phi_{u,i}(x, y)$, $\phi_{v,i}(x, y)$) are obtained using experimental data as explained in Sect. 4.11. Then, time dependent coefficients are computed by projecting the original PDE system onto the POD basis.

First, a set of snapshots (experimental data) must be constructed. This is a critical point in the POD technique. In order to obtain an accurate reduced-order model, the snapshots must be representative of the system behavior. Unfortunately, there is no systematic approach to decide the conditions that better represent the system behavior. However, the idea is to capture as much information as possible from a limited set of snapshots that may be obtained either through simulation of the original model or through appropriate experimental setups.

In this case all the snapshots are obtained from simulation of system (6.46)–(6.48) using the *complete* finite element model previously developed. The first set of snapshots aims at capturing the front-type behavior, so that the simulation starts with initial conditions of the form (6.86)–(6.87). The simulation runs till $t = 200$ and one snapshot is taken each $\Delta t = 10$ (21 snapshots). A second set is computed to capture the spiral behavior, such behavior is induced by resetting the upper half plane at a given instant. Snapshots are taken each $\Delta t = 10$ till $t = 200$ (also 21 snapshots).

Once the snapshots are available they are used to construct the kernel $\mathcal{K}(x, y, x', y')$ as in Eq. (4.125). In fact two kernels ($\mathcal{K}_u(x, y, x', y')$ and $\mathcal{K}_v(x, y, x', y')$) are constructed from the snapshots of the state variables u and v, respectively. The adaptation of Eq. (4.125) to the FHN system is now presented for the sake of clarity:

$$\mathcal{K}_u = \frac{1}{k}\sum_{i=1}^{k} u_i u_i^T; \qquad \mathcal{K}_v = \frac{1}{k}\sum_{i=1}^{k} v_i v_i^T.$$

Then the basis functions are computed by solving the integral eigenvalue problem (4.124) which, for the case of the FHN system, reads as:

$$\lambda_{u,i}\phi_{u,i}(x, y) = \iint_{\Omega} \mathcal{K}_u(x, y, x', y')\phi_{u,i}(x', y')\mathrm{d}x\mathrm{d}y',$$

$$\lambda_{v,i}\phi_{v,i}(x, y) = \iint_{\Omega} \mathcal{K}_v(x, y, x', y')\phi_{v,i}(x', y')\mathrm{d}x\mathrm{d}y'$$

The mass matrix obtained from the application of the finite element method can be exploited to numerically compute the previous spatial integrals (see section 4.3.5). As a result of this step, two basis sets ($\Phi_u = [\phi_{u,1}, \phi_{u,2}, \ldots, \phi_{u,p_u}]$ and $\Phi_v = [\phi_{v,1}, \phi_{v,2}, \ldots, \phi_{v,p_v}]$) are obtained.

At this point the basis functions are available and the only information required to recover the dependent variables u and v are the time dependent modes m_u and m_v. To this end, the original PDE system (6.46)–(6.48) is projected onto the basis functions. Projection is carried out by multiplying the PDE system with the basis functions and integrating the result over the spatial domain $\mathcal{V}$. Let us now describe the projection procedure for the dependent variable u. The same steps are followed for variable v:

$$\iint_{\Omega} \phi_{u,i}\frac{\partial u}{\partial t}\mathrm{d}x\mathrm{d}y = \iint_{\Omega} \phi_{u,i}\kappa\Delta u\mathrm{d}x\mathrm{d}y + \iint_{\Omega} \phi_{u,i}g(u) - v\mathrm{d}x\mathrm{d}y \tag{6.90}$$

where Δ represents the Laplacian operator.

Using the series expansion (6.88) the LHS term in (6.90) can be rewritten as:

$$\iint_{\Omega} \phi_{u,i} \frac{\partial u}{\partial t} \mathrm{d}x\mathrm{d}y = \iint_{\Omega} \phi_{u,i} \frac{\partial \sum_{j=1}^{p_u} m_{u,j}\phi_{u,j}}{\partial t} \mathrm{d}x\mathrm{d}y = \sum_{j=1}^{p_u} \frac{\mathrm{d}m_{u,i}}{\mathrm{d}t} \iint_{\Omega} \phi_{u,i}\phi_{u,j}$$

Since the basis functions are orthogonal

$$\iint_{\Omega} \phi_{u,i} \frac{\partial u}{\partial t} \mathrm{d}x\mathrm{d}y = \frac{\mathrm{d}m_{u,i}}{\mathrm{d}t}$$

and extending this expression for $i = 1, \ldots, p_u$

$$\iint_{\Omega} \phi_u \frac{\partial u}{\partial t} \mathrm{d}x\mathrm{d}y = \frac{\mathrm{d}\mathbf{m}_u}{\mathrm{d}t} \tag{6.91}$$

where $\mathbf{m}_u = [m_{u,1}, m_{u,2}, \ldots, m_{u,p_u}]^T$.

Now, the first term on the RHS of Eq. (6.90) can be rewritten as:

$$\kappa \iint_{\Omega} \phi_{u,i} \Delta u \mathrm{d}x\mathrm{d}y = \kappa \iint_{\Omega} \nabla(\phi_{u,i}\nabla u)\mathrm{d}x\mathrm{d}y - \kappa \iint_{\Omega} \nabla\phi_{u,i}\nabla u \mathrm{d}x\mathrm{d}y$$

where ∇ represents the gradient operator. Applying Green's theorem

$$\kappa \iint_{\Omega} \phi_{u,i} \Delta u \mathrm{d}x\mathrm{d}y = \kappa \iint_{\partial\Omega} \mathbf{n}\phi_{u,i}\nabla u \mathrm{d}x\mathrm{d}y - \kappa \iint_{\Omega} \nabla\phi_{u,i}\nabla u \mathrm{d}x\mathrm{d}y$$

Using expression (6.48) in the boundary term we obtain:

$$\kappa \iint_{\Omega} \phi_{u,i} \Delta u \mathrm{d}x\mathrm{d}y = -\kappa \iint_{\Omega} \nabla\phi_{u,i}\nabla u \mathrm{d}x\mathrm{d}y$$

Expanding u in a truncated Fourier series

$$\begin{aligned}
\kappa \iint_{\Omega} \phi_{u,i} \Delta u \mathrm{d}x\mathrm{d}y &= -\kappa \iint_{\Omega} \nabla\phi_{u,i}\nabla \sum_{j=1}^{p_u} m_{u,j}\phi_{u,j} \mathrm{d}x\mathrm{d}y \\
&= -\kappa \sum_{j=1}^{p_u} m_{u,j} \iint_{\Omega} \nabla\phi_{u,i}\nabla\phi_{u,j} \mathrm{d}x\mathrm{d}y
\end{aligned}$$

This latter equation can be expressed in matrix form

$$\kappa \iint_{\Omega} \phi_{u,i} \Delta u \mathrm{d}x\mathrm{d}y = -\kappa \mathscr{A} \mathbf{m}_u \tag{6.92}$$

where the elements of matrix $\mathscr{A}$ are given by:

$$\mathscr{A}_{i,j} = \iint_{\Omega} \nabla\phi_{u,i} \nabla\phi_{u,j} \mathrm{d}x\mathrm{d}y.$$

Finally, denoting by $\mathscr{F}$ the second term of RHS of Eq. (6.90), i.e. $\mathscr{F} = \iint_{\Omega} \phi_{u,i} g(u) - v \mathrm{d}x\mathrm{d}y$ and using expressions (6.91) and (6.92)

$$\frac{\mathrm{d}m_u}{\mathrm{d}t} = \mathscr{A} m_u + \mathscr{F}, \tag{6.93}$$

The same steps are followed in the case of dependent variable v to obtain:

$$\frac{\mathrm{d}\mathbf{m}_v}{\mathrm{d}t} = \mathscr{G} \tag{6.94}$$

where

$$\mathscr{G} = \int_{\mathscr{V}} \phi_v \varepsilon(\beta u - \gamma v + \delta) \mathrm{d}x\mathrm{d}y$$

The number of basis functions employed in the projection will determine the dimension of the reduced-order model. Several trial and error tests show that 85 and 28 basis functions for the projection of dependent variables u and v, respectively, is enough to accurately represent the system behavior.

Initial conditions are also projected as follows:

$$m_{u0} = \int_{\mathscr{V}} \Phi_u u_0 d\xi \tag{6.95}$$

$$m_{v0} = \int_{\mathscr{V}} \Phi_v v_0 d\xi \tag{6.96}$$

As a result a system with 113 ODEs (more than 40 times smaller than the original FE system) is obtained.

The solution of (6.93)–(6.96) can be computed with a standard initial value problem solver (see Chap. 2).

The time evolution of the first four time dependent coefficients for this problem is represented in Fig. 6.31. The vertical continuous line indicates the time at which the front is broken, i.e., when the system begins to behave as a wandering spiral. Continuous lines correspond to the projection of the FEM solution onto the basis

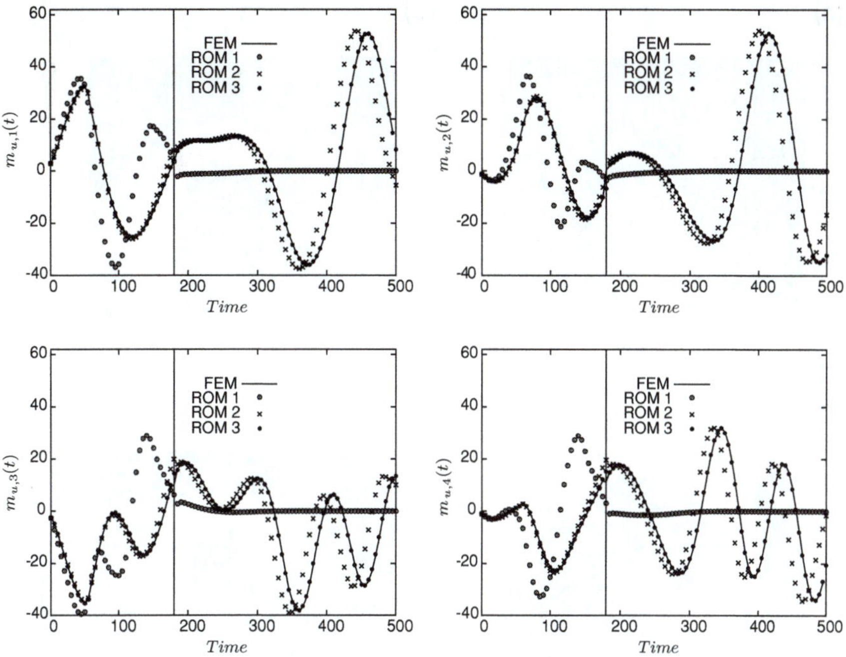

Fig. 6.31 Numerical solution, in terms of the first four time dependent coefficients, of the FitzHugh-Nagumo problem

functions, i.e., they are the time dependent coefficients of the FEM solution. Marks correspond to the results of three ROMs. The difference between the ROMs is the number of basis functions used in the projection. In this sense, in ROM 1, 30 and 17 basis functions are used for the projection of u and v, respectively, i.e. $p_u = 30$, $p_v = 17$. In ROM 2 $p_u = 40$, $p_v = 18$ while in ROM 3 $p_u = 80$, $p_v = 30$. As apparent in the figure, ROM1 can only capture the front behavior in a qualitative way. When the perturbation is introduced in the system to induce the wandering spiral ($t = 180$), the behavior of ROM 1 is completely different from the FEM solution. On the other hand, ROM 2 is able to capture in a qualitative manner the behavior of the FEM solution. The spiral part of the behavior seems to be reproduced quite well but it seems that the coefficients of ROM 2 are ahead of the FEM ones. Finally, ROM 3 is able to capture in a quantitative manner the FEM solution and as shown in the figure, the coefficients evolution coincides with the FEM ones.

Figure 6.32 shows the spatial distribution of dependent variable u at different times as predicted by ROM 3, in agreement with the behavior shown in Fig. 6.30.

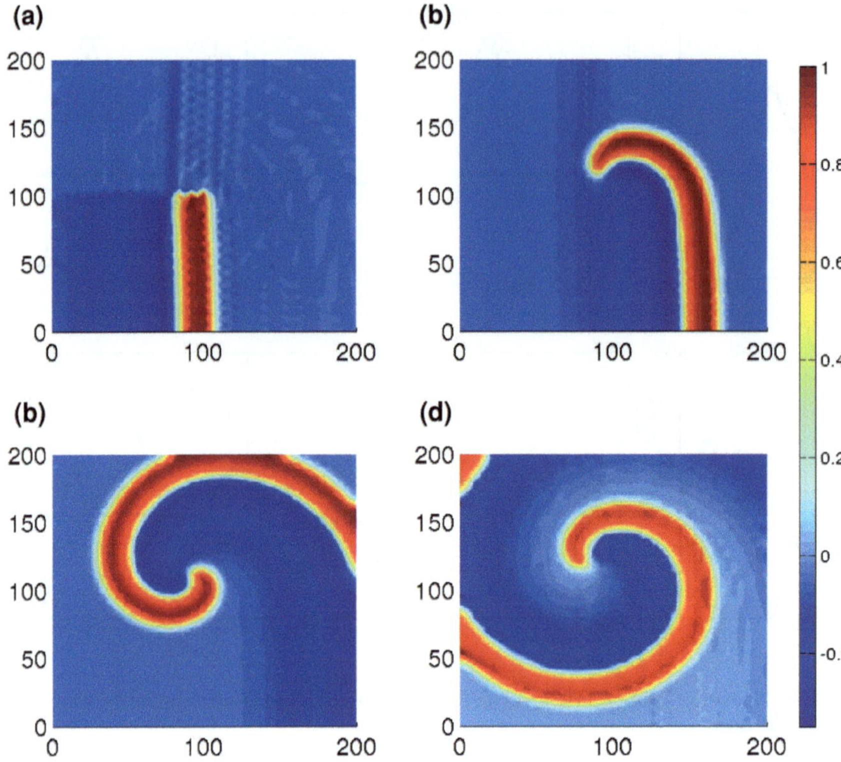

Fig. 6.32 Solution, in terms of variable u, of the FitzHugh-Nagumo problem using the POD technique (ROM 3). Plotting times are (a) $t = 0$, (b) $t = 105$, (c) $t = 215$, (d) $t = 320$ after resetting the upper half plane

6.3 Solution of PDEs on Time-Varying Domains

In the previous two sections, we have covered the basic concepts involved in the solution of PDEs on 2D spatial domains using either FDs or FEs. In the present section, we now introduce the situation where the spatial domain definition can change in time, i.e., the boundaries of a 1D spatial domain moves (or the shape of a 2D domain changes, but we will not consider this case in this introductory text). The approach that we will follow is based on a transformation of the spatial coordinates, so as to define a new, fixed, system of coordinates where the problem is easier to solve. The approach is introduced and illustrated with a concrete example of freeze drying (or lyophilization) in the food processing industry. Freeze-drying is a dehydration process used to preserve food, which consists in first freezing the material and then reducing the surrounding pressure to allow the frozen water in the material to sublimate directly from the solid phase to the gas phase.

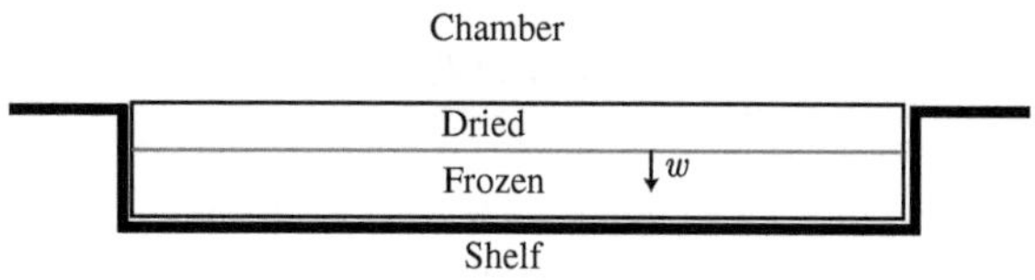

Fig. 6.33 Schematic representation of the freeze-drying chamber. *Grey line* indicates the front position at a given time and w denotes its velocity. The product is heated using the shelf. Chamber pressure and shelf temperature are controlled

6.3.1 The Freeze-Drying Model

At the beginning of the process the product is completely frozen (one phase). When the process starts ice sublimates and, as a consequence, a dried phase appears at the top of the product and increases as the process evolves (see Fig. 6.33). The interphase between both phases is known as the front. In our particular example, product height is much smaller than its width and length, therefore, the original 3D domain can be approximated by an equivalent 1D domain taking only product height into account.

The temperature in the dried and frozen zones is described using the Fourier equation:

$$\rho_{dr} c_{p,dr} \frac{\partial T_{dr}}{\partial t} = \kappa_{dr} \frac{\partial^2 T_{dr}}{\partial z^2} \tag{6.97}$$

$$\rho_{fr} c_{p,fr} \frac{\partial T_{fr}}{\partial t} = \kappa_{fr} \frac{\partial^2 T_{fr}}{\partial z^2} \tag{6.98}$$

where subindices dr and fr refer to dried and frozen regions, respectively. T is the temperature and z is the spatial coordinate. Parameters ρ, c_p and κ represent the density, the specific heat and heat conductivity, respectively.

Denoting the front position by $S(t)$, the spatial domain for Eq. (6.97) is $0 \leq z \leq S(t)$ while for Eq. (6.98) is $S(t) \leq z \leq L$ with L being the product height.

In order to solve Eqs. (6.97)–(6.98) boundary conditions are required. At $z = 0$ radiation is considered as the main heat transfer phenomena:

$$\kappa_{dr} \left. \frac{\partial T_{dr}}{\partial z} \right|_{z=0} = \sigma e_p f_p (T_{ch}^4 - T_{dr}|_{z=0}^4) \tag{6.99}$$

where σ is the well known Stefan-Boltzmann constant, e_p is the emissivity, f_p is a geometric correction factor and T_{ch} is the chamber temperature at the top of the sample.

Boundary conditions at $z = L$ are considered as a combination of radiation, convection and gas conduction which depends on the chamber pressure [12, 13]

$$\kappa_{fr} \left. \frac{\partial T_{fr}}{\partial z} \right|_{z=L} = h_L(T_{sh} - T_{fr}|_{z=L}) \tag{6.100}$$

In this equation T_{sh} is the shelf temperature. The heat transfer coefficient h_L containing the chamber pressure dependency takes the form [12]:

$$h_L = h_{L,1} + \frac{h_{L,2} P_c}{1 + \dfrac{P_c}{34.4}} \tag{6.101}$$

where $h_{L,1}$ and $h_{L,2}$ are given constant parameters and P_c represents the chamber pressure.

The third boundary condition required to solve the freeze-drying equations is located at $z = S(t)$, i.e. the front. Dirichlet boundary conditions are considered here:

$$T_{dr}|_{z=S(t)} = T_{\text{front}}; \qquad T_{fr}|_{z=S(t)} = T_{\text{front}} \tag{6.102}$$

The front temperature (T_{front}) is another process variable. In order to compute it, we use a combination between Darcy's law [14], which describes the movement of water vapor through a porous media, and Clausius-Clapeyron equation [15, 16], describing the relation between pressure and temperature in a phase change process.
From Darcy's equation we have, after a finite differences discretization:

$$P_{front} = P_c + \frac{(\rho_{fr} - \rho_{dr})w}{K_d} \tag{6.103}$$

where P_c is the chamber pressure, $w = \frac{dS}{dt}$ represents the front velocity while K_d denotes the mass resistance which is computed through

$$K_d = \frac{1}{k_1 P_c + k_2 S} \tag{6.104}$$

The front pressure is used in the Clapeyron equation to obtain the temperature:

$$T_{\text{front}} = \frac{1}{\dfrac{1}{273.11} - K_{\text{clap}} \log\left(\dfrac{P_{f\text{front}}}{611.72}\right)} \tag{6.105}$$

Finally, the description of the front motion is obtained by a heat balance at the interface. This balance considers the heat fluxes from the dried region to the interface and from the interface to the frozen region as well as the heat absorbed due to sublimation. Such balance results into the Stefan equation [17, 18]

$$(\rho_{fr} - \rho_{dr})\Delta H_s w = \left(\kappa_{fr} \left. \frac{\partial T_{fr}}{\partial z} \right|_{z=x^+} - \kappa_{dr} \left. \frac{\partial T_{dr}}{\partial z} \right|_{z=x^-} \right) \tag{6.106}$$

Table 6.1 Parameters employed in the freeze-drying model

Parameter	Value	Units
ρ_{dr}	200.31	$kg \cdot m^{-3}$
ρ_{fr}	1001.6	$kg \cdot m^{-3}$
$c_{p,dr}$	1254	$J \cdot kg^{-1} \cdot K^{-1}$
$c_{p,fr}$	1818.8	$J \cdot kg^{-1} \cdot K^{-1}$
κ_{dr}	0.0129	$W \cdot K^{-1} \cdot m^{-1}$
κ_{fr}	2.4	$W \cdot K^{-1} \cdot m^{-1}$
L	0.02	m
σ	5.6704×10^{-8}	$W \cdot m^{-2} \cdot K^{-4}$
e_p	0.78	–
f_p	0.99	–
K_{clap}	1.6548×10^{-4}	K^{-1}
$h_{L,1}$	3.85	$W \cdot m^{-2} \cdot K^{-1}$
$h_{L,2}$	0.352	$W \cdot m^{-2} \cdot K^{-1} \cdot Pa^{-1}$
k_1	4.75×10^3	$m \cdot s^{-1}$
k_2	6.051×10^7	s^{-1}
ΔH_s	2791.2×10^3	$J \cdot kg^{-1}$

where ΔH_s is the sublimation heat. The parameter values are given in Table 6.1.

Note that the front boundary is moving, therefore we have two moving domains. When dealing with this kind of problems it is common to use complex techniques such as the *Arbitrary Lagrangian-Eulerian* (ALE) method [17] for which specialized software is required. In this method the spatial grid moves with arbitrary motion. Details about this technique can be found in [17, 19, 20]. In order to avoid the use of such complex techniques and specialized software, a change of coordinates (Landau transform) is proposed here.

6.3.2 The Landau Transform

The Landau transform [21, 22] consists of a change of coordinates which allows to define fixed spatial domains in the transformed coordinates.

6.3.2.1 Transformation in the Dried Region

In this phase the following transformation is defined:

$$z \rightsquigarrow \xi \backslash z = S(t)\xi; \qquad \text{or} \quad \xi = \frac{z}{S(t)} \tag{6.107}$$

Note that $\xi \in [0, 1]$. The time coordinate is not transformed although we will rename it to avoid confusion $t \rightsquigarrow \theta \backslash t = \theta$. In the new coordinate system, the temperature will be denoted by R, this is: $T_{dr}(t,z) \rightsquigarrow R_{dr}(\theta,\xi)$. The following expressions for the partial derivatives can be derived from Eq. (6.107):

$$\frac{\partial \xi}{\partial \theta} = -\frac{zS_\theta}{S^2} = -\frac{\xi w}{S}; \quad \frac{\partial \xi}{\partial z} = \frac{1}{S}; \quad \frac{\partial^2 \xi}{\partial z^2} = \frac{1}{S^2} \tag{6.108}$$

The different time and spatial derivatives in Eq. (6.97) are rewritten, using the previous relations, as follows

$$\begin{aligned}
\frac{\partial T_{dr}}{\partial t} &= \frac{\partial R_{dr}}{\partial \theta}\frac{\partial \theta}{\partial t} + \frac{\partial R_{dr}}{\partial \xi}\frac{\partial \xi}{\partial t} = \frac{\partial R_{dr}}{\partial \theta} - \frac{\xi w}{S}\frac{\partial R_{dr}}{\partial \xi}\\
\frac{\partial T_{dr}}{\partial z} &= \frac{\partial R_{dr}}{\partial \theta}\frac{\partial \theta}{\partial z} + \frac{\partial R_{dr}}{\partial \xi}\frac{\partial \xi}{\partial z} = \frac{1}{S}\frac{\partial R_{dr}}{\partial \xi}\\
\frac{\partial^2 T_{dr}}{\partial z^2} &= \frac{1}{S^2}\frac{\partial^2 R_{dr}}{\partial \xi^2}
\end{aligned}$$

Therefore an expression equivalent to Eq. (6.97) is found in the new coordinate system where the spatial domain is fixed:

$$\rho_{dr} c_{p,dr}\frac{\partial R_{dr}}{\partial \theta} = \frac{\kappa_{dr}}{S^2}\frac{\partial^2 R_{dr}}{\partial \xi^2} + \rho_{dr} c_{p,dr}\frac{\xi w}{S}\frac{\partial R_{dr}}{\partial \xi}$$

Dividing the equation by $\rho_{dr} c_{p,dr}$

$$\frac{\partial R_{dr}}{\partial \theta} = \frac{\alpha_{dr}}{S^2}\frac{\partial^2 R_{dr}}{\partial \xi^2} + \frac{\xi w}{S}\frac{\partial R_{dr}}{\partial \xi} \tag{6.109}$$

where $\alpha_{dr} = \dfrac{\kappa_{dr}}{\rho_{dr} c_{p,dr}}$. Note that the price to pay to fix the domain is the addition of a *fictitious* negative convection term ($\frac{\xi w}{S}\frac{\partial R_{dr}}{\partial \xi}$). Furthermore there is an indeterminate form at $S = 0$, thus the simulation must start at $S = \varepsilon > 0$ where ε must be small enough to obtain a good approximation to the real solution.

6.3.2.2 Transformation in the Frozen Region

In this phase the transformation is defined as:

$$z \rightsquigarrow \eta \backslash z = \eta(L - S) + S; \qquad \text{or} \quad \eta = \frac{z - S}{L - S} \tag{6.110}$$

Note that $\eta \in [0, 1]$. As in the case of the dried region, the temperature will be denoted by R, this is, $T_{fr}(t,z) \rightsquigarrow R_{fr}(\theta,\eta)$. The following expressions for the partial derivatives can be derived from Eq. (6.110):

$$\frac{\partial \eta}{\partial \theta} = -w\frac{1-\eta}{L-S}; \qquad \frac{\partial \eta}{\partial z} = \frac{1}{L-S}; \qquad \frac{\partial^2 \eta}{\partial z^2} = \frac{1}{(L-S)^2} \tag{6.111}$$

Again, time and spatial derivatives in Eq. (6.98) can be rewritten in the new coordinate system with fixed domain as:

$$\frac{\partial T_{fr}}{\partial t} = \frac{\partial R_{fr}}{\partial \theta}\frac{\partial \theta}{\partial t} + \frac{\partial R_{fr}}{\partial \eta}\frac{\partial \eta}{\partial t} = \frac{\partial R_{fr}}{\partial \theta} - w\frac{1-\eta}{L-S}\frac{\partial R_{fr}}{\partial \eta}$$

$$\frac{\partial T_{fr}}{\partial z} = \frac{\partial R_{fr}}{\partial \theta}\frac{\partial \theta}{\partial z} + \frac{\partial R_{fr}}{\partial \eta}\frac{\partial \eta}{\partial z} = \frac{1}{L-S}\frac{\partial R_{fr}}{\partial \eta}$$

$$\frac{\partial^2 T_{fr}}{\partial z^2} = \frac{1}{(L-S)^2}\frac{\partial^2 R_{fr}}{\partial \eta^2}$$

Substituting the previous expressions into Eq. (6.98):

$$\rho_{fr}c_{p,fr}\frac{\partial R_{fr}}{\partial \theta} = \frac{\kappa_{fr}}{(L-S)^2}\frac{\partial^2 R_{fr}}{\partial \eta^2} + \rho_{fr}c_{p,fr}w\frac{1-\eta}{L-S}\frac{\partial R_{fr}}{\partial \eta}$$

and, dividing the equation by $\rho_{fr}c_{p,fr}$

$$\frac{\partial R_{fr}}{\partial \theta} = \frac{\alpha_{fr}}{(L-S)^2}\frac{\partial^2 R_{fr}}{\partial \eta^2} + w\frac{1-\eta}{L-S}\frac{\partial R_{dr}}{\partial \eta} \tag{6.112}$$

where $\alpha_{fr} = \dfrac{\kappa_{fr}}{\rho_{fr}c_{p,fr}}$. As in the previous case, a *fictitious* convection term appears as a consequence of the application of the Landau transform. The indeterminate form appears now at $S = L$ so that the simulation has to be halted before arriving to this point.

The transformation must be also applied to the expressions in the boundary and the front.

6.3.2.3 Transformation in the Boundaries and the Front

Boundary conditions at the top and the bottom of the sample—Eqs. (6.99) and (6.100)—are transformed into:

$$\left.\frac{\kappa_{dr}}{S}\frac{\partial R_{dr}}{\partial \xi}\right|_{\xi=0} = \sigma e_p f_p(R_{ch}^4 - R_{dr}|_{z=0}^4) \tag{6.113}$$

$$\left.\frac{\kappa_{fr}}{L-S}\frac{\partial R_{fr}}{\partial \eta}\right|_{\eta=1} = h_L(R_{sh} - R_{fr}|_{\eta=1}) \tag{6.114}$$

while Stefan equation in the new coordinate system reads as follows:

$$(\rho_{fr} - \rho_{dr})\Delta H_s w = \left(\frac{\kappa_{fr}}{L-S} \left.\frac{\partial R_{fr}}{\partial \eta}\right|_{\eta=0} - \frac{\kappa_{dr}}{S} \left.\frac{\partial R_{dr}}{\partial \xi}\right|_{\xi=1} \right) \tag{6.115}$$

With this transformation, we therefore have arrived at an equivalent system formed by two fixed spatial domains (dried and frozen) linked by Eq. (6.115).

6.3.3 The Finite Element Representation

In this section the FEM developed in Chap. 4 is used to compute the numerical solution of the freeze-drying model. The general form for the finite element representation of a given nonlinear diffusion-convection system using lagrangian elements is presented in Sect. 4.3 and the result is given in Eq. (4.76). The main points to note with respect to this formulation are:

- Two PDEs are considered.
- The freeze-drying model is linear, therefore the nonlinear part in Eq. (4.76) is avoided.
- The "trick" described in Sects. 3.10 and 4.11.2 to approximate Dirichlet boundary conditions (defined as an algebraic relation) by Robin boundary conditions (defined as an ODE) is used here to implement the boundary conditions.

The FEM version of the freeze-drying model is, therefore

$$\mathbf{M}\frac{\mathrm{d}\mathbf{R}_{dr}}{\mathrm{d}t} = \left(\alpha_{1,dr}\mathbf{D}_1 + \alpha_{2,dr}\mathbf{D}_2^{\mathrm{int}}\right)\mathbf{R}_{dr} + \mathbf{g}_{dr} \tag{6.116}$$

$$\mathbf{M}\frac{\mathrm{d}\mathbf{R}_{fr}}{\mathrm{d}t} = \left(\alpha_{1,fr}\mathbf{D}_1 + \alpha_{2,fr}\mathbf{D}_2^{\mathrm{int}}\right)\mathbf{R}_{fr} + \mathbf{g}_{fr} \tag{6.117}$$

where

$$\alpha_{1,dr} = \frac{w\xi}{S}; \quad \alpha_{2,dr} = \frac{k_{dr}}{\rho_{dr}c_{p,dr}S^2}; \quad \alpha_{1,fr} = \frac{w(1-\eta}{L-S}; \quad \alpha_{2,fr} = \frac{k_{fr}}{\rho_{fr}c_{p,fr}(L-S)^2};$$

$$\mathbf{g}_{dr} = \begin{bmatrix} \sigma e_p f_p \dfrac{T_{ch}^4 - R_{dr}(1)^4}{S\rho_{dr}c_{p,dr}} \\ 0 \\ \vdots \\ 0 \\ \dfrac{1}{\varepsilon}(T_{front} - R_{dr}(n_\xi)) \end{bmatrix}; \quad \mathbf{g}_{fr} = \begin{bmatrix} \dfrac{1}{\varepsilon}(T_{front} - R_{fr}(1)) \\ 0 \\ \vdots \\ 0 \\ h_L \dfrac{T_{sh} - R_{fr}(n_\eta)}{(L-S)\rho_{fr}c_{p,fr}} \end{bmatrix}; \tag{6.118}$$

and n_ξ and n_η are the number of discretization points in the dried and frozen regions, respectively. The first point in $\mathbf{g}_{dr}$ as well as the last point in $\mathbf{g}_{fr}$ correspond to Robin type boundary conditions so they enter naturally in the formulation. On the other hand, the last point in $\mathbf{g}_{dr}$ as well as the first point in $\mathbf{g}_{fr}$ correspond with Dirichlet boundary conditions and they are therefore multiplied by the inverse of a close to zero factor (ε).

For the sake of simplicity, notation for matrices **M**, $\mathbf{D}_1$ and $\mathbf{D}_2^{\text{int}}$ is the same in Eqs. (6.116) and (6.117) since the same discretization scheme was applied in the dried and frozen regions. However, we should keep in mind that depending on the problem, different discretization schemes could be employed for both regions.

The RHS of Eqs. (6.116) and (6.117) is coded in the MATLAB function `ode_fd`. We can note the following details:

1. The input parameters of the function are: the independent variable `t`, the dependent variable `y` (which contains the dependent states `Rdr`, `Rfr` and `x`) and a set of constants (that can be matrices, vectors or scalars).
2. The first block of the code focuses on the separate the dependent variables grouped in the input vector `y` and computing the gradients of the temperature related variables. Such gradients are computed using matrices `Adv_op_dr` = `Adv_op_fr` = $\mathbf{M}_1^{-1}\mathbf{D}_1$.
3. The second block of the program applies the algebraic relations (6.101), (6.103) and (6.105) in order to compute the heat transfer coefficient and the front temperature. The latter equation is coded in function `funcclapeyron`. Then the front velocity is computed using Stefan equation (6.106).
4. After this, boundary conditions are computed according to Eq. (6.118). Note that, at the end of this block boundary conditions are multiplied by the inverse of the mass matrices corresponding to both the dried and frozen regions.
5. Finally, Eqs. (6.116) and (6.117) are implemented. As a matter of fact, Eqs. (6.116) and (6.117) are multiplied by the inverse of the mass matrix before the implementation. Another option (already used in this book) is to pass the mass matrix to the IVP solver using function `odeset`.

```
function dy = ode_fd(t, y, Adv_op_dr, Adv_op_fr, Lap_op_dr,...
                     Lap_op_fr, inv_MMdr, inv_MMfr, ze, ye,...
                     ndz, ndy, L, alpha, Kfr, Kdr, sigma, ep,...
                     fp, Tcl, rhodr, cpdr, rhofr, cpfr, kd1,...
                     kd2, hL1, hL2, hL3, Pc, Tcr)

% The states
Rdr     = y(1:ndz);
Rfr     = y(1+ndz:ndz+ndy);
x       = y(ndz+ndy+1);

% States gradients
Rdrz = Adv_op_dr*Rdr;
Rfry = Adv_op_fr*Rfr;

% Diffusivities
difdr = Kdr/(rhodr*cpdr);
diffr = Kfr/(rhofr*cpfr);
```

```
% Heat transfer coefficient
hL   = hL1 + hL2*Pc/(1+Pc/hL3);

% Permeability
Kd =  1/(kd1*Pc+kd2*x);

% Front pressure and temperature
Pfront = Pc + ((rhofr-rhodr)*w)/(Kd);
Tfront = funcclapeyron(Pfront,'T');

% Front velocity
w = alpha*(Kfr/(L-x)*Rfry(1) - Kdr/x*Rdrz(ndz));

% Boundary conditions
Gdr        = zeros(ndz,1);
Gdr(1)     = sigma*ep*fp*(Tcl^4-Rdr(1)^4)/(x*rhodr*cpdr);
Gdr(ndz)  = 1e6*(Tfront-Rdr(ndz));
Gfr        = zeros(ndy,1);
Gfr(1)     = 1e6*(Tfront-Rfr(1));
Gfr(ndy)  = hL*(Tcr-Rfr(ndy))/((L-x)*rhofr*cpfr);
BVdr       = inv_MMdr*Gdr;
BVfr       = inv_MMfr*Gfr;

% ODEs
dRdr = w/x*ze.*Rdrz + (difdr/x^2*Lap_op_dr)*Rdr + BVdr;
dRfr = w/(L-x)*(1-ye).*Rfry+ (diffr/(L-x)^2*Lap_op_fr)*Rfr+ BVfr;
dx    = w;
dy    = [dRdr; dRfr; dx];
```

Function ode_fd Function to evaluate of the RHS of the Freeze-Drying problem—Eqs. (6.116) and (6.117)—after the application of the Landau transform.

As mentioned before, Clausius-Clapeyron equation is coded in function `funcclapeyron`. Two input arguments are used in this function: the second one is a string variable indicating the output parameter (`T` for temperature and `P` for pressure), the first parameter is the value of the pressure if temperature is the output parameter of the value of temperature if the pressure is the output parameter.

```
function varargout = funcclapeyron(varargin)

K = 8314/(2791.2e3*18);

% Output
switch varargin{2}
    case {'T'}
        % Output parameter is T. Input parameter is P
        P = varargin{1};
        T = 1./( 1/273.11  - K*log(P/611.73));
        varargout{1} = T;
    case {'P'}
        % Output parameter is P. Input parameter is T
        T = varargin{1};
        P = 611.73*exp(1/K*(1/273.11 -1./T));
        varargout{1} = P;
end
```

Function funcclapeyron Function to implement Clausius-Clapeyron equation (6.105).

The main file to solve the Freeze-Drying problem in MATLAB is coded in `fd_model`.

1. As usual, the script begins with the definition of the model parameters.
2. After this, control inputs, time span and initial conditions are defined. Note that the initial conditions for temperature must be defined in the coordinate system resulting from the Landau transform.
3. Then the mass matrix as well as first and second order differentiation matrices are computed with the MATLAB function `matfem`, included in the companion sofware.
4. The integration is performed using the IVP solver `ode15s`.
5. Finally, the transformation is undone in order to recover the original temperature variable and the solution is plotted.

```
clear all
clc

% Fixed parameters
% Dried region parameters
ep    = 0.78;          % Emisivity on the product top
fp    = 0.98;          % Geometrical correction factor on the top
rhodr = 200.31;        % Density
cpdr  = 1254;          % Specific heat
Kdr   = 0.0129;        % Heat conductivity in the dried region
ndz   = 13;            % Dried region discretization points
% Frozen region parameters
Kfr   = 2.4;               % Heat conductivity
rhofr = 1001.6;            % Density
cpfr  = 1818.8;            % Specific heat
ndy   = 13;                % Dried region discretization points
% Other parameters
L       = 5.75e-3;                 % Sample length
DHs     = 2791.2e3;                % Sublimation heat
sigma   = 5.6704e-8;               % Stefan-Boltzmann constant
alpha   = 1/((rhofr-rhodr)*DHs);
Tcl     = 20+273.15;               % Chamber temperature
kd1    = 4.75e3;   % Parameter in the mass resistance equation
kd2    = 6.051e7;  % Parameter in the mass resistance equation
hL1    = 3.85;     % Parameter in the heat transfer coefficient
hL2    = 0.352;    % Parameter in the heat transfer coefficient
hL3      = 34;     % Parameter in the heat transfer coefficient

% Control variables
Pc  = 20;            % Chamber pressure
Tcr = 273.15+20;     % Shelf temperature

% Time data
tf   = 26000;                  % final time
tl   = linspace(0,tf,200);

% Initial conditions (read experiments)
T_ini         = 244*ones(25,1);  % Dried region initial temperature
x0            = L*0.015;         % Initial front position (x0>0)
xi            = linspace(0,L,length(T_ini));
```

```
ze          = linspace(0,1,ndz)';   % Dryed region (Landau)
                                    % transformed coordinates
ye          = linspace(0,1,ndy)';   % Frozen region (Landau)
                                    % transformed coordinates
% Initial conditions (adaptation to Landau)
xi_ndz = linspace(0,x0,ndz);
xi_ndy = linspace(x0,L,ndy);
Rdr0   = interp1(xi, T_ini, xi_ndz, 'pchip')';
Rfr0   = interp1(xi, T_ini, xi_ndy, 'pchip')';
y0     = [Rdr0; Rfr0; x0];
clear T0 T_ini

% FEM matrices
[MMdr, DMdr, CMdr] = matfem (ze,'neu','neu','MM','DM','CM');
[MMfr, DMfr, CMfr] = matfem (ye,'neu','neu','MM','DM','CM');
inv_MMdr = MMdr\eye(size(MMdr));
inv_MMfr = MMfr\eye(size(MMfr));
% Spatial operators
Lap_op_dr = -inv_MMdr*DMdr;
Adv_op_dr = inv_MMdr*CMdr;
Lap_op_fr = -inv_MMfr*DMfr;
Adv_op_fr = inv_MMfr*CMfr;


%% Problem integration with ode15s
% Call the integrator
atol       = 1e-6;
rtol       = 1e-6;
optionspd = odeset('AbsTol',atol,'RelTol',rtol);
[tpd,ypd] = ode15s(@ode_fd, tl, y0, optionspd, Adv_op_dr,...
                   Adv_op_fr, Lap_op_dr, Lap_op_fr, inv_MMdr,...
                   inv_MMfr, ze, ye, ndz, ndy, L, alpha, Kfr,...
                   Kdr, sigma, ep, fp, Tcl, rhodr, cpdr, rhofr,...
                   cpfr, kd1, kd2, hL1, hL2, hL3, Pc, Tcr);

% Store the data
Rdr      = ypd(:,1:ndz)';
Rfr      = ypd(:,1+ndz:ndz+ndy)';
xx       = ypd(:,ndz+ndy+1)';

% Transformed spatial coordinates
ntpd   = length(tl);
zdr    = zeros(ndz,ntpd);
zfr    = zeros(ndy,ntpd);
for ii = 1:length(xx)
    zdr(:,ii) = xx(ii)*ze;
    zfr(:,ii) = (L-xx(ii))*ye+xx(ii);
end

% Computation of the temperature (undo the landau transform)
ndxi   = 11;
pts    = linspace(0,L,ndxi);
zz     = [zdr;zfr(2:end,:)];
RR     = [Rdr;Rfr(2:end,:)];
Tpd    = zeros(ndxi,size(zz,2));
for jj = 1 : size(zz,2)
    Tpd(:,jj) = interp1(zz(:,jj),RR(:,jj),pts,'pchip');
end
```

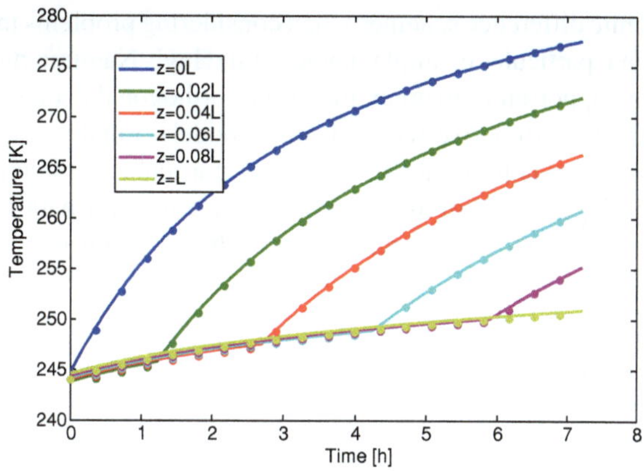

Fig. 6.34 Comparison in terms of the product temperature between the ALE technique (*continuous lines*) and the Landau transform (*marks*)

```
% Plot the solution
plot(tl(1:10:end)/3600,Tpd(1:2:end,1:10:end))
xlabel('Time [h]','Fontsize',16)
ylabel('Temperature [K]','Fontsize',16)
```

Script fd_model Main program for the solution of the Freeze-Drying problem using the Landau transform.

Figure 6.34 shows the comparison between the results (evolution of the product temperature at different spatial locations) obtained with an advanced simulation technique (ALE) [17] implemented in a specialized software (COMSOL, http://www.comsol.com/) and the Landau transform. The values of the control inputs (chamber temperature and pressure) are $T_{sh} = 293.15$K and $P_c = 20$Pa. As shown in the figure, the different approaches produce the same numerical results confirming the validity of the approach based on Landau transformation. The latter has the advantage of simplicity. Additional applications of the Landau transformation in combination with finite difference approximation can be found in [22].

6.4 Summary

Whereas the previous chapters are exclusively dedicated to lumped systems (systems of dimension 0 described by ODEs) and distributed parameter systems in one spatial dimension, this chapter touches upon the important class of problems in more space dimensions, as well as problems with time-varying spatial domains. Both are difficult topics and the ambition of this chapter is just to give a foretaste of possible numerical approaches. Finite difference schemes on simple 2D domains, such as squares, rectangles or more generally convex quadrilaterals, are first introduced, including several examples such as the heat equation, Graetz problem, a tubular chemical reactor, and Burgers equation. Finite element methods, which have more potential than finite difference schemes when considering problems in 2D, are then discussed based on a particular example, namely FitzHugh-Nagumo model. This example also gives the opportunity to apply the proper orthogonal decomposition method (presented in Chap. 4) to derive reduced-order models. Finally, the problematic of time-varying domains is introduced via another particular application example related to freeze drying. The main idea here is to use a transformation so as to convert the original problem into a conventional one with a time-invariant domain.

References

1. Graetz L (1883) Ueber die wärmeleitungsfähigkeit von flüssigkeiten. Annalen der Physik und Chemie 18:79
2. Pozrikidis C (2005) Introduction to finite and spectral element methods using matlab. Chapman and Hall/CRC, Boca Raton
3. Tabata M (1986) A theoretical and computational study of upwind-type finite element methods. Stud Math Appl 18:319–356
4. Kurganov A, Tadmor E (2000) New high-resolution central schemes for nonlinear conservation laws and convection-diffusion equations. J Comput Phys 160:241–282
5. FitzHugh R (1961) Impulses and physiological states in theoretical models of nerve membrane. Biophys J 1:445–466
6. Nagumo J, Arimoto S, Yoshizawa Y (1962) Active pulse transmission line simulating nerve axon. Proc Inst Radio Eng 50:2061–2070
7. Hodgkin AL, Huxley AF (1952) A quantitative description of membrane current and its application to conduction and excitation in nerve. J Physiol 117:500–544
8. Murray JD (2002) Mathematical biology II: spatial models and biomedical applications, 3rd edn. Springer, Berlin
9. Keener JP (2004) The topology of defibrillation. J Theor Biol 203:459–473
10. Argentina M, Coullet P, Krinsky V (2000) Head-on collisions of waves in an excitable fitzhugh-nagumo system: a transition from wave annihilation to classical wave behavior. J Theor Biol 205:47–52
11. Vilas C, García MR, Banga JR, Alonso AA (2008) Robust feed-back control of travelling waves in a class of reaction-diffusion distributed biological systems. Physica D Nonlinear Phenom 237(18):2353–2364
12. Pikal MJ (2000) Heat and mass transfer in low pressure gases: applications to freeze drying. Drugs Pharm Sci 102(2):611–686
13. Trelea IC, Passot S, Fonseca F, Marin M (2007) An interactive tool for the optimization of freeze-drying cycles based on quality criteria. Drying Technol 25(5):741–751
14. Dagan G (1979) The generalization of darcy's law for nonuniform flows. Water Resour Res 15(1):1–7
15. Velasco S, Roman FL, White JA (2009) On the clausius-clapeyron vapor pressure equation. J Chem Educ 86(1):106–111
16. Velardi SA, Rasetto V, Barresi AA (2008) Dynamic parameters estimation method: advanced manometric temperature measurement approach for freeze-drying monitoring of pharmaceutical solutions. Ind Eng Chem Res 47:8445–8457
17. Mascarenhas WJ, Akay HU, Pikal MJ (1997) A computational model for finite element analysis of the freeze-drying process. Comput Methods Appl Mech Eng 148:105–124
18. Crank J (1984) Free and moving boundary problems. Clarendon Press, Oxford
19. Belytschko T, Kennedy JM (1978) Computer models for subassembly simulation. Nucl Eng Des 49:17–38

20. Donea J (1983) Arbitrary Lagrangian-Eulerian finite element methods. In: Belytschko T, Hughes TJR (eds) Computational methods for transient analysis. North-Holland, Amsterdam, pp 473–516
21. Landau HG (1950) Heat conduction in a melting solid. Q Appl Mech Math 8:81–94
22. Illingworth TC, Golosnoy IO (2005) Numerical solutions of diffusion-controlled moving boundary problems which conserve solute. J Comput Phys 209:207–225

Further Readings

23. LeVeque RJ (1990) Numerical methods for conservation laws. Lectures in mathematics. ETH-Zurich. Birkhauser-Verlag, Basel
24. LeVeque RJ (2002) Finite volume methods for hyperbolic problems. Cambridge University Press, Cambridge
25. Reddy JN (1993) Introduction to the finite element method, 2nd edn. McGraw Hill, New York

Index

A. Vande Wouwer et al., *Simulation of ODE/PDE Models with MATLAB®, OCTAVE and SCILAB*, DOI: 10.1007/978-3-319-06790-2,

MIX
Papier aus verantwortungsvollen Quellen
Paper from responsible sources
FSC® C105338

If you have any concerns about our products,
you can contact us on
ProductSafety@springernature.com

In case Publisher is established outside the EU,
the EU authorized representative is:
Springer Nature Customer Service Center GmbH
Europaplatz 3, 69115 Heidelberg, Germany

Printed by Libri Plureos GmbH
in Hamburg, Germany